Nucleosides as Biological Probes

NUCLEOSIDES AS BIOLOGICAL PROBES

ROBERT J. SUHADOLNIK

Department of Biochemistry
Temple University
School of Medicine

A WILEY-INTERSCIENCE PUBLICATION

JOHN WILEY & SONS
New York • Chichester • Brisbane • Toronto

Library of Congress Cataloging in Publication Data

Suhadolnik, Robert J 1925–
Nucleosides as biological probes.

"A Wiley-Interscience publication."
Includes index.
1. Nucleosides—Physiological effect.
2. Nucleosides—Metabolism. I. Title.

QP625.N88S93 574.8'76'028 79-10719
ISBN 0-471-05317-1

Printed in the United States of America

10 9 8 7 6 5 4 3 2 1

To the many scientists whose spirit of generosity and cooperation was extended to me in the preparation of this book

Preface

In this book a detailed, up-to-date emphasis is placed on the application of the 70 known naturally occurring nucleoside analogs as biochemical probes in cellular reactions. Nine years ago, when my book, *Nucleoside Antibiotics*, was published, our knowledge of the nucleoside analogs as biochemical probes was not as clearly defined. Much unexpected information has been obtained concerning the simple and complex enzyme reactions, macromolecular syntheses, viral replication, and cell wall formation. This vast accumulation of knowledge has made it possible to complete the difficult transition of taking nucleoside analogs from the laboratory to their use in the treatment of human diseases.

Many interesting and exciting developments have been revealed during these last 9 years in terms of unique structures and physical, chemical, biosynthetic, pharmacological, and biochemical properties. Nine years ago the 30 known naturally occurring nucleoside analogs were diverse both in their structure and their activity against animals, plants, bacteria, and viruses. They had found valuable use in the elucidation of the complex steps involved in reading the genetic message on the ribosomes for protein, RNA and DNA synthesis and in the regulation of purine and pyrimidine nucleotide biosynthesis, enzyme reactions, chitin synthesis, and subcellular organization. Many of the 33 newly discovered analogs have found continued use in the above cellular processes, but also have expanded properties such that they are not only antagonists of purine and pyrimidine nucleotides but exert other physiological actions on humans.

The recently discovered analog tunicamycin, which is a structural analog of UDP-*N*-acetylglucosamine and dolichol, has been especially useful in studies related to glycoprotein and peptidoglycan synthesis in animals, plants, bacteria, and viruses. Similarly, puromycin, which has been so valuable as a codon-independent functional analog of aminoacyl-tRNA is now used as the *p*-azido photoaffinity label to form covalent bonds with the ribosomal proteins to more clearly define the exact protein with which puromycin reacts.

Cordycepin, which was originally reported to be a false feedback inhibitor of purine nucleotide biosynthesis as well as RNA synthesis, has been of great value as a biochemical probe in the synthesis of poly(A) and eventually the formation of functional, cytoplasmic mRNA. Three of the newly discovered analogs are extremely useful immunosuppressive agents (coformycin, 2′-deoxycoformycin, and bredinin). Coformycin and 2′-deoxycoformycin are also potent inhibitors of adenosine deminase. As such, they prevent normal mammalian and tumor cells from deaminating adenosine analogs (i.e., cordycepin, formycin, and ara-A). This allows for a marked increase in the concentration of the 5′-triphosphate of the analogs such that the tumor cell is bathed in much higher concentrations of the analogs for prolonged periods of time. The net result is a marked increase in the therapeutic efficiency of the drug. A new concept of inhibition by adenosine analogs via a "nucleotide independent" mechanism has been recently introduced. Ara-A and cordycepin bind irreversibly by first order kinetics to S-adenosylhomocysteine hydrolase, which is a "target" enzyme for these adenosine analogs. The result is a "suicide" inactivation (see Chapter 4).

Eritadenine, an adenine analog, is not a true nucleoside analog. However, it is reviewed here because it has the property of decreasing plasma cholesterol. Clitidine, one of two pyridine nucleoside analogs, has powerful hypoesthetic and hyperemic activity in mammals. Two new analogs, thuringiensin and agrocin 84, are elaborated as the nucleotides and exert their inhibitory activity as the nucleotides. Since 1970, new data have been accumulated showing that crotonoside and showdomycin can distinguish the inducible constitutive binding sites on the bacterial cell wall. Decoyinine, the ketohexose purine analog, is used to induce sporulation. One of the new analogs of guanosine, 2′-amino-2′-deoxyguanosine, has also been discovered. Bredinin, a nucleoside analog of AICA, in addition to its immunosuppressive properties and inhibitor of RNA and protein synthesis, has potent antiarthritic antirheumatoid activity.

Another important finding since 1970 is the report of the pyrrolopyrimidine ring in Q base of tRNA of mammalian tissue. Prior to this, the pyrrolopyrimidine aglycon was found in the nucleoside analogs, tubercidin, toyocamycin, and sangivamycin. The biosynthesis in the eukaryote and the prokaryote uses guanine as the carbon–nitrogen skeleton.

Finally, in 1970 only 5-azacytidine had found limited use in the treatment of human leukemia. None of the nucleoside analogs had FDA clearance. Today, tubercidin and ara-A have been cleared by the FDA for the treatment of human basal cell carcinoma, herpes simplex keratitis, and herpes simplex encephalitis. In terms of structures, the carbohydrate moiety of the octosyl acids and the ezomycin complex is a new type trans-fused anhydrooctouronic acid.

This volume on the nucleosides as biological probes is structured on the basis of the biological role that most accurately describes each nucleoside/nucleotide analog in cellular processes. Emphasis is also placed on the pleotropic effect of each analog to assist the reader in the multifunction activity of the analogs.

For uniformity, the newly discovered nucleoside analogs are presented according to the following plan: introductory comments; Discovery, Isolation, and Production; Physical and Chemical Properties; Structural Elucidation; Chemical Synthesis; Synthesis of Analogs; Inhibition of Growth; Biosynthesis; Biochemical Properties; Summary; and References. Sections concerning those nucleoside analogs reviewed in *Nucleoside Antibiotics* (Wiley, 1970) contain only the Synthesis; Biosynthesis; Biochemical Properties; Summary; and References. The reader is referred to this edition for supplemental reading.

Finally, I have carefully read and reviewed more than 4000 publications related to the naturally occurring nucleoside/nucleotide analogs presented in this revised edition. The preparation of this book was made possible by the numerous reprints, preprints, and personal detailed data made available to me by the many scientists throughout the world who have been actively engaged in research with these nucleosides and those who have given generously of their valuable time in editing the chapters sent to them.

ROBERT J. SUHADOLNIK

Philadelphia, Pennsylvania
June 1979

Acknowledgments

I should especially like to thank the following colleagues for their valuable editorial comments, suggestions, and criticisms: Drs. R. Agarwal, A. Argoudelis, O. Bârzu, L. Bennett, Jr., G. Bettinger, A. Bloch, V. Bohr, R. Borchardt, B. Brdar, A. Bruzel, C. Cheung, S. Cohen, B. Cooperman, J. Coward, F. Cramer, J. Darnell, E. De Clercq, P. Doetsch, J. Doskočil, J. Drach, K. Eckardt, A. Elbein, N. Fedorinchik, J. Fox, M. Franze-deFernandez, E. Freese, G. Gentry, A. Glazko, G. Gentry, G. Gutowski, H. Hadler, R. Hamill, R. Hamilton, L. Hanka, T. Haskell, A. Holý, M. Hori, J. Horwitz, L. Hurley, M. Ikehara, K. Isono, K. Iwai, D. Johnson, G. Just, T. Kalman, A. Kerr, T. Kishi, W. Klassen, H. Kneifel, W. König, J. Lampen, L. Lerner, W. Lennarz, F. Lichtenthaler, D. Lichtenwalner, C. Lum, R. Majima-Tsutsumi, J. Moffatt, K. Moldave, S. Morimoto, J. Mosca, W. Müller, R. Nagarajan, S. Nakamura, T. Nakanishi, J. Nevins, R. Parks, S. Pestka, H. Pitot, W. Plunkett, C. Pootjes, M. Robins, R. Robins, K. Sakaguchi, T. Sakagami, K. Sakata, K. Šebesta, J. Schell, V. Shchepetil'nikova, H. Seto, C. Smith, F. Šorm, A. Takatsuki, J. Tkacz, S. Tokuda, L. Townsend, J. Veselý, R. Vince, D. Visser, K. Watanabe, M. Weigele, R. Whitley, J. Wu, I. Zaenen, H. Zähner, J. Žemlička.

I also wish to express my sincerest appreciation to Nancy L. Reichenbach for her valuable assistance and patience in the assembling and organization of the chapters, tables, permissions, drawing of the structures of the analogs, and many of the figures. Without these many hours of dedication, the final publication would have been greatly delayed.

R.J.S.

Contents

Chapter 1 Inhibition of Cell Wall Synthesis, Cell Wall Receptors and Transport, Viral Coat Formation, Fungi, and Yeast **1**

1.1 Tunicamycin and the Streptovirudins, 4
1.2 Sinefungin, 19
1.3 The Ezomycins, 23
1.4 The Polyoxins, 30
1.5 Thraustomycin, 38
1.6 Amipurimycin, 40
1.7 Nikkomycin, 43
1.8 Septacidin, 45
1.9 Platenocidin, 46
1.10 Agrocin 84, 47
1.11 Showdomycin, 52
1.12 Crotonoside, 60

Chapter 2 Inhibition of Protein Synthesis **71**

2.1 The Aminoacylaminohexosylcytosine Analogs, 73
2.2 The Purine Nucleoside Analogs, 89
2.3 The Clindamycin Ribonucleotides, 104

Chapter 3 Inhibition of RNA Synthesis **113**

3.1 Cordycepin, 118
3.2 5-Azacytidine, 135
3.3 2′-Aminoguanosine, 145
3.4 3′-Aminoadenosine, 146
3.5 Aristeromycin, 147
3.6 Bredinin, 149

3.7 Pentopyranines, 154
3.8 Pyrrolopyrimidine Nucleoside Analogs, 158
3.9 Pyrazolopyrimidine Nucleoside Analogs, 169
3.10 Thuringiensin, 183
3.11 Puromycin Aminonucleoside, 195

Chapter 4 Inhibition of DNA Synthesis, Viruses, and Neoplastic Tissue **215**

4.1 9-β-D-Arabinofuranosyladenine, 217
4.2 1-β-D-Arabinofuranosylthymine, 233
4.3 Oxazinomycin, 236
4.4 1-Methylpseudouridine, 241
4.5 Nebularine, 244
4.6 5,6-Dihydro-5-azathymidine, 245

Chapter 5 Inhibition of Adenosine Deaminase and Immunosuppressive Activity of Nucleoside Analogs **258**

5.1 Coformycin, 259
5.2 2′-Deoxycoformycin, 260
5.3 Biochemical Properties, 262

Chapter 6 Inhibition of Purine and Pyrimidine Interconversions **279**

6.1 Psicofuranine and Decoyinine, 279
6.2 Pyrazofurin, 281

Chapter 7 Hyperesthetic and Hyperemic Nucleosides **292**

7.1 Clitidine, 292

Chapter 8 Inhibition of Cyclic-AMP Phosphodiesterase **295**

8.1 Octosyl Acids, 295

Chapter 9 Induction of Hypocholesterolemia **298**

9.1 Eritadenine, 298

Chapter 10 Naturally Occurring Nucleosides with Limited Biological Activity **311**

10.1 Herbicidins A and B, 311
10.2 5′-*O*-Glycosyl-*ribo*-nucleosides, 312
10.3 Raphanatin and 6-Benzylamino-7-β-D-glucopyranosylpurine, 313

Author Index **317**

Subject Index **327**

Nucleosides as Biological Probes

Chapter 1 Inhibition of Cell Wall Synthesis, Cell Wall Receptors and Transport, Viral Coat Formation, Fungi, and Yeast

1.1 TUNICAMYCIN AND THE STREPTOVIRUDINS 4

Discovery, Production, and Isolation 5
Physical and Chemical Properties 6
Structural Elucidation 6
Inhibition of Growth 7
Biochemical Properties 7
Effect on Eukaryotes 7
Effect on Viral Coat Protein Synthesis 9
Inhibition of Glycosylation of Interferon by TM 12
Formation of Fungal Multinuclear Giant Cells by TM 12
Inhibition of the Synthesis of Yeast Glycoproteins 13
Effect on Bacteria 13
Teichoic Acid Biosynthesis 15
Inhibition of *N*-Acetylglucosamine-lipid Formation in Plants 16
Inhibition of the Conversion of Procollagen to Collagen and Secretion of IgA and IgE and Serum Proteins 16
Inhibition of Biosynthesis of Acid Mucopolysaccharides 16

1.2 SINEFUNGIN 19

Discovery, Production, and Isolation 20
Physical and Chemical Properties 20
Structural Elucidation 20
Inhibition of Growth 20
Biosynthesis of Sinefungin 21
Biochemical Properties 22
Inhibition of Methyltransferases 22

1.3 THE EZOMYCINS 23

Discovery, Production, and Isolation 23
Physical and Chemical Properties 24

Structural Elucidation 24
Inhibition of Growth 29
Biochemical Properties 29

1.4 THE POLYOXINS 30

Total Chemical Synthesis of Polyoxin J 31
Polyoxin Analogs 32
Biosynthesis of the Polyoxins and the Octosyl Acids 32
Biochemical Properties 35
Antifungal Insecticidal Activity 35
Inhibition of Chitin Synthetase 35
The Study of Hyphal Growth with Polyoxin D 37

1.5 THRAUSTOMYCIN 38

Discovery, Production, and Isolation 38
Physical and Chemical Properties 38
Structural Elucidation 38
Biochemical Properties 40

1.6 AMIPURIMYCIN 40

Discovery, Production, and Isolation 41
Physical and Chemical Properties 41
Structural Elucidation 41
Inhibition of Growth and Biochemical Properties 42

1.7 NIKKOMYCIN 43

Discovery, Production, and Isolation 43
Physical and Chemical Properties 43
Structural Elucidation 44
Inhibition of Growth 44
Biochemical Properties 44

1.8 SEPTACIDIN 45

1.9 PLATENOCIDIN 46

Discovery, Production, and Isolation 46
Physical and Chemical Properties and Structural Elucidation 46
Biochemical Properties 47

1.10 AGROCIN 84 47

Discovery, Production, and Isolation 47
Physical and Chemical Properties 48
Structural Elucidation 48
Inhibition of Growth 49
Biochemical Properties 49
 Plasmid Required for Virulence of *A. tumefaciens* 50

1.11 SHOWDOMYCIN 52

Chemical Synthesis of Showdomycin and Showdomycin Analogs 53
Enzymatic Transformation of Showdomycin to Isoshowdomycin 53
Phosphorylation of Showdomycin to Showdomycin 5′-Monophosphate 53
Biosynthesis of Showdomycin 55
Biochemical Properties 56
 Inhibitory Effects of Showdomycin: Permeability and Alkylation Properties 56
 Inhibition of Thymidylate Synthetase and Pseudouridylate Synthetase 59
 Radiosensitization of *E. coli* B/r by Showdomycin 59
 Oxidative Phosphorylation and Carcinogenesis 59

1.12 CROTONOSIDE 60

Biochemical Properties 61
 Effect on Glutamine Dehydrogenase and cAMP 61
 Allosteric Modifier Activities with Isocitrate Dehydrogenase 62

SUMMARY 62

REFERENCES 63

This chapter reviews the eukaryotes, prokaryotes, and viruses as they are affected by the naturally occurring nucleoside/nucleotide analogs that inhibit (i) cell wall synthesis (i.e., tunicamycin and the streptovirudins), (ii) fungi (i.e., sinefungin, the ezomycins, the polyoxins, thraustomycin, amipurimycin, nikkomycin, and septacidin), (iii) yeast (i.e., platenocidin, and sinefungin) and (iv) cell wall receptors and transport (i.e., agrocin 84, showdomycin, and crotonoside). The polyoxins, showdomycin, crotonoside, and septacidin, have been reviewed in detail (Suhadolnik, 1970) and are discussed here only in terms of studies related to their biological properties since 1970. The remaining nucleoside analogs are reviewed in detail with respect to discovery, inhibition of growth, chemical, physical, and biochemical properties, and biosynthesis.

1.1 TUNICAMYCIN AND THE STREPTOVIRUDINS

Some 60–70 naturally occurring nucleoside analogs have been widely used as biochemical probes for many complex cellular reactions. An equally valuable analog is tunicamycin (TM) (Fig. 1.1), discovered by Takatsuki et al. (1971), which is a structural analog of UDP-*N*-acetylglucosamine and dolichol. Takatsuki and Tamura (1971a, b) were the first to demonstrate that TM selectively inhibits the incorporation of sugars into acid-insoluble compounds and the first to show that TM inhibits the multiplication of Newcastle disease virus. Takatsuki and coworkers were also the first to report that TM induces morphological changes in bacteria and yeast (Takatsuki et al., 1971, 1972). TM prevents glycosylation of proteins by specifically inhibiting the formation of the lipid-linked sugar intermediate. TM was first isolated by Takatsuki et al. (1971) from the mycelium and the culture filtrates of *Streptomyces lysosuperficus*. Ito et al. (1977) and Takatsuki et al. (1977a) have shown that TM is not a single compound but, rather, a mixture of homologous antibiotics. There appear to be at least

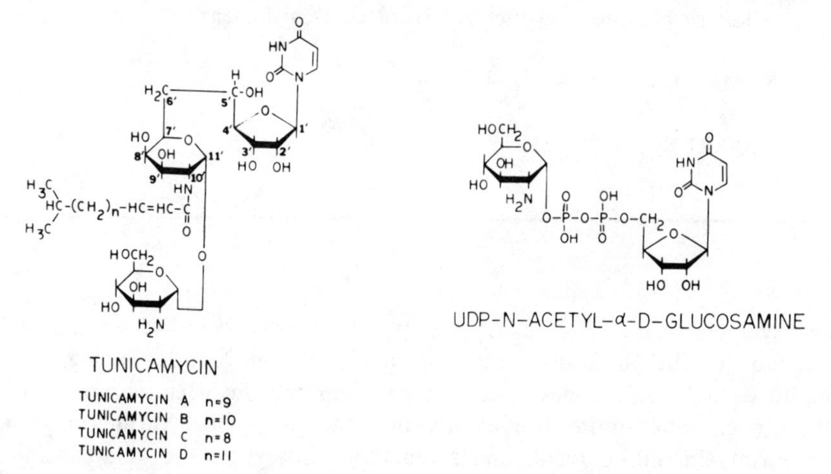

Figure 1.1 Structure of the tunicamycin complex and comparison to UDP-*N*-acetyl-α-D-glucosamine.

tunicamycins A, B, C, and D (Fig. 1.1). Based on the elegant degradations and physical data obtained by Ito et al. (1977), the structure has been assigned for the main components of the TM complex. Each contains uracil, a fatty acid component that is a trans α,β-unsaturated branched chain acid, and *N*-acetylglucosamine. Although TM and the polyoxins (see page 30) are structural analogs of UDP-*N*-acetylglucosamine, the polyoxins, without the unsaturated branched chain acid, are limited to the competitive inhibition of chitin synthetase. The structure of TM has two features, that is, the nucleoside part of TM is an analog of the nucleoside diphosphate sugars, and the unsaturated acid part is an analog of dolichol. It is these structural features of TM that make it an analog of dolichyl-*N*-acetyl-glucosaminyl pyrophosphate which inhibits the transfer of *N*-acetylglucosamine from the lipid saccharide required for the assembly of the oligosaccharide component of glycoproteins and peptidoglycan. A new C_{11} aminodeoxydialdose, named tunicamine, has also been isolated and characterized (Ito et al., 1978). Hydrolysis of TM with 3 *N* HCl, 3 h, at 100°C cleaves TM to tunicaminyl uracil, D-glucosamine, and several unsaturated fatty acids. Isopentadecanoic acid is present in addition to the fatty acid already mentioned (Takatsuki et al., 1977a). TM, which derives its name from the latin, *tunica*, meaning coat, inhibits gram-positive bacteria (especially *Bacillus* species), yeast, fungi, protozoa, plants, envelope viruses, and mammalian cells in culture (Takatsuki et al., 1971). Because of its ability to inhibit the synthesis of *N*-acetylglucosaminyl-lipids in prokaryotes, and eukaryotes, tunicamycin is one of the most useful biochemical probes for elucidation of the complex reactions involved in the assembly of glycoproteins and cell walls. Tunicamycin differs from other glycolipid antibiotics such as diumycin, macarbomycin, moenomycin, and prasinomycin in that it does not contain phosphorus; however, TM is capable of inhibiting the transferases involved in glycolipid synthesis in prokaryotes, eukaryotes, and viral coat formation. Although the streptovirudins differ chemically from TM, their primary action also involves the inhibition of glycoprotein synthesis (J. S. Tkacz, personal communication). The antibiotic mycospocidin, which was isolated by Nakamura et al. (1957), is in the same class as the TM and streptovirudin groups (Eckardt et al., 1973; K. Eckardt, personal communication; J. Tkacz, personal communication). For reviews, see Lambert et al. (1977), Kuo and Lampen (1972), and Waechter and Lennarz (1976).

Discovery, Production, and Isolation

Takatsuki et al. (1971) were the first to report on the isolation and production of TM from *S. lysosuperficus*. TM is isolated from either the cells or the culture filtrates. The medium, growth conditions, and isolation of TM have been described in detail (Takatsuki et al., 1971).

Physical and Chemical Properties

Tunicamycin is a white crystalline powder; mol wt ~ 870; mp 234–235°C; $[\alpha]_D^{20}$ +52° (*c* 0.5, pyridine); ultraviolet spectral properties: $\lambda_{max}^{methanol}$ 250 nm ($E_{1\,cm}^{1\,percent}$ 230), 260 nm ($E_{1\,cm}^{1\,percent}$ 110); the nmr spectrum of TM has been reported; treatment with periodate results in a rapid loss of antiviral activity (Takatsuki et al., 1971).

Structural Elucidation

Takatsuki et al. (1977a) and Ito et al. (1978) have reported on the structure of some members of the TM complex (Fig. 1.1). Hydrolysis of TM in 3 *N* HCl, 100°C, 3 h results in the isolation of four major *trans*-α,β-unsaturated *iso* acids (Takatsuki et al., 1979; Ito et al., 1978) and tunicaminyl uracil, 1-β-D-(10′-amino-6′,10′-dideoxy-L-galacto-D-allo-undecadialdo-7′,11′-pyranose-1′,4′-furanosyl)uracil. The ultraviolet spectral properties are as follows: $\lambda_{max}^{H_2O}$ 261 nm (ϵ = 10,700); mass spectra and nmr have been described (Takatsuki et al., 1977a). Ito et al. (1978) have proposed the name tunicamine for the dialdose with 11 carbons attached to uracil. Crystalline D-

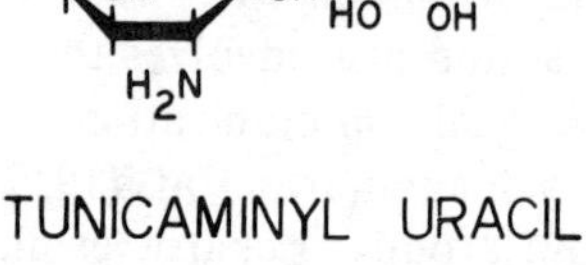

TUNICAMINYL URACIL

glucosamine is isolated from the hydrolyzed TM and the α,α-linkage between C-11 of tunicamine and C-1″ of *N*-acetylglucosamine has been assigned on the basis of the high molecular rotation of TM and their spectrum; the permethylated derivative of TM complex showed peaks at *m*/*e* 970, 984, 998, and 1012, which might be assigned to M^+ ion peaks of permethylated tunicamycins C, A, B, and D, respectively (Takatsuki et al., 1977a). The major structural differences between the streptovirudin complex (series II) and the tunicamycin complex are in their fatty acid chains (J. S. Tkacz, personal communication; Eckardt et al., 1975).

Inhibition of Growth

Tunicamycin is active against plant and animal RNA and DNA viruses, gram-positive bacteria, yeast, and fungi (Takatsuki et al., 1971; Takatsuki and Tamura, 1971a). *Bacillus* species are the most sensitive bacteria studied. Tunicamycin induces various morphological changes among the organisms sensitive to the antibiotic without inhibiting protein, RNA, or DNA synthesis.

Biochemical Properties

Effect on Eukaryotes. Tunicamycin has been used as a biological probe to study the assembly, secretion and function of glycoproteins in mammalian cells, bacteria and viruses (Lucas et al., 1975; Chen and Lennarz, 1976). Elbein and coworkers (Heifetz et al., *Biochemistry,* **18,** 2186, 1979; ref. added in pg. proof) have reported that TM is a noncompetitive inhibitor of GlcNAc 1-phosphate transferase (Fig. 1.2), inhibits the reverse reaction and is a substrate-product transition-state analog. The synthesis of the polypeptide chain of ovalbumin has been reported in great detail by Shimke and coworkers (Palmiter et al., 1971). Keiley et al. (1976) documented that nascent chains of ovalbumin are still attached to tRNA when the carbohydrate chain of mannose and *N*-acetylglucosamine units are added.

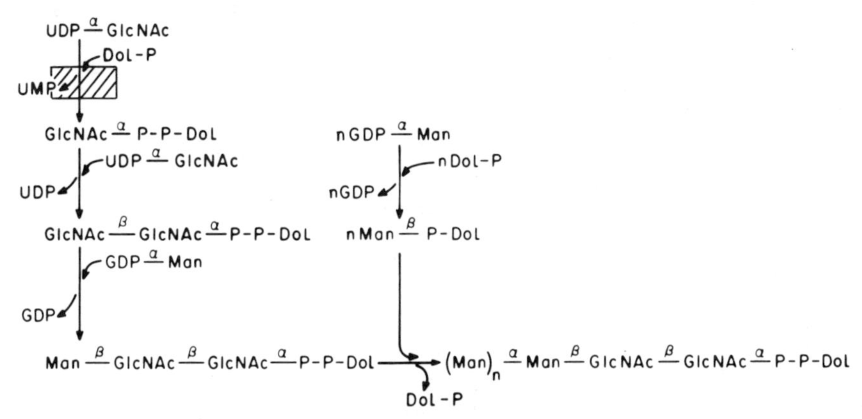

Figure 1.2 Lipid-linked sugars in glycoprotein synthesis. Postulated sequence in assembly of the oligosaccharide-lipid based on preparations of liver, myeloma, pancreas, lymphocyte, and oviduct membrane. The known site of inhibition by tunicamycin is shown by the crosshatched bar. Modified from Waechter and Lennarz, 1976.

Takatsuki and Tamura (1971b) showed that tunicamycin preferentially inhibits the incorporation of glucosamine in glycoproteins of chick embryo fibroblasts. Takatsuki et al. (1975) reported that in chick embryo microsomes, TM inhibits the production of an *N*-acetylglucosamine-lipid. Tkacz and Lampen (1975) also reported that in calf liver rough microsomes, TM inhibits the production of dolichyl *N*-acetylglucosaminylpyrophosphate but does not affect the synthesis of mannolipid (Fig. 1.3). Subsequently, Struck and Lennarz (1977) used TM in three different types of experiments on glycoprotein synthesis to show that lipid-linked sugars participate in the assembly of the oligosaccharide side chain of ovalbumin. From their data they concluded that tunicamycin: (i) inhibits the transfer of *N*-acetylglucosamine from UDP-*N*-acetylglucosamine to form the glycosylated ovalbumin (Fig. 1.4), (ii) does not inhibit the synthesis of phosphoryldolichol or protein synthesis (Fig. 1.3), (iii) does not inhibit elongation of preexisting *N*-acetylglucosamine-lipids to oligosaccharide-lipid or the transfer of oligosaccharide from oligosaccharide-lipid to protein, (iv) appears to block the synthesis of *N*-acetylglucosaminylpyrophosphorylpolyisoprenol, (v) does not block other *N*-acetylglucosaminyl transferases present in oviduct membrane preparations and (vi) cannot utilize this mannosyl donor for glycoprotein synthesis. Kinetic studies on the inhibition of the synthesis of *N*-acetylglucosaminylpyrophosphorylpolyisoprenol indicate that TM competes with UDP-*N*-acetylglucosamine (Takatsuki et al., 1976;

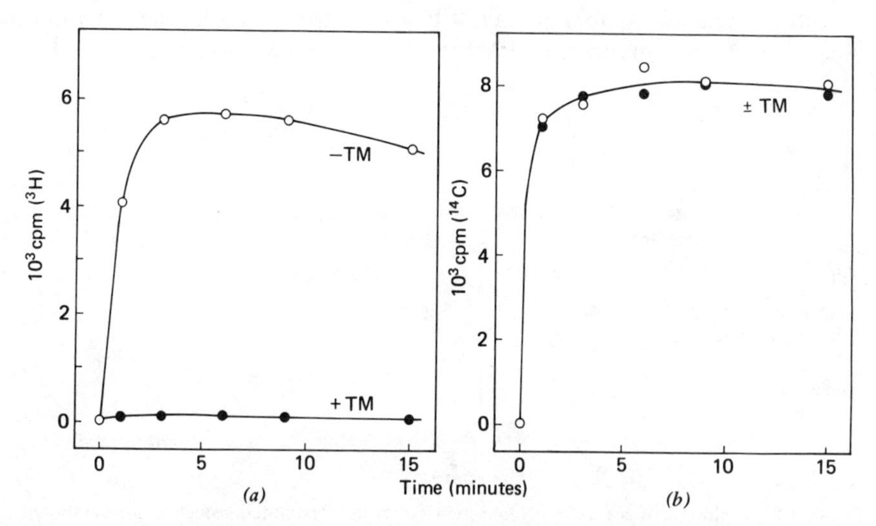

Figure 1.3 Incorporation of radioactivity from (*a*) UDP-[³H]GlcNAc and (*b*) GDP-[¹⁴C]Man into CM-soluble products in the presence (●) or absence (○) of tunicamycin (5 μg/ml). Reprinted with permission from Tkacz and Lampen, 1975.

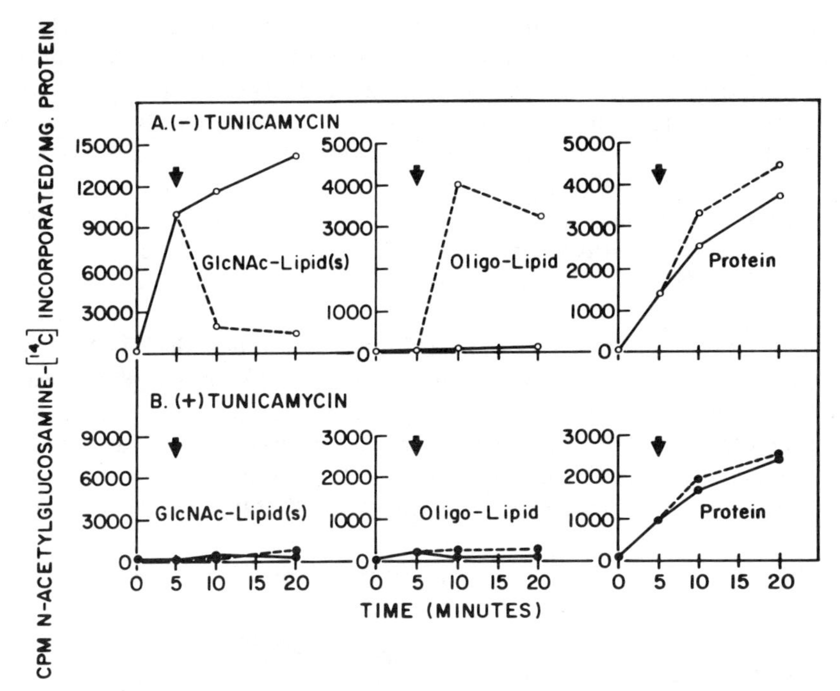

Figure 1.4 Effect of tunicamycin on *N*-acetylglucosamine incorporation from UDP-*N*-acetylglucosamine by oviduct membranes. At 5 min, unlabled GDP-mannose (---) or water (——) was added. Reprinted with permission from Struck and Lennarz, 1977.

Takatsuki, 1978). However, tunicamycin does not merely act as an analog of UDP-*N*-acetylglucosamine because TM does not inhibit transfer of *N*-acetylglucosamine from UDP-*N*-acetylglucosamine to dolichylpyrophosphoryl-*N*-acetylglucosamine (Lehle and Tanner, 1976; Struck and Lennarz, 1977) nor does it inhibit chitin synthesis (Kuo and Lampen, 1974). These results suggest an important role of the lipid moiety of TM in its action. It appears that the saccharide-lipids participate in (i) the assembly of core-type oligosaccharide of membrane glycoproteins of the oviduct and (ii) the synthesis of the carbohydrate chain of the major secretory glycoproteins. Struck and Lennarz (1977) conclude that it is not unreasonable that the synthesis of the hydrophobic saccharide-lipid, a membrane-dependent synthesis, is closely related to translation of the polypeptide that occurs on the ribosome-associated membrane.

Effect on Viral Coat Protein Synthesis. A number of laboratories have suggested that viral glycoproteins are essential for the formation of infectious virus (Duda and Schlesinger, 1975; Kaluza et al., 1972, 1973;

Klenk et al., 1972; Leavitt et al., 1977; Scholtissek et al., 1974, 1975; Schwarz and Klenk, 1974). These findings are based on the inhibition of glycosylation of glycoproteins by 2-deoxy-D-glucose and D-glucosamine. However, because other cellular reactions are also affected, these data do not allow unequivocal interpretations. This experimental difficulty has now been overcome with the discovery of TM. Takatsuki and Tamura (1971a, b) were the first to show that TM preferentially blocks the multiplication of Newcastle disease virus when added during the viral multiplication cycle in cultured chick embryo fibroblasts (Fig. 1.5). Viral RNA synthesis proceeds normally in the presence of tunicamycin.

Furthermore, Takatsuki et al. (1977c) demonstrated that SV40-transformed cells show a higher sensitivity to TM than do normal cells in terms of three parameters: cell growth, incorporation of *N*-acetylglucosamine into glycoproteins, and the formation of lipid-linked *N*-acetylglucosamine. The inhibition of cell growth was the most striking. Takatsuki et al. (1977c) suggested that the increased hypersensitivity of SV40-transformed cells to TM

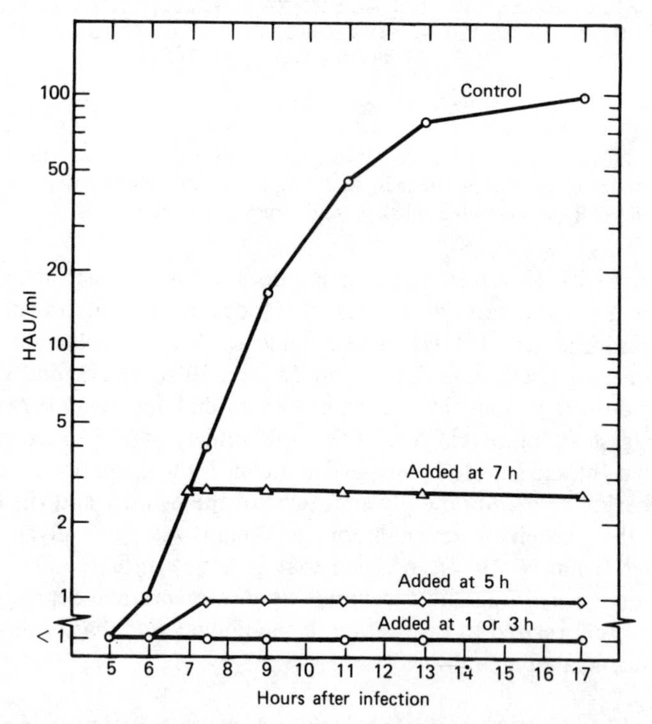

Figure 1.5 Anti-Newcastle disease virus activity of tunicamycin added at various times after the infection. Virus production is expressed as hemagglutinin units (HAU) per milliliter. Reprinted with permission from Takatsuki and Tamura, 1971b.

may indicate TM hypersensitive glycoprotein(s) that play an important role in cell growth. The use of TM may help elucidate the function of the glycoproteins with respect to cell division. This will contribute greatly to our understanding of the complex process of viral transformation and cell growth. Of extreme interest is a recent report by Takatsuki and Tamura (1978), which is an extension of their earlier report (Takatsuki and Tamura, 1972) that phosphatidylcholine, lauric acid, and monoglyceride of lauric acid effectively reverse the antiviral activity of TM (at minimal inhibitory concentrations) against Newcastle disease virus multiplication. The authors suggest that the recovery of virus multiplication by lipids is due to the interaction of TM with the lipophilic membrane components.

The notion that TM interferes with the formation of cell-surface glycoproteins has been demonstrated by Duksin and Bornstein (1977a, b). They showed that virally transformed mouse (3T3) cells on treatment with TM caused detachment and death of simian virus 40- and polyoma-transformed cells. Analysis of labeled proteins showed that a high molecular weight protein, presumably related to fibronectin, is markedly reduced in cells treated with TM. These data and the studies of Olden et al. (1977) suggest that TM interferes with the function of the cell-surface glycoproteins. The inhibition of viral glycoproteins by TM has been demonstrated by Nakamura and Compans (1978), who showed that TM completely inhibited the incorporation of D-glucosamine.

Tunicamycin has also been used to study the synthesis of Semliki Forest virus (SFV) membrane proteins. Translation studies have shown that the synthesis of membrane ectoproteins, like glycoproteins, of SVS and SFV occurs simultaneously with their insertion into membranes, their glycosylation, and their proteolytic processing (Katz et al., 1977; Rothman and Lodish, 1977). By using tunicamycin, Garoff and Schwarz (1978) have demonstrated by means of SDS gel electrophoresis that glycosylation of viral proteins is not necessary for membrane insertion and cleavage of SFV membrane proteins. The nonglycosylated proteins are present in viral proteins of control and tunicamycin-treated cells, whereas no bands that correspond to the glycosylated proteins are in the tunicamycin-treated cells.

Gibson et al. (1979) investigated the role of glycosylation in the assembly and infectivity of several isolates of VSV. At 38°, TM prevented the assembly of VSV (San Juan and Orsay) and in both cases the nonglycosylated viral glycoprotein aggregated in the cell and failed to migrate to the cell surface. At 30° in the presence of TM, the assembly of VSV (San Juan) was still markedly inhibited, but VSV (Orsay) was only inhibited 30 to 50%. The nonglycosylated G protein of VSV (Orsay) migrated to the cell surface and was incorporated into fully infectious virions. These data show that (i) glycosylation of the viral G protein plays a very important role in the assembly of VSV, but not in its infectivity, (ii) nonglycosylated G protein is temperature sensitive, and (iii) glycosylation of the G protein may influence its folding and acquisition of a stable conformation.

Inhibition of Glycosylation of Interferon by TM. The biological significance of the carbohydrate moieties on interferon has been lacking. To evaluate the effect of glycosylation, Sulkowski and coworkers (Mizrahi et al., 1978) have described the properties of human immune interferon prepared in the presence and absence of TM. Their data show that human interferon undergoes glycosylation; TM inhibits this process. Glycosylation is *not* necessary for secretion of interferon by the cell. However, most important is that the nonglycosylated TM is able to express its antiviral properties. Therefore, the antiviral activity of the interferon does not seem to depend on the *N*-glycosidically linked carbohydrates.

Fujisawa et al. (1978) also reported that TM blocks glycosylation of interferon production by L cells induced by Newcastle disease virus. The nonglycosylated interferon of smaller size (T interferon, 18,000 daltons) has full antiviral activity. However, the molecular characteristics of glycosylated interferon and interferon produced in the presence of TM are markedly different. For example, normal L cell interferon is stable against heating at 70°C, 1 h in SDS and a reducing agent, whereas interferon produced in the presence of TM is inactivated under these same conditions.

Another group of antibiotics that have potent antiviral activity against the replication of envelope viruses is the streptovirudin complex (Eckardt et al., 1973). The streptovirudins show antiviral activity against RNA and DNA viruses cultured in chick embryo cells (Tonew et al., 1975; Eckardt et al., 1975; Thrum et al., 1975); at 2.5–20 μg/ml, there is a 100% plaque reduction. It is likely that the streptovirudins are directed primarily against glycoprotein synthesis, for these antibiotics are structurally related to tunicamycin and like tunicamycin are potent inhibitors of dolichylpyrophosphorylacetylglucosamine synthesis (Tkacz, unpublished results).

Formation of Fungal Multinuclear Giant Cells by TM. Takatsuki et al. (1972) reported that TM causes morphological changes in bacteria and yeasts. Takatsuki and coworkers (Katoh et al., 1976, 1978) investigated the effect of TM on *Penicillium* and *Aspergillus*. TM inhibits the germination and cell division conidia of these fungi. The conidia of these fungi incubated with TM increased in cell mass and formed multinuclear giant cells (10–100 μm in diameter). Removal of TM restored the giant cells to normal mycelial form. Polyoxin D also induces morphological changes in fungi (Endo et al., 1970; Bartnicki-Garcia and Lippman, 1972); however, unlike TM, the polyoxin D-treated cells are osmotically labile. It appears that TM selectively inhibits the synthesis of the cell surface. This inhibition is independent of the type of cell (Katoh et al., 1976). Katoh et al. (1978) showed, by scanning electron microscopy, that there is a lack of network structure of cell surface, thick cell walls, and a lack of internal septa separating individual nuclei of TM-treated fungi. These findings suggest that the inhibition of germination and cell division by TM is due to the inhibition and perturbation of normal cell wall synthesis.

Inhibition of the Synthesis of Yeast Glycoproteins. The *Saccharomyces* have been studied extensively with tunicamycin (Tkacz et al., 1977; Kuo and Lampen, 1972, 1974; Hasilik and Tanner, 1976). TM acts as an inhibitor of glycoprotein synthesis in yeast. The synthesis of active invertase, acid phosphatase and carboxypeptidase Y is inhibited by TM. Phosphatidyl choline and phosphatidyl-L-serine reverse the activity of TM against invertase (Kuo and Lampen, 1976). The syntheses of α-glucosidase and alkaline phosphatase (nonglycosylated proteins) are not inhibited by tunicamycin. TM does not affect glucosamine metabolism. Kuo and Lampen (1976) have shown that TM inhibits the incorporation of glucosamine into mannan peptides of protoplasts prepared from *Saccharomyces*. Chitin synthesis is not inhibited. Lehle and Tanner (1976) have shown that TM inhibits the synthesis of dolichylpyrophosphorylacetylglucosamine. They used membrane preparations of *Saccharomyces cerevisiae*. TM does not inhibit the transfer of a second residue of *N*-acetylglucosamine to this precursor. Therefore, TM inhibits the translocase reaction and not the *N*-acetylglucosaminyl transferase.

Effect on Bacteria. Takatsuki et al. (1971) and Takatsuki and Tamura (1971b) reported that TM induces morphological changes such as elongation, conversion from rod form to cocci form, and enlargement of microbial cells. Although RNA and DNA degradation occurred in *Bacillus subtilis*, the incorporation of [^{14}C]glucosamine was inhibited most extensively. The mode of inhibition by TM must be directed on the cell surfaces. Tamura et al. (1976), using cellfree extracts of *Micrococcus lysodeikticus* (*luteus*), showed that TM specifically inhibits peptidoglycan synthesis by blocking the translocase reaction (Fig. 1.6). This finding was subsequently confirmed by Ward (1977). TM inhibits formation of lipid pyrophosphoryl *N*-acetylmuramylpentapeptide, but does not affect the transfer of *N*-acetylglucosamine from UDP-*N*-acetylglucosamine to lipid pyrophosphoryl *N*-acetylmuramyl (pentapeptide)-*N*-acetylglucosamine (Fig. 1.6).

Lambert et al. (1977) demonstrated that *Bacillus licheniformis* membranes can also incorporate *N*-acetylglucosamine from UDP-*N*-acetylglucosamine into a lipid and nonlipid fraction. Therefore, in the absence of

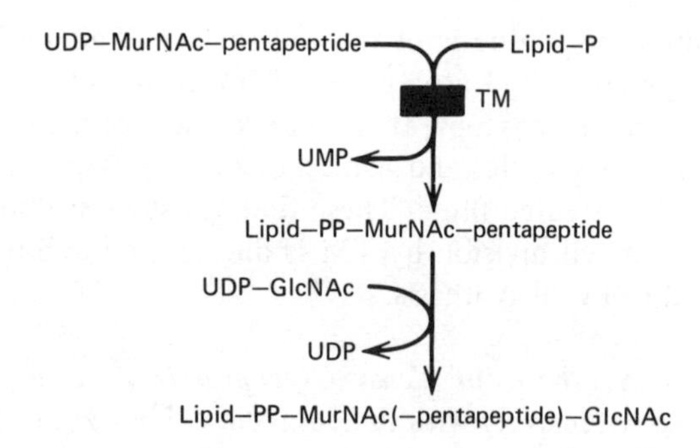

Figure 1.6 Inhibition of the translocase reaction in *M. lysodeikticus* (*luteus*) by tunicamycin (Tamura et al., 1976).

UDP-*N*-acetylmuramyl-pentapeptide, membranes are capable of transferring one *N*-acetylglucosamine from UDP-*N*-acetylglucosamine to undecaprenyl phosphate by means of transphosphorylation. Bettinger et al. (1977a, b) studied the effect of TM with isolated membranes of *B. subtilis*. The addition of *N*-acetylglucosamine from UDP-*N*-acetylglucosamine to undecaprenyl phosphate occurs in isolated membrane of *B. subtilis* in two different steps.

Bettinger and Young (1975) reported that TM inhibits peptide synthesis in *B. subtilis* about 10%. While this percentage is low, it is consistent with the findings of Ward (1977) and Tamura et al. (1976).

Bettinger and Young (1975, 1977b) and Takatsuki et al. (1978a) showed that with *B. subtilis* membrane TM can inhibit the reversible formation of undecaprenyl-*N*-acetylglucosaminyl-lipid by the release of UMP or UDP in the following lipid-linked reactions of unknown function where lipid-PP is undecaprenyl pyrophosphate and lipid-P is undecaprenyl phosphate:

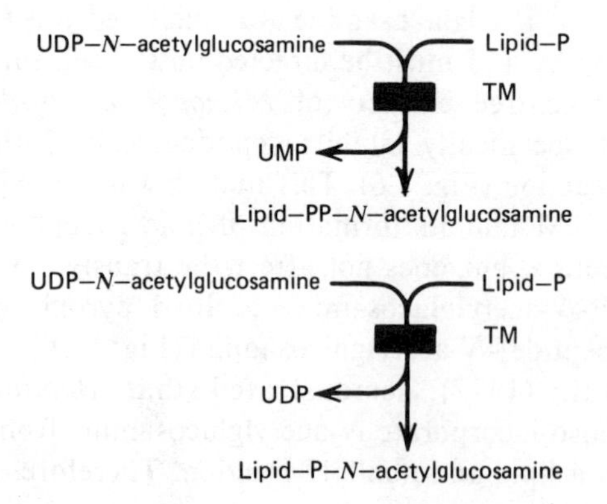

TM inhibits the UMP and UDP lipid formation/depolymerization reactions by 74 and 78%, respectively (Bettinger, personal communication).

Teichoic Acid Biosynthesis. Because of the role of teichoic acid in cell wall formation, Bracha and Glaser (1976), Hancock et al. (1976), Weston and Perkins (1977), and Wyke and Ward (1977) studied the effect of tunicamycin on the lipophilic compound that serves as an intermediate in the synthesis of the linkage region between teichoic acid and peptidoglycan in membrane preparations from several organisms. Baddiley and coworkers (McArthur et al., 1978) have used TM to elucidate the sequence of synthesis of lipid intermediates **I**, **II**, and **III** in the biosynthesis of the linkage unit between teichoic acid and peptidoglycan. TM inhibits the synthesis of lipid intermediate **I**, hence of lipid intermediates **II** and **III**. The assembly of the poly(*N*-acetylglucosamine phosphate) teichoic acid-linkage unit in *Micrococcus varians* ATCC 29750 is shown in Fig. 1.7.

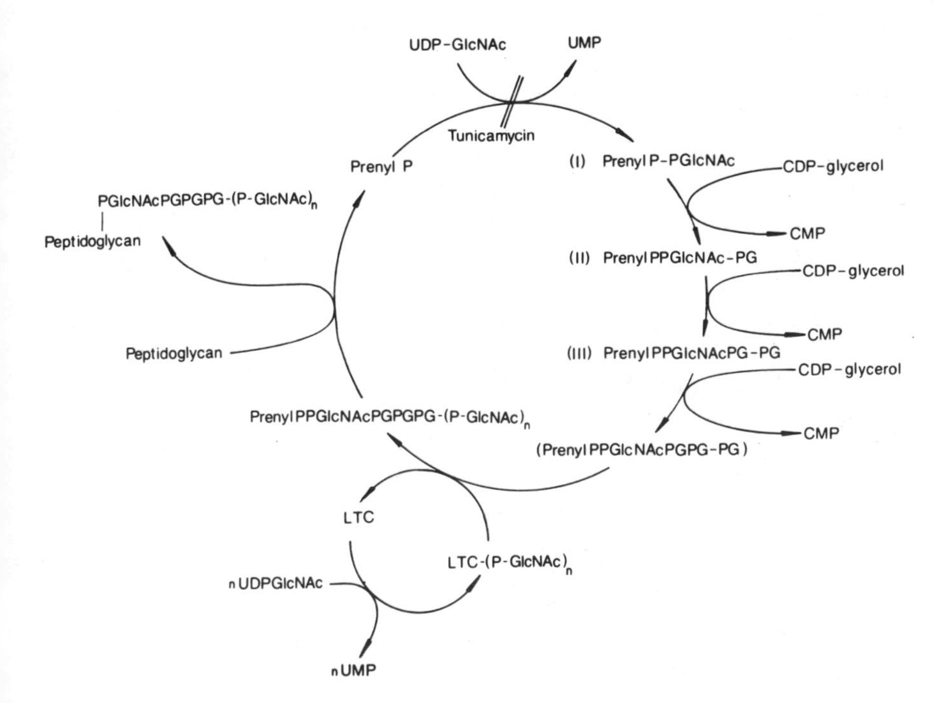

Figure 1.7 Proposed scheme for the synthesis of teichoic acid, poly(*N*-acetylglucosamine phosphate), and the linkage unit that attaches it to peptidoglycan in *Micrococcus varians* ATCC29750. PG = phosphoglycerol; P-GlcNAc = *N*-acetylglucosamine-1-phosphate; LTC = lipoteichoic acid carrier. Reprinted with permission from Roberts et al., 1979.

Inhibition of N-Acetylglucosamine-lipid Formation in Plants. The synthesis of the lipid-linked oligosaccharides of mannose and *N*-acetylglucosamine in plants appears to be similar to that in the mammalian systems (Forsee and Elbein, 1975; Forsee et al., 1976). Proof that plants form *N*-acetylglucosaminepyrophosphorylpolyprenol that eventually forms glycoproteins was demonstrated by Ericson et al. (1977) using particulate extracts from cotton fibers and mung bean seedlings. They showed that TM (5 μg/ml) blocked the incorporation of *N*-acetylglucosamine into the *N*-acetylglucosaminepyrophosphorylpolyprenol.

Inhibition of the Conversion of Procollagen to Collagen and Secretion of IgA and IgE and Serum Proteins. Duksin and Bornstein (1977b) reported that TM inhibits the conversion of procollagen to collagen. TM inhibits the incorporation of mannose by 90%, whereas general protein and collagen synthesis is decreased only 10–20%.

Hickman and coworkers (Hickman et al., 1977; Hickman and Kornfeld, 1978) found that TM inhibited the secretion of IgA, IgM, and IgE by various plasma cell lines. In contrast, the secretion of IgG was not significantly inhibited even though TM completely prevented the glycosylation of the IgG heavy chains. In contrast, TM produced only a small inhibition of secretion of normally nonglycosylated λ_2 light chains by a variant light chain-secreting mouse plasmacytoma. Hickman and Kornfeld (1978) have demonstrated that TM produced an 81% inhibition of IgM secretion by MOPC 104E plasma cells. TM did not inhibit the initial rate of synthesis of intracellular IgM. TM produced a 64% inhibition of IgA secretion. With IgG, TM inhibited the incorporation of D[^{14}C]glucosamine into newly synthesized IgG, but only produced a 28% inhibition of IgG secretion. These data indicate that immunoglobin secretion produced by TM depends on the immunoglobin class produced by plasma cells.

As is reported above, tunicamycin preferentially blocks the glycosylation of newly synthesized proteins (Struck and Lennarz, 1977; Kuo and Lampen, 1974; Schwarz et al., 1976). However, despite the inhibition of glycosylation of the serum proteins, rat liver transferrin, and the apoprotein B chain of chick liver VLDL by tunicamycin, the secretion of the unglycosylated form of these proteins is virtually unimpaired (Struck et al., 1978). Similarly, the secretion of the ovalbumin by hen oviduct is not inhibited by TM (Keller and Swank, 1978). Therefore, no generalization can be made about the role of carbohydrate chains in secretion.

Inhibition of Biosynthesis of Acid Mucopolysaccharides. The mucopolysaccharides are important in cell physiology (Brimacombe and Webber, 1964). The biosynthesis of chondroitin sulfate and heparin proceeds by a stepwise transfer of monosaccharides from their nucleotide sugars (Lindahl, 1972; Rodén et al., 1972; Richmond et al., 1973; Silbert and Repucci, 1976). When TM is added to chick embryo fibroblast cultures, there is a marked inhibition of $H_2{}^{35}SO_4$ into acid mucopolysaccharides (Takatsuki et al., 1977b), but there is no indication if this is a direct or indirect effect.

Although there was a marked inhibition of glycosaminoglycan synthesis by TM, Takatsuki et al. (1977b) did not investigate the effect of TM on the individual glycosaminoglycans. The involvement of lipid–saccharide intermediates for the biosynthesis of individual glycosaminoglycans has now been reported by Hart and Lennarz (1978). Cornea explants were treated with D-[6-^{3}H]glucosamine and $^{35}SO_4{}^{2-}$ in the presence and absence of TM. The amount of radioactivity in each class of glycosaminoglycan was determined (Fig. 1.8). Whereas TM essentially abolishes the biosynthesis of the polysaccharide chain of corneal keratan sulfate, the synthesis of the polysaccharide chains of chondroitin sulfate, heparan sulfate, and hyaluronic acid is unaffected.

Takatsuki et al. (1978b) observed that TM does not affect the biosynthesis of hyaluronic acid, but strongly inhibits the biosynthesis of heparan sulfate in cultures of chick embryo fibroblasts (Table 1.1). The explanation for the discrepancies awaits further studies. Participation of lipid-linked

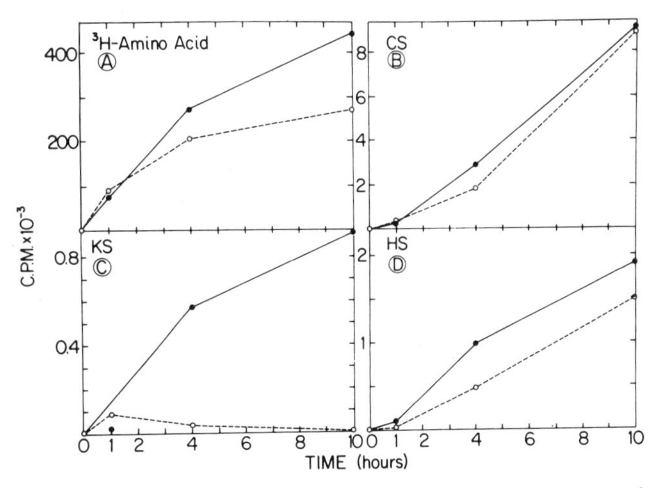

Figure 1.8 Time course of the incorporation of [^{3}H]amino acids or $^{35}SO_4{}^{2-}$ and D-[6-^{3}H]glucosamine into proteins and glycosaminoglycans of chick corneas in the presence or absence of tunicamycin. Only the data for ^{3}H are shown. (*A*) incorporation of [^{3}H]amino acids into protein; (*B*) incorporation of [^{3}H]glucosamine into chondroitin sulfates (CS); (*C*) incorporation of [^{3}H]glucosamine into high-sulfated keratan sulfate (KS) fractions; (*D*) incorporation of [^{3}H]glucosamine into heparan sulfates (HS): with tunicamycin (○), without tunicamycin (●). Reprinted with permission from Hart and Lennarz, 1978.

Table 1.1 Enzymatic Analysis of Glycosaminoglycan (GAG) by Culture of Chick Embryo Fibroblasts (CEF) in the Presence of Tunicamycin (TM)

		Total		HA[a]			CA[a]			HS[a]		
	TM (μg/ml)	cpm	Percent Inh.	cpm	Percent Comp.	Percent Inh.	cpm	Percent Comp.	Percent Inh.	cpm	Percent Comp.	Percent Inh.
Expt. 1	Control	184,208	—	78,475	42.6	—	56,067	30.4	—	49,666	27.0	—
	1.0	113,600	61.7	69,069	60.8	12.0	32,799	28.9	41.5	11,752	10.3	86.3
Expt. 2	Control	29,471	—	7,191	24.4	—	12,495	42.4	—	9,784	33.2	—
	0.1	18,039	38.8	7,089	39.3	1.5	6,440	35.7	48.5	4,510	25.0	53.9
Expt. 3	Control	10,563	—	2,958	28.0	—	4,162	39.4	—	3,507	33.2	—
	0.1	7,912	25.1	3,244	41.0	–10.3	3,020	38.2	26.2	1,598	20.2	—

Reprinted with permission from Takatsuki et al., 1978b.

[a] ^{3}H-GlcN-labeled GAG biosynthesized by CEF were prepared and analyzed by enzymatic digestion as described in the text. Distribution of ^{3}H among GAG was expressed as counts per minute (cpm) and percent composition (comp.). Effect of TM on the biosynthesis was expressed as percent inhibition (inh.). HA, hyaluronic acid; CS, chondroitin sulfate; HS, heparan sulfate.

intermediates in the biosynthesis of some mucopolysaccharides is strongly suggested by the inhibition with TM (Takatsuki et al., 1977b, 1978b; Hart and Lennarz, 1978) and also by the formation of lipid-linked glucuronic acid (Turco and Heath, 1977; Hopwood and Dorfman, 1977; Takatsuki et al., 1978b).

1.2 SINEFUNGIN

In the search for analogs with antifungal activity, a new strain of *Streptomyces griseolus* (NRRL 3739) was isolated from a soil sample collected on the Ivory Coast region of Africa. This strain produces the antifungal nucleoside analogs, sinefungin (A9145), A9145A, and A9145C (Fig. 1.9) (Boeck et al., 1973; Hamill and Hoehn, 1973; Gordee and Butler, 1973; Hamill et al., 1977; Nagarajan et al., 1977). These nucleosides are unique in that they contain adenine, a ribofuranosyl moiety similar to *S*-adenosylmethionine (Nagarajan et al., 1977), and exert their action by inhibiting methyltransferase reactions (Fuller and Nagarajan, 1978). They are potent inhibitors of fungi, plant disease fungi, parasites, viruses, and cancer (Nagarajan, personal communication). A9145E and A9145H are produced from sinefungin (Nagarajan et al., 1977).

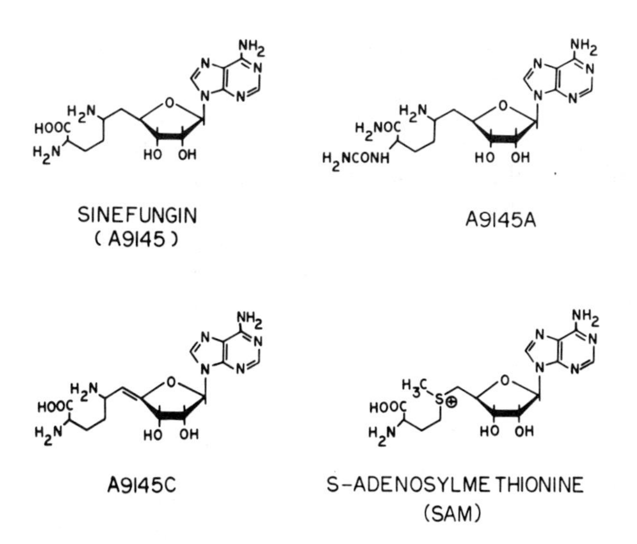

Figure 1.9 Structures of sinefungin (A9145), A9145A, and A9145C and comparison with *S*-adenosylmethionine (Nagarajan et al., 1977).

Discovery, Production, and Isolation

Sinefungin is produced by a new strain of *S. griseolus* in the medium described (Boeck et al., 1973; Hamill and Hoehn, 1973). *Saccharomyces pastorianus* (ATCC 2366) is the test organism in a turbidimetric assay. Addition of adenine increased the yield of sinefungin (Boeck et al., 1973). The isolation procedure has been described (Hamill and Hoehn, 1973; Hamill et al., 1977). Sinefungin has been crystallized from acetone–water and crystallizes with 1 mole of acetone of crystallization (Nagarajan, personal communication).

Physical and Chemical Properties

The molecular formula for sinefungin is $C_{15}H_{23}N_7O_5$; mol wt 381; mp 190–192°C; $[\alpha]_D^{25}$ +13.4° (*c* 1.0, 0.001 *M* KOH); ultraviolet spectral properties: $\lambda_{max}^{H_2O}$ 260 nm (ϵ = 16,000), 208 nm (ϵ = 22,000). No shift in acid or base is observed. Potentiometric titration (80% Methyl Cellosolve): initial pH 9.5, pK_a's at 2.5, 3.3, 8.1, and 10.1. Infrared spectrum (mull): 3380, 3200 (OH, NH_2), 2800–2600 (COOH), 1615 (C=N, C=C) (Nagarajan, personal communication).

Structural Elucidation

The nmr and cmr spectra of sinefungin suggest the presence of adenine and sugar residues. Sinefungin is slowly converted to A9145E and A9145H (Fig. 1.10) in 0.05 *N* NH_4OH or NaOH (pH 8–11) at room temperature and A9145C is converted to an analogous product under the same conditions. The stereochemistry of the furanose ring of A9145E has been established by nmr as the β-D-ribofuranosyl moiety (Nagarajan et al., 1977). Sinefungin and A9145C, but not A9145A, are substrates for L-α-amino acid oxidase (Hamill et al., 1977). The CD of a 1:2 Cu^{2+} complex with sinefungin also suggests the L-configuration at C-9′ (Nagarajan et al., 1977).

Inhibition of Growth

Sinefungin inhibits *Ceratostomella ulmi*, *Helminthosporium sativum*, and *Penicillium expansum*. It is a strong inhibitor of *Saccharomyces pastorianus* and *Candida tropicalis*. It also inhibits *Candida albicans* (Gordee and Butler, 1973). Efficacy of sinefungin was demonstrated in the kidney and intestinal tract of mice colonized with *C. albicans* infections. The reduction of yeast cell count decreased from 5 × 10^5/g feces to zero in 7 days (Turner et al., 1977). Sinefungin is as active as amphotericin B against *C. albicans*

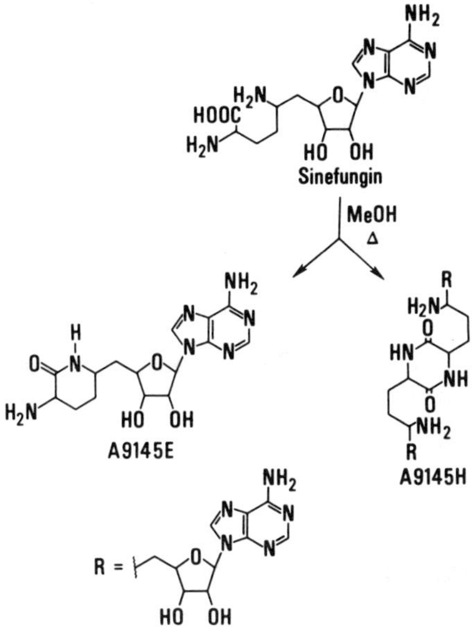

Figure 1.10 Conversion of sinefungin to A9145E and A9145H. Reprinted with permission from Nagarajan et al., 1977.

infections in mice. In terms of its activity against foliar plant diseases, as a spray it inhibits powdery mildew (*Erysiphe polygoni*), bean rust (*Uromyces phaseoli* var. *typico*), antracnose (*Collectotrichum lagenarium*), crown gall (*Agrobacterium tumefaciens*), and bacterial blight (*Xanthomonas phaseoli* var. *sojensis*). Sinefungin is effective against infections of *Trypanosoma rhodesiense*, *Trypanosoma gambiense*, and *Trypanosoma congolense*. The acute toxicity (LD_{50}) by subcutaneous injection in mice is 185 mg/kg. Sinefungin also inhibits Newcastle disease virus and vaccinia virus (Nagarajan et al., 1977; Pugh et al., 1978) (see page 23).

Biosynthesis of Sinefungin

The analog sinefungin contains an ornithine residue linked to C-5′ of adenosine. The addition of [8-^{14} C]adenosine and [U-^{14}C]ornithine to cultures of *S. griseolus* results in a 52.8 and 18.6% incorporation of ^{14}C into sinefungin, respectively (Berry and Abbott, 1978). The labeling studies suggest that a preformed adenine derivative and ornithine or a derivative of ornithine are close precursors of sinefungin. A suggested biosynthetic mechanism is shown in Fig. 1.11. The ^{14}C incorporated into sinefungin reaches a maximum and then decreases sharply. The decrease of ^{14}C in sinefungin is

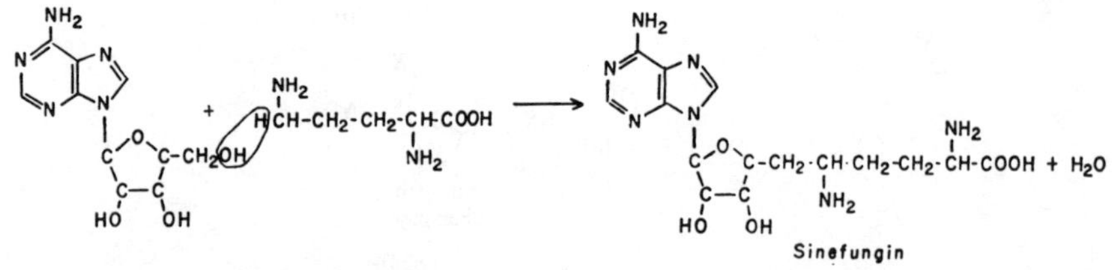

Figure 1.11 Proposed mechanism for the biosynthesis of sinefungin. Reprinted with permission from Berry and Abbott, 1978.

accompanied by an increase of ^{14}C in the closely related antibiotic, factor C. Therefore, the loss of ^{14}C from sinefungin does not appear to be due to a conversion of sinefungin to factor C.

Biochemical Properties

Inhibition of Methyltransferases. Of the four nucleoside analogs, A9145C is the most toxic against *C. albicans* (Turner et al., 1977). A9145A and A9145C are inactive *in vitro*, but are active *in vivo*. A9145E can be converted to sinefungin *in vivo*. Antibiotic activity *in vivo* was reversed by *S*-adenosylmethionine (SAM) in the culture medium. Analog C inhibited the

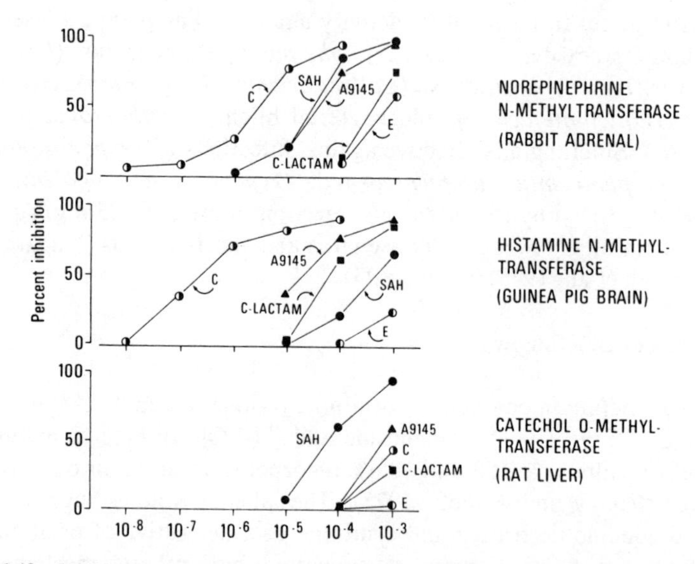

Figure 1.12 Inhibition of mammalian methyltransferases by *S*-adenosylhomocysteine (SAH) (●), sinefungin (A9145) (▲), A9145C (◑), A9145E (◐), and C-lactam (■). Reprinted with permission from Fuller and Nagarajan, 1978.

SAM mediated transmethylation as evidenced by the decreased incorporation of [^{14}C]-(CH_3)methionine into ergosterol. The selective inhibition of mammalian transferases by sinefungin, A9145C, A9145E, and C-lactam is shown in Fig. 1.12.

Sinefungin and A9145C inhibit Newcastle disease virus and are competitive inhibitors of vaccinia virion mRNA (guanine-7)-methyltransferase and vaccinia virion (nucleoside-2′)-methyltransferase. The inhibitory constants for sinefungin and A9145C were substantially less than that for *S*-adenosyl-L-homocysteine (Pugh et al., 1978).

1.3 THE EZOMYCINS

The ezomycin complex is a new group of eight antifungal bicyclic anhydrooctose uronic acid nucleoside analogs with a pseudouridine-like C—C structure. They are produced by a strain of *Streptomyces*. Their discovery was made by Takaoka et al. (1971) while screening for antibiotics against stem rot of the kidney bean plant, *Phaseolus vulgaris* L. As a result of the very elegant studies of Sakata, Sakurai, and Tamura, the physicochemical properties, structure, and inhibitory properties of eight ezomycin antibiotics are known. Ezomycins A_1 and A_2 are N-nucleosides in which the anomeric carbon of the sugar moiety is attached to N-1 of cytosine; ezomycins B_1, B_2, C_1, C_2, D_1 and D_2 are C-nucleosides in which the anomeric carbon of the sugar moiety is attached to C-5 of uracil (Fig. 1.13). The latter ezomycins are the first pseudouridine-type C-nucleoside analogs to be identified. The ezomycins are further divided into the L-cystathionine-minus (A_2, B_2, C_2, and D_2) and the L-cystathionine-containing (A_1, B_1, C_1, and D_1) analogs. Another property of the ezomycins is the difference in the α- or β-linkage between the sugar and aglycon. Ezomycins A_1, A_2, B_1, and B_2 have the β-linkage; ezomycins C_1 and C_2 have the α-linkage. The anomeric linkage in ezomycins D_1 and D_2 has not yet been established.

The octose derivative containing the bicyclic system in the ezomycins is similar to the octosyl moiety of octosyl acids A, B, and C (Isono et al., 1975a) (page 296). The amino sugars in the ezomycins are similar to the amino sugars found in puromycin, 3′-amino-3′-deoxyadenosine, the polyoxins, and aminoacyl aminohexosyl pyrimidine antibiotics. The ezomycins have limited antifungal activity against the phytopathogenic fungi.

Discovery, Production, and Isolation

Takaoka et al. (1971) first reported on the isolation of the *Streptomyces* that produce ezomycin A_1 and B_1. The composition of the culture medium is

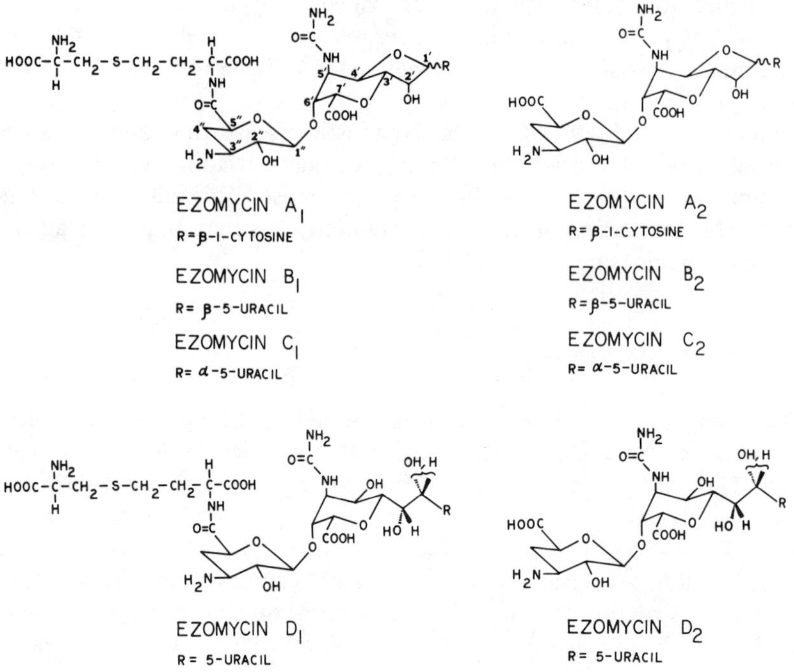

EZOMYCIN A_1
R = β-1-CYTOSINE

EZOMYCIN B_1
R = β-5-URACIL

EZOMYCIN C_1
R = α-5-URACIL

EZOMYCIN A_2
R = β-1-CYTOSINE

EZOMYCIN B_2
R = β-5-URACIL

EZOMYCIN C_2
R = α-5-URACIL

EZOMYCIN D_1
R = 5-URACIL

EZOMYCIN D_2
R = 5-URACIL

Figure 1.13 Structures of the ezomycins.

described by Sakata et al. (1973). The isolation of the eight ezomycins is shown in Fig. 1.14. Crystallization is accomplished from water.

Physical and Chemical Properties

The molecular formula, molecular weights, specific rotations and the ultraviolet spectral properties of the ezomycins are summarized in Table 1.2 (Sakata et al., 1974a, 1975a, 1975b, 1977a). The melting points of all ezomycins are above 200°C (with decomposition).

Structural Elucidation

By chemical degradations and ultraviolet, infrared, nmr, ^{13}C nmr and mass spectrometry, Sakata et al. (1973, 1974a, 1974b, 1974c, 1975b) were able to propose the structures for ezomycin A_1 and A_2 as shown in Fig. 1.13.

Treatment of ezomycin A_1 (**1**) (Scheme 1.1) with Dowex 50 (H^+) afforded L-cystathionine (**2**) and ezomycin A_2 (**3**). Acid hydrolysis of ezomycin A_2 (**3**) gave rise to the aminodeoxyuronic acid, ezoaminuroic acid (**4**), which has

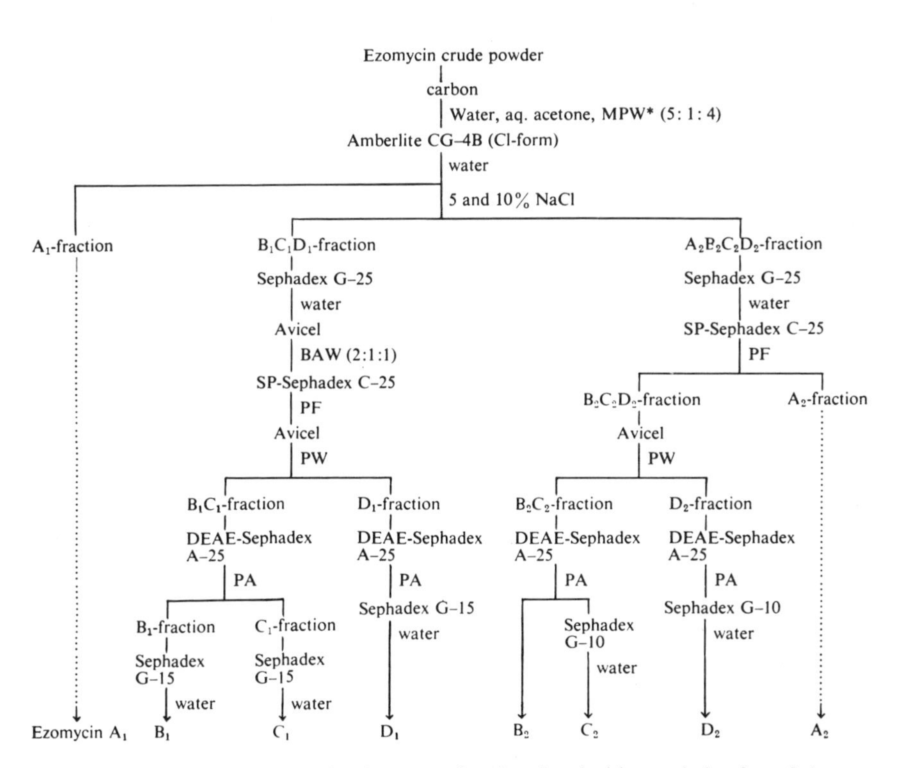

Figure 1.14 Isolation procedure for the ezomycins. Reprinted with permission from Sakata et al., 1977a.

Table 1.2 Physical and Chemical Properties of the Ezomycins

Ezomycin	Molecular Formula	Molecular Weight	$[\alpha]^{22}$ (°)	(c, H_2O)	λ_{max}^{pH} (ϵ)	λ_{max}^{pH} (ϵ)
A_1	$C_{26}H_{38}N_8O_{15}S \cdot H_2O$	752.7	+13.5	(1.05)	278 (10,500)[a]	271 (7900)[b]
A_2	$C_{19}H_{26}N_6O_{12} \cdot H_2O$	548.5	+44.4	(1.03)[c]	278 (11,100)[a]	271 (7100)[b]
B_1	$C_{26}H_{37}N_7O_{16}S \cdot H_2O$	753.7	−5.5	(0.83)	262 (7000)	286 (6100)
B_2	$C_{19}H_{25}N_5O_{13} \cdot H_2O$	549.5	+10.8	(1.02)[c]	261.5 (6300)	286 (4900)
C_1	$C_{26}H_{37}N_7O_{16}S \cdot H_2O$	753	−76.4	(0.94)	263 (6400)	287 (4900)
C_2	$C_{19}H_{25}N_5O_{13} \cdot H_2O$	549	−100	(0.52)[d]	263.5 (6200)	287 (4800)
D_1	$C_{26}H_{39}N_7O_{17}S$	753	−62.1	(0.84)	263.5 (2900)	287 (5200)
D_2	$C_{19}H_{27}N_5O_{14}$	549	+58.8	(0.78)[d]	264 (6700)	287 (4900)

From Sakata et al., 1975a, b, 1977a.
[a] Measured in 0.05 *N* HCl (Sakata et al., 1975a).
[b] Measured in 0.05 *N* NaOH (Sakata et al., 1975a).
[c] Measured at 16°C in 0.2 *N* NaOH (Sakata et al., 1977a).
[d] Measured at 18°C in 0.5 *N* NH_4OH (Sakata et al., 1977a).

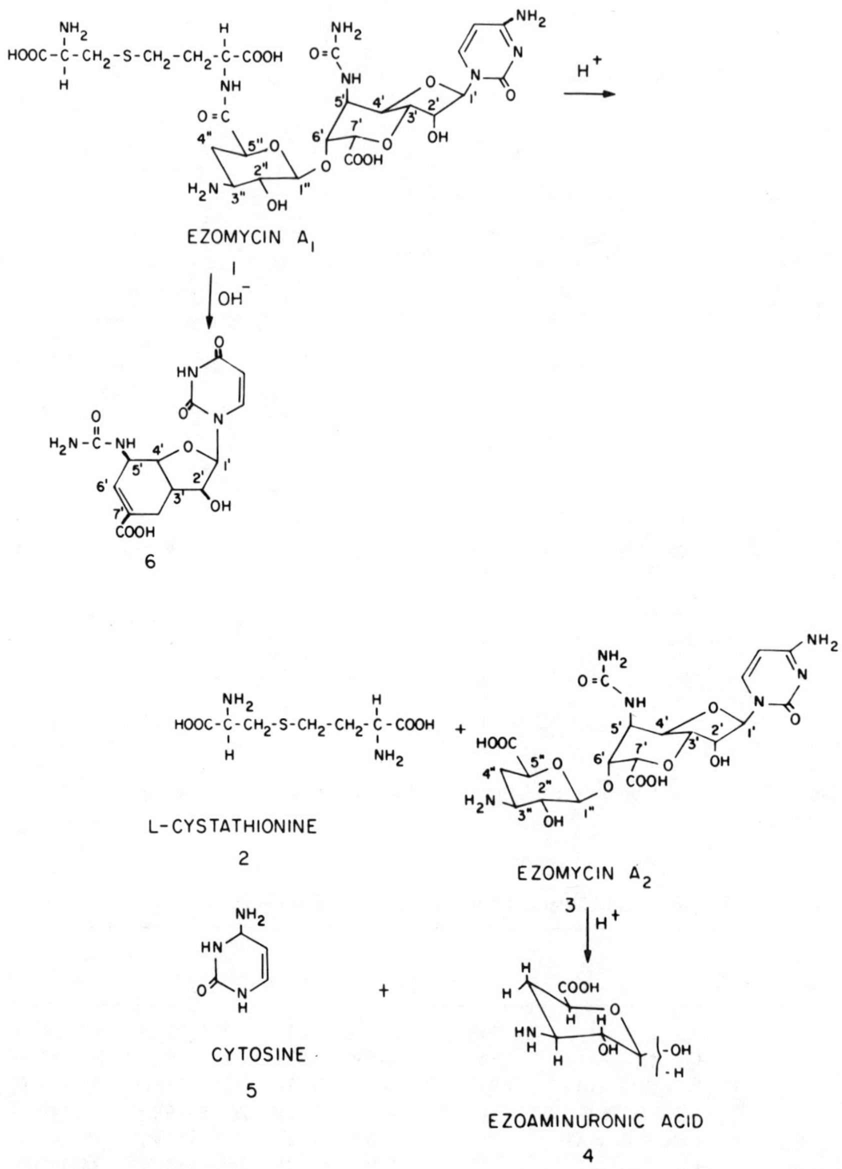

EZOMYCIN A_1

1

L-CYSTATHIONINE

2

EZOMYCIN A_2

3

CYTOSINE

5

EZOAMINURONIC ACID

4

6

Scheme 1.1 Degradation of ezomycin A_1. Reprinted with permission from Sakata et al., 1974a, 1976.

the structure 3-amino-3,4-dideoxy-D-xylohexopyranuroic acid, plus cytosine (**5**) (Sakata et al., 1974a, 1974b). This is the first example of naturally occurring 3-amino-3-deoxyhexuronic acids.

When ezomycin A_1 is treated with 1.5 *N* NaOH at 90°C, anhydrodeaminonucleoside (**6**) forms (Sakata et al., 1976). The ureido group of an *O*-carbamoyl function in anhydrodeamino nucleoside (**6**) was detected by a positive *p*-dimethylaminobenzaldehyde test (Sakata et al., 1975a). Further proof for the structure of ezomycin A_1 was obtained by periodate oxidation (Scheme 1.2). The deoxypyranose moiety of ezomycin A_1 and A_2 is indicated by the large coupling constant ($J = 11$) of the high field signals (δ = 1.55 and 2.33) of their nmr (Sakata et al., 1974b, 1975a).

Periodate oxidation of ezomycin A_1 (**1**) gave rise to the lactam amino hemiacetal (**7**) and nucleoside (**8**), which by means of the nmr and the CD spectra is assigned the structure 1-(3′,7′-anhydro-5′-deoxy-5′-ureido-D-threo-β-*D*-allooctofuranosyluronic acid) cytosine (Sakata et al., 1974a). The nmr spectra of **8** suggest that the furanose ring is fixed in the C-2′ exo or C-3′ endo conformation. H-3′ and H-4′ in the fused ring are in the 1,2-trans-diaxial relationship. This conclusion is based on their large coupling constant ($J_{3',4'} = 11$). Reduction of **7** with $NaBH_4$ gave *N*-(2′,4′-dihydroxybutyryl)-L-cystathionine sulfoxide (**9**). Acid hydrolysis of **9** gave cystathionine sulfoxide (**10**) and α-hydroxybutyro-lactone (**11**).

The absolute structures of ezomycins A_1, A_2, B_1, B_2, C_1, and C_2 were established by the ^{13}C nmr spectrum of the periodate oxidation product of ezomycin A_1 or A_2 (Fig. 1.15) (Sakata et al., 1977a, b, c; Sakata and Uzawa, 1977). These are the first reports on the complete ^{13}C nmr spectra of bicyclic anhydrooctose uronic acid nucleosides. Ezomycins A_1 or A_2, B_1 or B_2, C_1 or C_2, and A_1 can be degraded to the bicyclic anhydrooctose uronic acid nucleosides (**12**, **13**, and **14**) and the anhydrodeaminonucleoside (**15**), respectively.

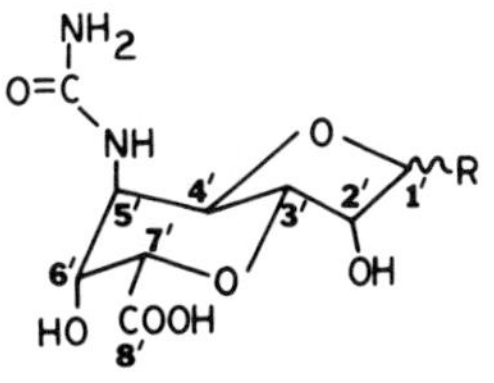

12 R=β-1-CYTOSINE
13 R=β-5-URACIL
14 R=α-5-URACIL

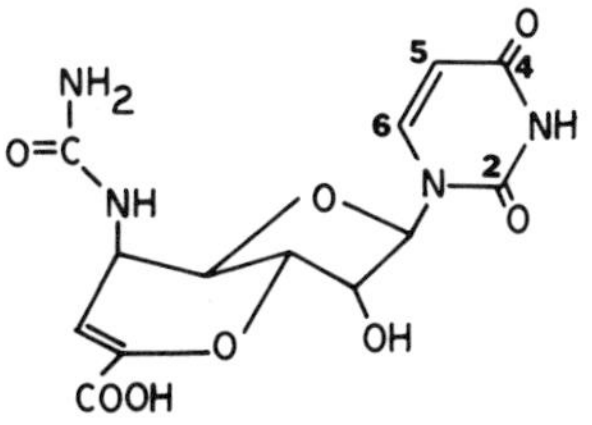

15

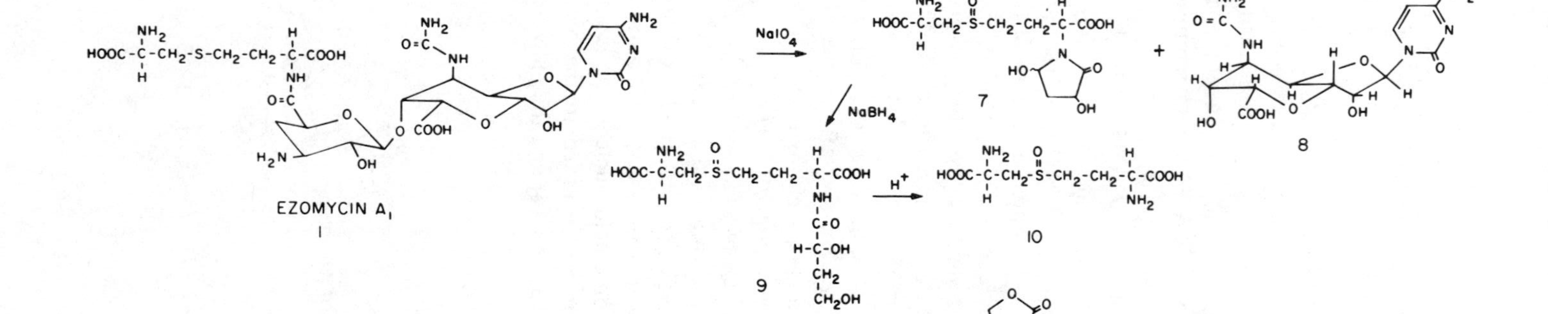

Scheme 1.2 Products of periodate, borohydride, and acid treatment of ezomycin A_1. Reprinted with permission from Sakata et al., 1974c.

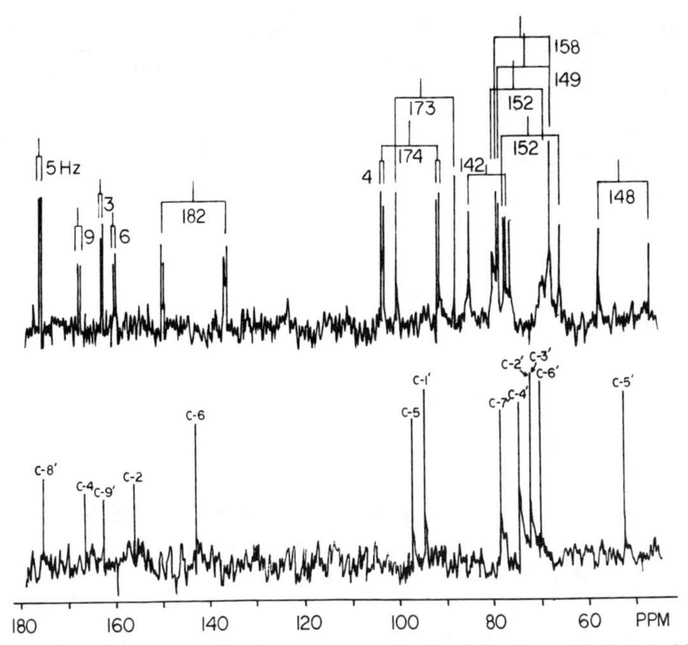

Figure 1.15 Carbon-13 nmr spectra of nucleoside (**8**) following periodate oxidation of ezomycin A_1 or A_2 (30 mg in 0.5 ml of 2% ND_3 in D_2O): (top) gated decoupled spectrum with NOE (1.2 × 20,000); (bottom) complete proton noise decoupled spectrum (2.0 s × 1000). From Sakata et al., 1977c. From *Organic Magnetic Resonance*, Vol. 10, December 1977. Published by Heyden & Son, Ltd., London.

Inhibition of Growth

The ezomycins are active against a limited species of phytopathogenic fungi. Ezomycin B_1 is most inhibitory. The fungi sensitive to the ezomycins are *Sclerotinia* and *Botrytis* sp. (Sakata et al., 1973, 1974c). The ezomycins do not inhibit yeast or bacteria.

Biochemical Properties

Although no studies have been reported on the biochemical properties of the ezomycins, the antifungal properties suggest that ezomycins A_1, B_1, and C_1 (the β-anomers) have biological activity. Ezomycin B_1 shows the strongest antifungal activity. When ezomycins B_1, C_1, and D_1 are incubated with *Sclerotinia sclerotiorum* for 14 days, there is no decrease in toxicity (Sakata et al., 1976). These results suggest that there is no interconversion between B_1, C_1, and D_1 during the bioassay. Because the naturally occurring C-nucleoside antibiotics, formycin, pyrazofurin, showdomycin, and oxazinomycin, are potent antibacterial and antineoplastic agents, the isolation of the ezomycins with the pseudouridine-type C-nucleoside linkage may further extend the inhibitory properties of these type nucleosides. The presence of the amino group on the 3-position of the hexuronic acid moiety of the ezomycins may place them in the class with gougerotin, blasticidin S, amicetin, and hikizimycin. Although the ezomycins have a uronic acid *N*-carbamoyl moiety, they do not inhibit chitin synthetase as do the polyoxins, which have the same type group (Isono et al., 1975a).

1.4 THE POLYOXINS

The polyoxin complex is a mixture of antifungal peptidylpyrimidine nucleoside antibiotics elaborated by *Streptomyces cacaoi* var. *asoensis*. The elegant work of Isono, Suzuki, and coworkers resulted in the isolation, physical properties, chemical degradations, structural elucidation, and biochemical properties of polyoxins A–N (Fig. 1.16) (for review, see Suhadolnik, 1970). Isono and coworkers (Uramoto et al., 1978) have reported on a newly isolated soil Streptomycetes, *S. piomogenus*, which produces polyoxins L and M. Apparently this strain lacks the enzyme(s) needed to incorporate the C-1 unit into C-5 of uracil (Isono and Suhadolnik, 1976) and the biosynthesis of the C-terminal polyoximic acid. They have isolated another active component. On the basis of degradation

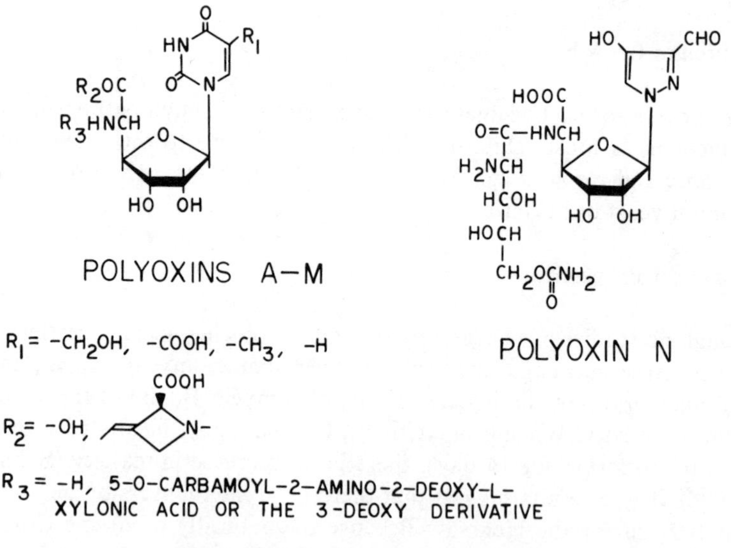

POLYOXINS A–M

POLYOXIN N

R_1 = $-CH_2OH$, $-COOH$, $-CH_3$, $-H$

R_2 = $-OH$,

R_3 = $-H$, 5-O-CARBAMOYL-2-AMINO-2-DEOXY-L-XYLONIC ACID OR THE 3-DEOXY DERIVATIVE

Figure 1.16 Structures of the polyoxins.

studies and spectral analyses, the new minor component, polyoxin N (Fig. 1.16), has the structure, 1-[5′-(5″-*O*-carbamoyl-2″-amino-2″-deoxy-L-xylonyl)-amino-5′-deoxy-D-allofuranosyl uronic acid]-3-formyl-4-hydroxypyrazole. Polyoxin N, as an *N*-nucleoside, is another example of a naturally occurring nucleoside analog having the pyrazole ring as its base. The aglycon in all the polyoxins contains either a pyrimidine or a pyrazole ring and the hexuronic acid 5-amino-5-deoxyallofuranose. Absolute proof of the nucleoside skeleton of the polyoxins has been reported by Moffatt and coworkers (Damodaran et al., 1971). Kuzuhara et al. (1973) reported on the total chemical synthesis of polyoxin J. In addition to the production of the polyoxins by *S. cacaoi*, Isono et al. (1975a) have also isolated octosyl acids A, B, and C. The structural similarity of the polyoxins to uridine diphospho-*N*-acetylglucosamine prompted Isono et al. (1969) to speculate that the polyoxins would act as competitive inhibitors for chitin synthesis. Endo and Misato (1969) subsequently reported that the polyoxins are powerful inhibitors of chitin synthesis in filamentous fungi.

Total Chemical Synthesis of Polyoxin J

Kuzuhara et al. (1973) reported on the total synthesis of polyoxin J (**I**) (Fig. 1.17). Compound **VI** (2-azido-3,4-di-*O*-benzyl-2-deoxy-1-*O*-trityl-L-xylitol) was converted to compound **VII** with $LiAlH_4$. Compound **VII** was benzoylated to form compound **VIII**, which on treatment in pyridine with *p*-

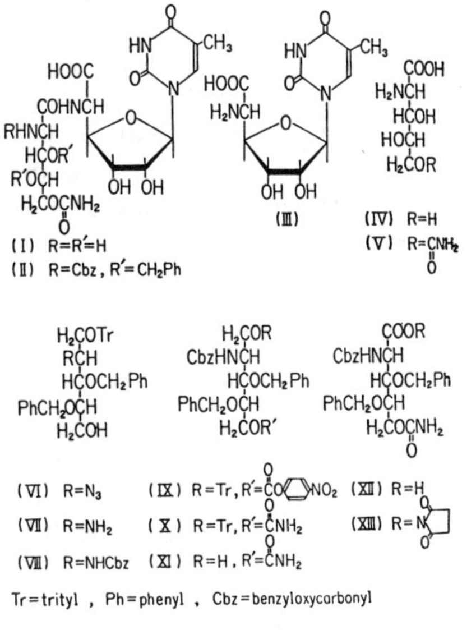

Figure 1.17 Total chemical synthesis of polyoxin J (**I**). Reprinted with permission from Kuzuhara et al., 1973.

nitrophenyl chloroformate gave derivative **IX**. Treatment of **IX** with methanolic ammonia resulted in a 5-*O*-carbamoyl derivative (**X**). Removal of the trityl group of **X** resulted in the isolation of **XI**, 3,4-di-*O*-benzyl-2-benzyloxycarbonylamino-5-*O*-carbamoyl-2-deoxy-L-xylitol. To synthesize **XII**, compound **XI** was treated with CrO_3 in aqueous sulfuric acid. The *N*-hydroxysuccinimide ester (**XIII**) was prepared by treating compound **XII** in the presence of dicyclohexylcarbodiimide and *N*-hydroxysuccinimide. Polyoxin J was synthesized by the addition of deoxypolyoxin C (**III**) to a solution of **XIII**. The coupled product, compound **II**, was hydrogenated with aqueous methanol and Pd–C. The overall yield on the basis of compound **XII** through the coupling reaction was 28%. Synthetic polyoxin J (**I**) showed equivalent inhibition with that of natural polyoxin J against several species of phytopathogenic fungi.

Polyoxin Analogs

In an attempt to determine the role of the 5′-amino group of the polyoxins, Isono et al. (1971) reported that the carboxyl group of C-5′ of the polyoxins is an absolute requirement for inhibition of chitin synthetase. In a more recent paper, Isono and coworkers (Azuma et al., 1977a) synthesized a number of aminoacyl derivatives of polyoxin C and polyoxin L. Antifungal activities were tested with several lines of phytopathogenic fungi. The α-L-amino group is absolutely essential for activity. Addition of an alkyl group to the 5′-amino group of polyoxin C and L alters the inhibitory activity. None of the aminoacyl derivatives reported shows a broader antimicrobial spectrum than do the natural polyoxins. Hori et al. (1971, 1974a, 1974b) have also studied the relation of the polyoxin structure to chitin synthetase. Similarly, Isono and coworkers (Shibuya et al., 1972) chemically modified the 5-carboxyuracil of polyoxins D, E, and F by treatment with sodium bisulfite. The dihydrouracil-6-sulfonates showed reduced antifungal activity. The adenine analogs of the polyoxins have been prepared by an improved transnucleosidation reaction (Azuma et al., 1977b).

Biosynthesis of the Polyoxins and the Octosyl Acids

Although thymine is the pyrimidine chromophore of polyoxin H, exogenously supplied thymine cannot be utilized by *S. cacaoi* for the biosynthesis of the thymine in the polyoxins. Furthermore, the synthesis of dTTP from [2-^{14}C]uracil by either 5-fluorouracil (an inhibitor of thymidylate synthetase) or 1-formylisoquinoline (an inhibitor of ribonucleotide reductase) is inhibited, whereas the incorporation of ^{14}C from [2-^{14}C]uracil into the thymine of polyoxin is not inhibited (Isono and Suhadolnik, 1976).

The methyl group of the thymine in polyoxin H is formed from C-3 of serine, which serves as the one-carbon donor; the methyl group of methionine is not utilized. The methyl group must add to the uracil chromophore at some step near the completion of the biosynthesis of polyoxin H (Scheme 1.3). Therefore, the thymine in polyoxin H can form by a pathway independent of thymidylate synthetase. Funayama and Isono (1976) and Isono et al. (1978) have demonstrated that C-3 of serine also serves as the C-1 precursor for the biosynthesis of the hydroxymethyl group at C-5 of uracil in polyoxins A, B, G, H, I, J, K, L, and M and for the 5-carboxyuracil-type polyoxins D, E, and F.

S. cacaoi can synthesize unnatural polyoxins as evidenced by the incorporation of 5-fluoro-, 5-bromo-, and 6-azauracil into the polyoxins (Isono et al., 1973; Isono and Suhadolnik, 1976). 5-Fluoropolyoxin L and 5-fluoropolyoxin C inhibit *Escherichia coli* and *Streptococcus faecalis*.

The biosynthesis of the carbon skeleton of polyoximic acid, 3-ethylidene-L-azetidine-2-carboxylic acid, in polyoxins A, F, H, and K has also been described (Isono et al., 1975b). This unique cyclic amino acid is the *C*-terminal amino acid of the peptide bond of the polyoxins. The $^{3}H/^{14}C$ ratio of the [1-^{14}C;4,5-^{3}H]isoleucine added to the polyoxin-producing *S. cacaoi* was 4.24. The theoretical ratio of the $^{3}H/^{14}C$ in the 3-ethylidene-L-azetidine-2-carboxylic acid when corrected for the loss of one tritium atom is 3.39; experimentally the ratio is 3.42. Therefore, four of the five hydrogens of the

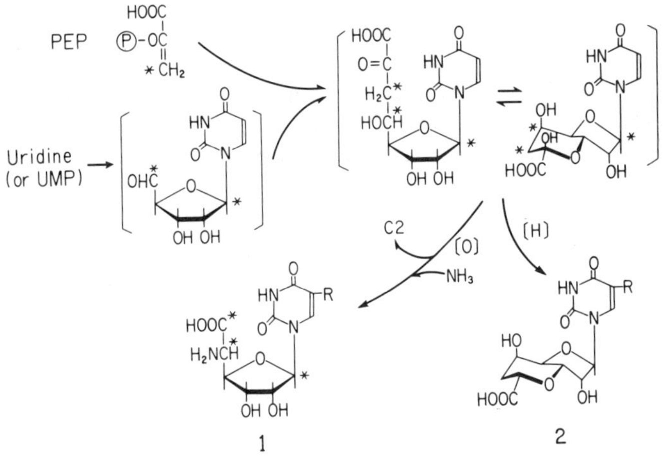

PEP HOOC Ⓟ-OC *CH₂ Uridine (or UMP) OHC* HN O N OH OH HOOC O=C H₂C* HOCH* OH OH OH HOOC C2 [O] NH₃ [H] HOOC* H₂NCH* R HO 1 2

Scheme 1.3 Proposed biosynthetic pathway for the nucleoside skeleton of the polyoxins. (R = $-CH_3$, $-CH_2OH$, $-COOH$) and the octosyl acids (R = $-COOH$, $-CH_2OH$). Asterisk shows the ^{13}C enrichment from D-[1-^{13}C]glucose. Incorporation stage of the one-carbon unit into C-7 is not known. Reprinted with permission from Isono et al., 1978.

ethyl group of isoleucine are retained in the formation of the ethylidene group, and the carbon–nitrogen skeleton of isoleucine is incorporated intact into 3-ethylidene-L-azetidine-2-carboxylic acid. The incorporation of L-isoleucine into 3-ethylidene-L-azetidine-2-carboxylic acid can be visualized as shown in Scheme 1.4.

The biosynthesis of 2-amino-2-deoxy-L-xylonic acid, which makes up the *N*-terminal amino acid of the peptide bond of the polyoxins, requires glutamate (Funayama and Isono, 1975). The δ-carboxyl group of glutamate is reduced to the hydroxymethyl group. Whereas there is a complete loss of ^{3}H on C-2 of [2-^{3}H]glutamate incorporated into this aminoaldonic acid, there is also a 65% loss of the tritium from C-2 of the glutamate that is incorporated into the cellular protein. Therefore, there must be a rapid equilibrium between glutamate and α-ketoglutarate.

Ammonium sulfate fractionation of cellfree extracts of *S. cacaoi* show that [^{14}C]carbamoyl phosphate is transferred by α-amino-δ-hydroxyvaleric acid carbamoyl transferase to C-4 of α-amino-δ-hydroxyvaleric acid (Scheme 1.5). The oxidation of C-3 and C-4 of glutamate to give 5-*O*-carbamoyl-2-amino-2-deoxy-L-xylonic acid is not known. 2-Amino-4,5-dihydroxyvaleric acid, 4-hydroxyglutarate, and 3,4-dihydroxyglutamic acid have been isolated from plants (Larsen, 1967; Virtanen and Ettala, 1957; Hatanaka, 1962; Brandner and Virtanen, 1963). [1-^{14}C]Polyoximic acid, 3-ethylidene-L-azetidine-2-carboxylic acid is efficiently incorporated into the polyoxins (Isono, unpublished data), whereas [^{14}C]polyoxamic acid is not incorporated.

Finally, Isono et al. (1978) have reported on the biosynthesis of the unique aminouronic acid, 5-amino-5-deoxy-D-allofuranoseuronic acid, in the polyoxins (Scheme 1.3). Phosphoenolpyruvate (PEP) condenses with uridine as the 5′-aldehyde, to form octofuranuloseuronic acid as the intermediate. Oxidative elimination of the two terminal carbon atoms forms the carbon

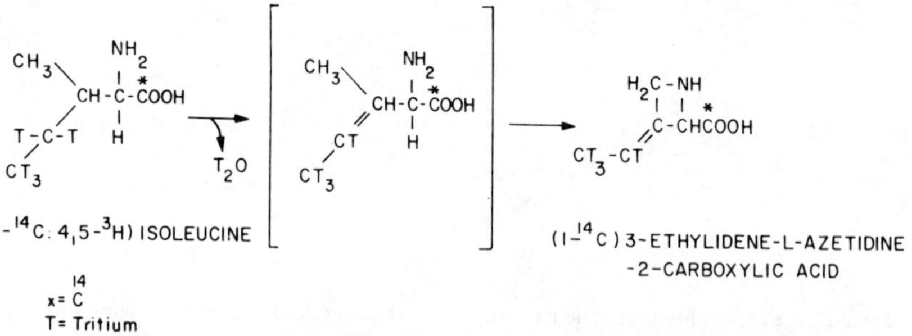

(1-^{14}C: 4,5-^{3}H) ISOLEUCINE

T_2O

(1-^{14}C) 3-ETHYLIDENE-L-AZETIDINE-2-CARBOXYLIC ACID

x = ^{14}C
T = Tritium

Scheme 1.4 Hypothetical biosynthetic pathway of 3-ethylidene-L-azetidine-2-carboxylic acid. Modified from Isono et al., 1975b, reprinted with permission.

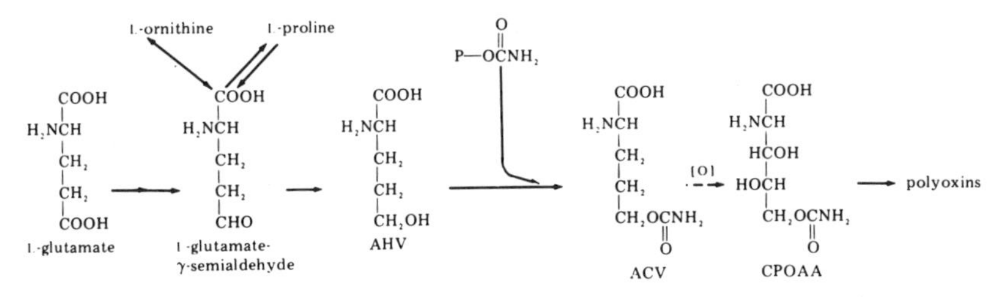

Scheme 1.5 Biosynthetic pathway for 5-*O*-carbamoyl-2-amino-2-deoxy-L-xylonic acid (carbamoylpolyoxamic acid, CPOAA). Reprinted with permission from Funayama and Isono, 1977.

skeleton of the nucleoside moiety of the polyoxins (Scheme 1.3, compound **1**). The octosyl acids (**2**) could be formed by reduction of C-7′ of the intermediate. The glycolytic pathway, and not the one-carbon pool, contributes to the biosynthesis of C-6′. The use of PEP for the formation of 3-deoxy-D-arabino-hepturosonate-7-phosphate, the precursor of the shikimate pathway, occurs by condensation of PEP with the anomeric carbon (Wood, 1972). With the polyoxin complex, PEP condenses with the C-5 aldehyde of uridine instead of the anomeric carbon of uridine.

Biochemical Properties

Antifungal Insecticidal Activity. The polyoxins have been used as practical fungicides in Japan. Minor structural changes of the polyoxins are important in their inhibitory properties. For example, polyoxin D, with its 5-carboxyuracil aglycon, has found wide application in the treatment of the *Pellicularia* disease of rice plants (Shibuya et al., 1972). It is only moderately inhibitory to *Alternaria kikuchiana*. Polyoxin L, with its uracil aglycon, has found application in the treatment of *Alternaria* disease of fruit trees (Isono et al., 1967). Therefore, there is a great difference in the sensitivity of phytopathogenic fungi to the polyoxins with the 5-carboxyuracil chromophore compared with the uracil chromophore. Polyoxins also affect the mycelia of *Alternaria kikuchiana* by the formation of interhyphae (Ishizaki et al., 1974, 1976). Polyoxin D also inhibits the germination of spores of *Trichoderma viride* (Benitez et al., 1976). The polyoxins are very effective insecticides against immature insects (Vardanis, 1977).

Inhibition of Chitin Synthetase. The first suggestion concerning the action of the polyoxins was in the reports by Isono and Suzuki (1968) and Eguchi et al. (1968) in which they showed that the polyoxins caused fungal cell walls to swell. Subsequently, other independent studies showed that the polyoxins are powerful competitive inhibitors of chitin UDP-*N*-acetylglucosaminotransferase (chitin synthetase) in cellfree extracts from *Piricularia oryzae*, *Sacchromyces cerevisiae*, and *Aspergillus flavus* (Keller and Cabib, 1971; Cabib and Keller, 1971; Endo and Misato, 1969; Gooday et al., 1976; Bartnicki-Garcia and Lippman, 1972; Ohta et al., 1970; Lopez-Romero and Ruiz-Herrera, 1976). Polyoxin D is a competitive inhibitor, $K_i = 6.5\ \mu M$. This compares with a K_m of 18 μM for UDP-*N*-acetylglucosamine with chitin synthetase (Lopez-Romero and Ruiz-Herrera, 1976). Keller and Cabib (1971) reported inhibition constants of chitin synthetase for polyoxins A, B, and L of 8 μM. Studies by Isono et al. (1973) with the unnatural polyoxin, 5-fluoropolyoxin L, have shown a K_i of 5.5 μM. This compares with a K_i of 7 μM for polyoxin L. The recent studies of Hori et al. (1971, 1974a, 1974b) using the polyoxins and polyoxin analogs have shown that the pK_i–pH plots for these competitive inhibitors can be divided as follows: the ionized group of the C-2″ position, the carbonyl oxygen atom at C-1″, and the carbamoyloxy group (Fig. 1.18).

In a polyoxin-resistant mutant of *A. kikuchiana*, Hori et al. (1974b) reported that cell wall synthesis was not inhibited by polyoxin B. However, with cellfree extracts chitin synthetase was equally sensitive to polyoxin B in the resistant strain as in the sensitive strain. It was found that carbon-14 labeled polyoxins A, B, C, and I show a decreased uptake by the polyoxin-resistant strain (Hori et al., 1975, 1976). Therefore, the resistance is not attributed to an increased concentration of UDP-*N*-acetylglucosamine, which is competitive against the polyoxins. Hori et al. (1977) extended our knowledge of the polyoxins by showing that the antifungal activity of polyoxin A against *A. kikuchiana* disappears when dipeptides are added. The dipeptides are competitive inhibitors of polyoxin uptake.

As is usually the case, there are several techniques that can be used to overcome the resistance of the drug to the inhibitor. With the polyoxins, this was accomplished first by Isono and coworkers, who reported on an efficient transnucleosidation reaction in which they prepared the adenine analog of the polyoxins (Azuma et al., 1976, 1977b). A second approach involved the biosynthesis of polyoxins in which the uracil aglycon was substituted with 5-fluorouracil (Isono et al., 1973). A third method to overcome drug resistance is the chemical transformation of the inhibitor (Brink et al., 1976). This was accomplished by reacting an amino acid sugar derivative with N-3′ of silylated uracil. Isono et al. (1972) have reported the successful transformation of the 5-carboxyl-uracil polyoxins by treatment with aqueous sodium bisulfite to give the corresponding 5-carboxyl-5,6-dihydrouracil-6-sulfonate derivative.

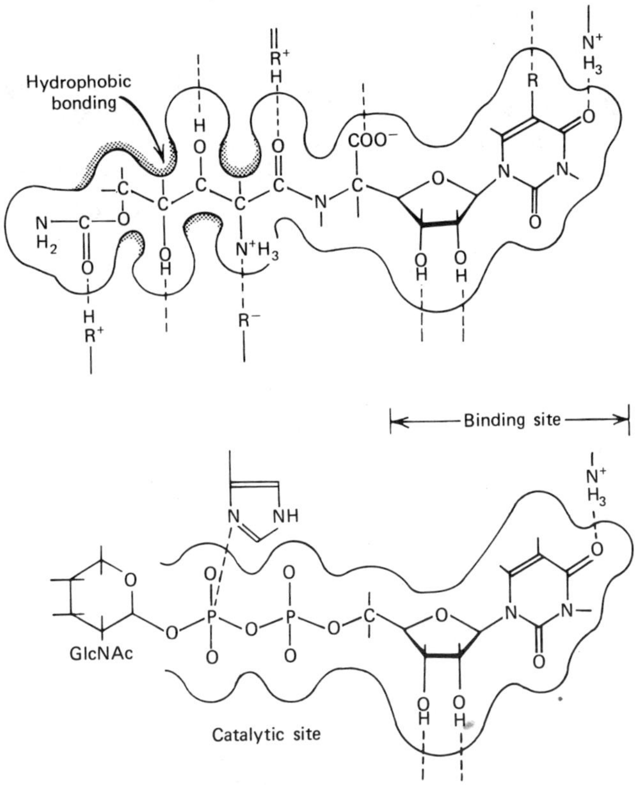

Figure 1.18 Proposed mechanism of interaction between polyoxin (top) or UDP-*N*-acetylglucosamine (bottom) and the active center of chitin synthetase. Reprinted with permission from Hori et al., 1974b.

The Study of Hyphal Growth with Polyoxin D. Nucleated protoplasts of *Schizophyllum commune* can regenerate their walls and 50% can revert to hyphal mode of growth. Of the three main wall components (*S*-glucan, *R*-glucan, and chitin), polyoxin D inhibited the synthesis of chitin and *R*-glucan (deVries and Wessels, 1975; van der Valk and Wessels, 1976). The reversion to hyphal growth was also inhibited. The polyoxins inhibit the mycelia of *A. kikuchiana* by the formation of interhyphae (Ishizaki et al., 1974, 1976). Polyoxin D also inhibits the germination of spores of *Trichoderma viride* and *A. kikuchiana* (Benitez et al., 1976). Mycelium growing in the presence of polyoxin D was irregular. The biosyntheses of β-(1,3)-glucan and chitin are inhibited.

1.5 THRAUSTOMYCIN

Kneifel et al. (1974) reported on the isolation of thraustomycin, an antifungal nucleoside analog, from *Streptomyces exfoliatus* (Fig. 1.19). Although the exact structure of thraustomycin has not been established, it is composed of equal amounts of adenine, L-leucine, and an unusual tetrahydroxymonocarboxylic acid that has properties related to a carbohydrate (Fig. 1.19). Thraustomycin is a potent inhibitor of fungi, especially *Mucor hiemalis* (+).

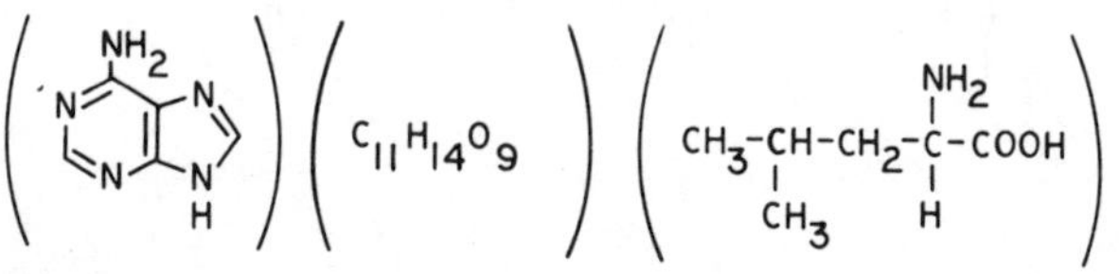

Figure 1.19 Structure of thraustomycin (Kneifel et al., 1974).

Discovery, Production, and Isolation

Kneifel et al. (1974) isolated thraustomycin from *S. exfoliatus*. The preparation of the spore solution and fermentation medium and isolation are described. A second substance, β-thraustomycin, a decomposition product of thraustomycin, was used for most of the structural studies.

Physical and Chemical Properties

The molecular formula for β-thraustomycin is $C_{16}H_{17}N_5O_8$; mol wt 407 (from mass spectral studies). Thraustomycin is a white substance that decomposes upon heating without melting; the ultraviolet spectral properties are: $\lambda_{max}^{pH\,7}$ 259 nm ($E_{1\,mg/ml}^{259}$ 25.0); thraustomycin is readily converted to the inactive β-thraustomycin (minus the L-leucine moiety) following treatment with dilute HCl or ammonia (Kneifel et al., 1974).

Structural Elucidation

Although the exact structures of thraustomycin and β-thraustomycin are not known, Kneifel et al. (1974) have obtained extensive information on thraustomycin by way of high resolution mass spectrometry of the TMS derivative of β-thraustomycin (Fig. 1.20). The corresponding fragmentation scheme is shown in Fig. 1.21. The TMS derivative of β-thraustomycin

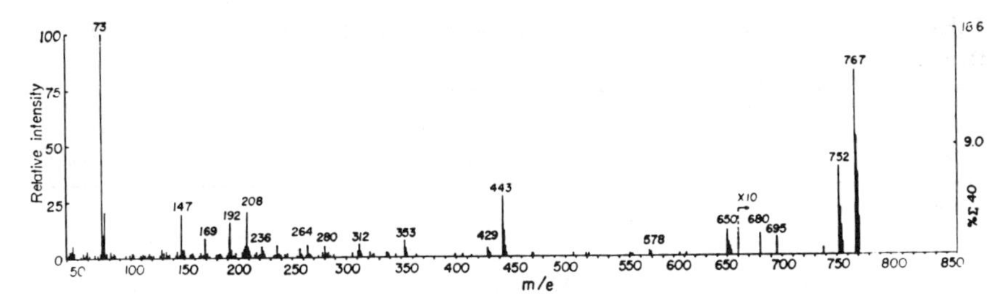

Figure 1.20 Mass spectrum of the TMS derivative of β-thraustomycin. Reprinted with permission from Kneifel et al., 1974.

shows a molecular ion at *m/e* 767. A difference of 203 mass units between the TMS derivatives of thraustomycin and β-thraustomycin implies the loss of a leucine residue with one TMS group. The presence of a carboxylic group was ascertained by the formation of the methyl ester by way of diazomethane. The mass spectrometry fragmentation patterns of the TMS derivatives of β-thraustomycin are very similar to those of psicofuranine and decoyinine. Acid hydrolysis of β-thraustomycin in 3 *N* HCl (2 h, 100°C) forms adenine and two isomeric component products, probably sugar derivatives. The molecular formula of the sugar component is $C_{11}H_{14}O_9$. The sugar has one carboxyl group, one glycosidic hydroxyl group, and three other hydroxyl groups. Kneifel et al. (1974) suggested that thraustomycin is an adenine nucleoside with a hydroxymethyl group on C-1′. The mass spectrometry data indicate that the adenine is substituted on the N-9 position and the leucine is connected to the sugar component.

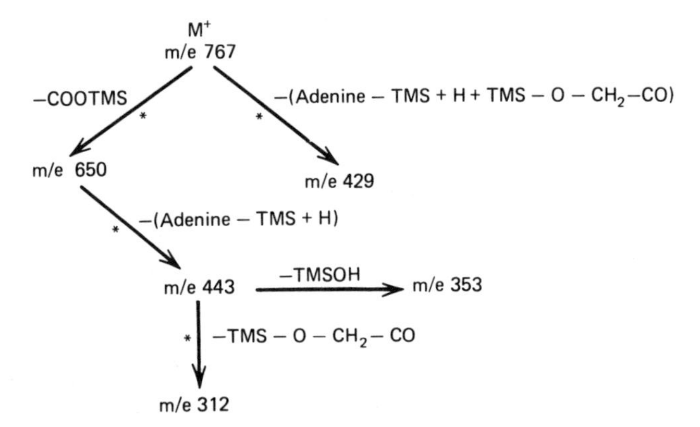

Figure 1.21 Fragmentation scheme of the TMS derivative of β-thraustomycin. Reprinted with permission from Kneifel et al., 1974.

Table 1.3 Minimal Inhibitory Concentration (MIC) of Thraustomycin in the Plate Diffusion Test

Strain	MIC (mg/ml)	Strain	MIC (mg/ml)
A. Bacteria			
		Mucor parvisporus	30~100
Escherichia coli	>100	*Mucor racemosus* (+)	10~30
Escherichia coli[a]	>100	*Mucor racemosus* (–)	10~30
Bacillus subtilis	>100	*Phycomyces blakesleeanus* (+)	10~30
Bacillus subtilis[a]	>100		
Pseudomonas saccharophilia	>100	*Phycomyces blakesleeanus* (–)	10~30
Streptomyces viridochromogenes	10~30	*Rhizopus circinans* (+)	10~30
		Rhizopus circinans (–)	10~30
		Zygorhynchus moelleri	30
B. Fungi			
1. Zygomycetes		2. Ascomycetes	
Blakeslea trispora (+)	3~10	*Aspergillus Melleus*	>100
Blakeslea trispora (–)	3~10	*Botrytis cinerea*	30~100
Mucor hiemalis (+)	0.1~0.3	*Candida vulgaris*	>100
Mucor hiemalis (–)	30~100	*Paecilomyces varioti*	>100
Mucor mucedo (+)	0.1~0.3	*Piricularia oryzae*	10~30
Mucor mucedo (–)	0.1~0.3	*Saccharomyes cerevisiae*	3
Mucor luteus (+)	10~30	*Spicaria* sp.	10
Mucor luteus (–)	1~4		

Reprinted with permission from Kneifel et al., 1974.
[a] Synthetic medium.

Biochemical Properties

Thraustomycin inhibits the germination of the fungus *M. hiemalis* (+). It shows no activity against bacteria (Kneifel et al., 1974) (Table 1.3).

1.6 AMIPURIMYCIN

In a continuation of the screening program for useful antifungal analogs, Takeda Chemical Industries isolated a new adenosine analog from a soil sample collected in Rae, Papua, New Guinea. The name amipurimycin was

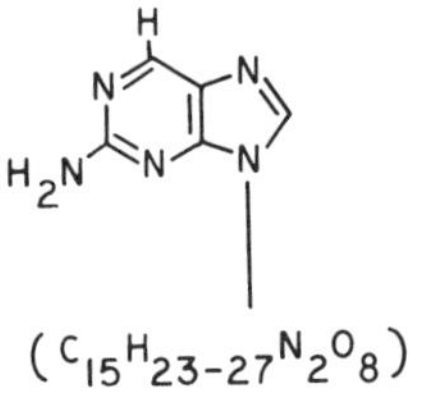

Figure 1.22 Structure of amipurimycin.

given to this analog because it contains the 2-aminopurine chromophore plus a $C_{15}H_{23\text{-}27}N_2O_8$ fragment (Fig. 1.22). This is the first naturally occurring nucleoside analog containing the aglycon, 2-aminopurine.

Discovery, Production, and Isolation

Amipurimycin is produced in the fermentation broth of *Streptomyces novoguineensis* T-36496 as described by Iwasa et al. (1977). The isolation as described by Harada and Kishi (1977) is based on combination of ion exchange and adsorption chromatography. The nucleoside crystallizes from ethanol–water as colorless prisms.

Physical and Chemical Properties

The molecular formula for amipurimycin is $C_{20}H_{27\text{-}31}N_7O_8$; mp 217°C (decomp.); $[\alpha]_D$ −3.2° (in H_2O); ultraviolet spectral properties: $\lambda_{max}^{H_2O}$ 218 nm ($E_{1\,cm}^{1\,percent}$ 429), 243 (116), 305 (130); $\lambda_{max}^{0.1\,N\,NaOH}$ 243 nm ($E_{1\,cm}^{1\,percent}$ 108), 305 (136); $\lambda_{max}^{0.1\,N\,HCl}$ 222 nm ($E_{1\,cm}^{1\,percent}$ 634), 244 (shoulder), 313 (78) (Fig. 1.23); the mobilities of amipurimycin by thin layer chromatography and paper electrophoresis have been reported (Harada and Kishi, 1977). The pK_a' values are estimated at 3.7 and 9.1. Amipurimycin is soluble in water, slightly soluble in dimethyl sulfoxide, alcohols, pyridine, and acetic acid, and insoluble in organic solvents. In the nmr spectrum, amipurimycin signals were observed at 8.38 and 8.36 (H-6, H-8) and 5.98 ppm (H-1). Almost identical signals were obtained with authentic 2-aminopurine riboside. The ^{13}C nmr spectrum of amipurimycin in deuterium oxide contained signals associated with the purine base at 159.9 ppm (s, C-2), 153.2 (s, C-4), 126.3 (s, C-5), 149.7 (d, C-6), and 142.8 (d, C-8).

Structural Elucidation

Although the total structure of amipurimycin is not established, all the physical and chemical data indicate that the chromophore is 2-aminopurine.

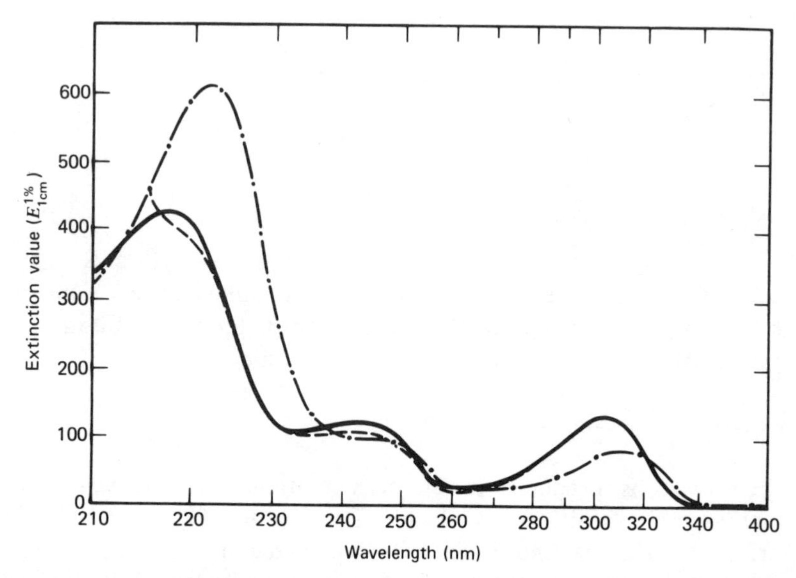

Figure 1.23 Ultraviolet spectrum of amipurimycin. (——) Buffer (pH 7), (- - -) 0.1 *N* NaOH, (—·—) 0.1 *N* HCl. Reprinted with permission from Harada and Kishi, 1977.

Amipurimycin appears to be similar to miharamycin A and B in that they both inhibit the growth of *P. oryzae in vitro* and *in vivo* (Tsuruoka et al., 1967; Shomura et al., 1967; Iwasa et al., 1977). However, the miharamycins differ in a Sakaguchi reaction, specific rotation, elemental analysis, molecular formula, and antibacterial action against pseudomonas (Iwasa et al., 1977). All the chemical, physical, and biological data suggest that amipurimycin is a new analog.

Inhibition of Growth and Biochemical Properties

Amipurimycin is effective against some pathogenic fungi. At 16 ppm it controls blast of rice plants. Although blasticidin S and kasugamycin also inhibit rice blast, Iwasa et al. (1977) found that the antimicrobial spectra of amipurimycin and these two antiblast antibiotics are not identical. They described in detail the taxonomical studies of the organism, control of blast of rice plant, and the antimicrobial spectrum of amipurimycin. Amipurimycin shows no antimicrobial activity against gram-positive and gram-negative bacteria, yeast, and saprophytic fungi. The toxicity of amipurimycin to the Kilifish was not observed at 10 ppm after 2 days; however, all fish tested died after 3 days. The LD_{50} values of amipurimycin administered i.v. in mice and rats were about 1–5 mg/kg body weight, and by oral administration they were 10–20 and 20–30 mg/kg of body weight. When tested with rabbits, there was no irritation to the cornea at 500 ppm, but a strong irritation was observed on the skin following daily application of 200 ppm for 10 days (Iwasa et al., 1977).

1.7 NIKKOMYCIN

Zähner and coworkers (Dähn et al., 1976) isolated nikkomycin (Fig. 1.24) from the fermentation broth of *Streptomyces tendae* Tü 901. The tentative structure of nikkomycin was identified by a combination of physical and chemical measurements of the product obtained from chemical degradation. Nikkomycin is a nucleoside-peptide consisting of uracil, a 5-aminohexuronic acid, and a new α-amino acid with a pyridine ring. Nikkomycin is structurally similar to the polyoxins (page 30), ablastimycin, bulgerin, and 24010 (Isono et al., 1969; Hashimoto et al., 1968; Shoji et al., 1970; Mizuno et al., 1971). Nikkomycin and clitidine (Fig. 7.1) are the first two naturally occurring nucleosides with the pyridine ring. Nikkomycin is a potent inhibitor of fungal chitin synthesis.

Discovery, Production, and Isolation

S. tendae Tü 901 was obtained from a soil sample from Nikko, Japan. Nikkomycin was isolated from the culture medium of *S. tendae* Tü 901 4–5 days after inoculation by a combination of anion and cation exchange chromatography and gel chromatography (Dähn et al., 1976).

Physical and Chemical Properties

The molecular formula for nikkomycin is $C_{20}H_{25}N_5O_{10}$; mol wt 495; ultraviolet spectral properties: $\lambda_{max}^{0.1\,N\,HCl}$ 205, 230 (sh), 282 nm; $\lambda_{max}^{0.1\,N\,NaOH}$ 241 and 306 nm (Fig. 1.25). Nikkomycin is soluble in water and pyridine and insolu-

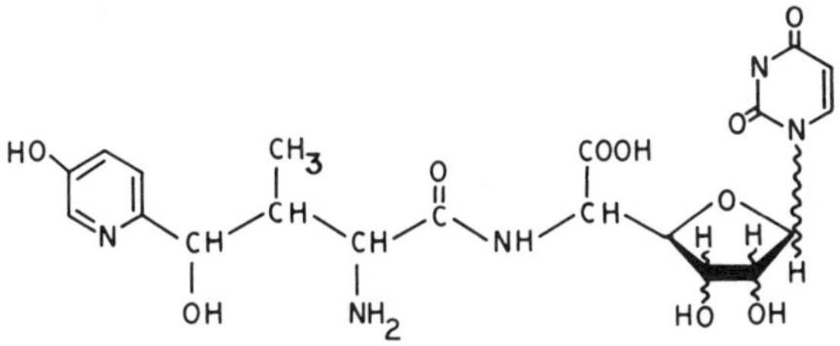

Figure 1.24 Structure of nikkomycin.

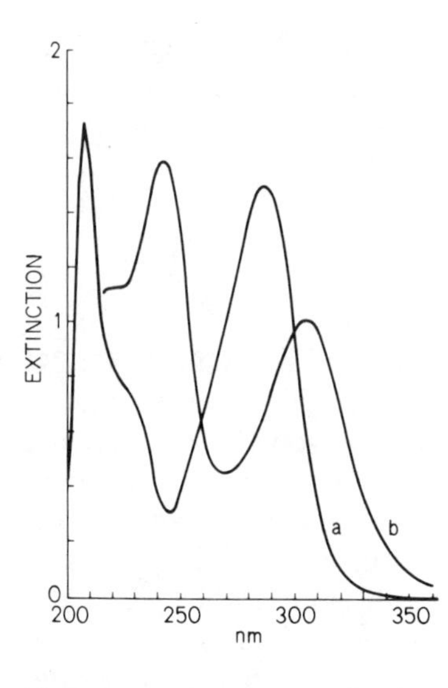

Figure 1.25 Ultraviolet spectrum of nikkomycin as determined in (*a*) 0.1 *N* HCl and (*b*) 0.1 *N* NaOH. Reprinted with permission from Dähn et al., 1976.

ble in organic solvents. It gives a positive reaction with ninhydrin and is stable between pH 3 and 9 at 0 or 20°C (Dähn et al., 1976).

Structural Elucidation

The acid degradation of nikkomycin is shown in Fig. 1.26. Hydrolysis with 6 *N* HCl, 110°C, 20 h gave compound **1** (the new α-amino acid containing the pyridine ring) plus uracil, 3-hydroxypyridine, and 3-hydroxypyridine-6-carboxylic acid. Hydrolysis in 1 *N* trifluoroacetic acid (F_3AcOH) (100°C, 1 h) produced compound **1** and compound **2**, 1-(5′-amino-5′-deoxyhexuronic acid) uracil. Mild acid hydrolysis (1.25 *N* HCl–methanol) gave compound **4**, a dipeptide ester. When the uracilaminohexuronic acid (compound **2**) is further hydrolyzed with 3 *N* HCl, the products are uracil plus compound **3**, 5-amino-5-deoxyhexuronic acid. The identification of compounds **1–4** was accomplished as the TFA and TMS derivatives by mass spectrometry (Dähn et al., 1976).

Inhibition of Growth

Of all the organisms tested for sensitivity to nikkomycin the *Zygomycetes* were the most sensitive. Nikkomycin did not inhibit the gram-negative or gram-positive bacteria (Dähn et al., 1976).

Biochemical Properties

The effect of nikkomycin on RNA, DNA, protein, and chitin synthesis has been studied (Dähn et al., 1976). When nikkomycin (0.5 μg/ml) is added to

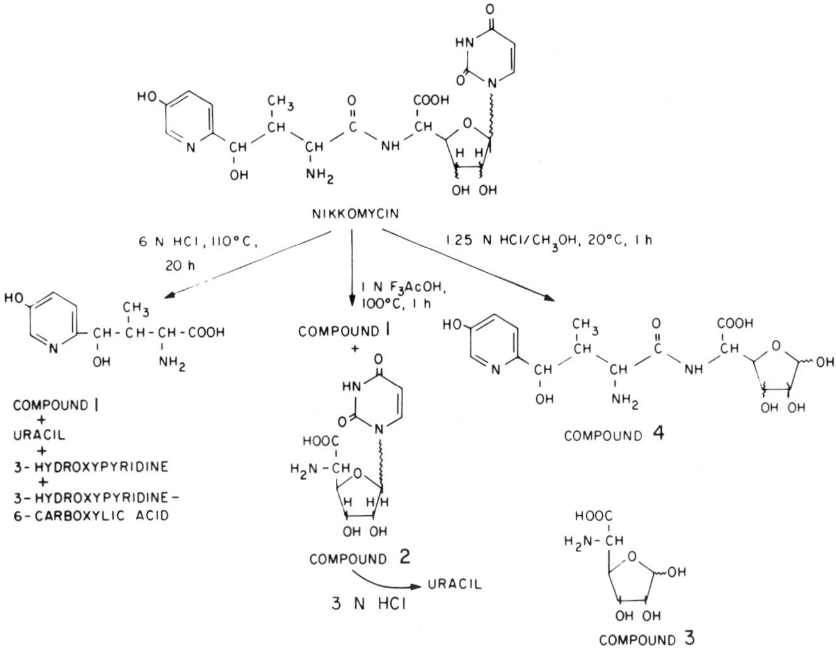

Figure 1.26 Acid degradation of nikkomycin. After Dähn et al., 1976.

cultures of *M. hiemalis* (+), there is a 63% inhibition of RNA synthesis; this compares with a 77% inhibition when cyclohexamide at 10 μg/ml is added. With protein synthesis, the incorporation of [^{14}C]isoleucine into the protein of *M. hiemalis* (+) is inhibited by 40%, whereas cyclohexamide inhibits protein synthesis by 80%. The inhibition of chitin synthesis by nikkomycin (0.5 μg/ml), as determined by the incorporation of [^{3}H]*N*-acetyl glucosamine into *M. hiemalis* (+) is very similar to that reported for the polyoxins (page 35). Cyclohexamide does not inhibit chitin synthesis. Nikkomycin can also serve as a negative allosteric effector, much the same way as *N*-acetylglucosamine. Although nikkomycin inhibits chitin synthesis and RNA synthesis, chitin synthesis is more sensitive.

1.8 SEPTACIDIN

Septacidin (Fig. 1.27) is not a true adenine N^9 riboside nucleoside analog. However, septacidin is included because it contains the 4-amino-substituted amino sugar, 4-aminoaldoheptose, which is covalently bound to the N^6-amino group of adenine. The products of hydrolysis of septacidin are adenine, glycine, a 4-aminoaldoheptose, and C_{12}–C_{18} fatty acids (mostly

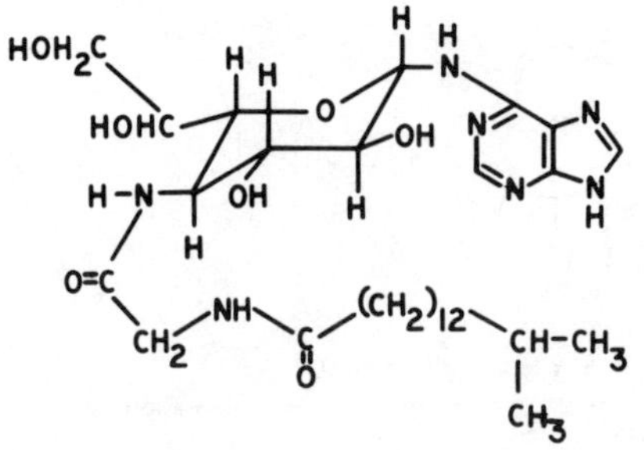

Figure 1.27 Structure of septacidin.

branched). Isopalmitic acid is the predominant fatty acid following hydrolysis. Septacidin does not have antibacterial activity, but it does inhibit filamentous fungi. The discovery, production, isolation, physical and chemical properties, and structural elucidation have been reviewed (Suhadolnik, 1970; Townsend, 1975).

Septacidin is cytotoxic against Earle's L cells in tissue culture (Deutcher et al., 1963). It is toxic to *Trichophyton mentagrophytes* and *Fusarium bulbigenum*. Because the 4-amino group of the sugar moiety exists as the amide, septacidin does not inhibit protein synthesis as does puromycin, gougerotin, and so on.

1.9 PLATENOCIDIN

Platenocidin is a naturally occurring nucleoside analog that inhibits certain species of yeast.

Discovery, Production, and Isolation

The 5-hydroxymethyluracil nucleoside analog platenocidin has been isolated from the culture broth of *Streptomyces* H 273 N-SY2, which has been classified as belonging to *S. platensis* (Honke et al., 1977). This is the same organism from which Argoudelis and Mizsak (1976) isolated 5,6-dihydro-5-azathymidine (antibiotic U-44590), 1-methylpseudouridine, and pseudouridine (see pages 241 and 245). The yield is 23 mg/12 liters culture medium.

Physical and Chemical Properties and Structural Elucidation

Elemental analysis gave C, 43.03%; H, 4.73%; and N, 15.65%. Mp 118–122°C; ultraviolet spectral properties: $\lambda_{max}^{pH\ 7.9\ H_2O}$ 255 nm ($E_{1\,cm}^{1\,percent}$ 219); $\lambda_{max}^{0.1\,N\,HCl}$ 260 nm ($E_{1\,cm}^{1\,percent}$ 257); $\lambda_{max}^{0.1\,N\,NaOH}$ 257 ($E_{1\,cm}^{1\,percent}$ 230); acid or

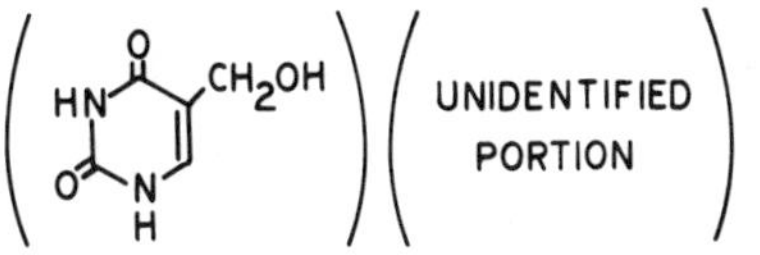

Figure 1.28 Structure of platenocidin.

alkaline hydrolysis gives 5-hydroxymethyluracil. Platenocidin gives positive ninhydrin, $KMnO_4$, α-naphthol phosphate, and Ehrlich reactions (Honke et al., 1977).

The physicochemical properties of platenocidin are very similar to those of polyoxins A, B, C, and I, which are known to contain 5-hydroxymethyluracil (Isono et al., 1965, 1967, 1969; Isono and Suzuki, 1966). The isolation of 5-hydroxymethyluracil suggests that polyoxin C may be a part of the platenocidin molecule (Fig. 1.28).

Biochemical Properties

The biological properties of platenocidin are markedly different from those of polyoxins A and B (Honke et al., 1977). Platenocidin shows inhibitory properties against certain species of yeasts, but not bacteria or pathogenic fungi. Polyoxins A and B inhibit phytopathogenic fungi, but not yeast; polyoxins C and I are inactive (Isono et al., 1965, 1967, 1969; Isono and Suzuki, 1967). Therefore, platenocidin is differentiated from the polyoxins.

1.10 AGROCIN 84

Agrocin 84 is a unique adenine nucleotide elaborated by the nonpathogenic strain *Agrobacterium radiobacter* var. *radiobacter* strain K-84 and pathogenic strain 396 (J. Schell, personal communication). The proposed partial structure of agrocin 84 is shown (Fig. 1.29) (Roberts et al., 1977; A. Kerr, personal communication). Thompson et al. (1977) confirmed the structure of agrocin 84 as reported by Roberts et al. (1977). This is the first naturally occurring adenosine derivative with a 6-*N*-phosphoramidate to which glucofuranose is covalently bound. The ribofuranosyl moiety of adenosine is replaced by the rare pentose, 3-deoxy-D-arabinofuranose. The biological control of crown gall operates through the production of agrocin 84. Most pathogens are inhibited by agrocin 84.

Discovery, Production, and Isolation

Agrocin 84 is produced by culturing a rough colony variant strain of *A. radiobacter* var. *radiobacter* in a defined liquid medium (Roberts et al.,

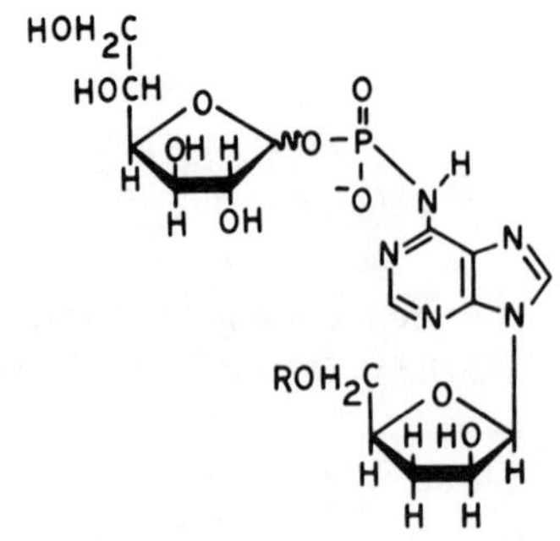

R=UNIDENTIFIED PHOSPHORYLATED CARBOHYDRATE

Figure 1.29 Partial structure of agrocin 84 (Roberts et al., 1977; Kerr, personal communication).

1977). Although agrocin 84 has not been crystallized, Kerr and coworkers (Roberts et al., 1977) report 10,000-fold purification of agrocin 84 by gradient elution techniques involving charcoal adsorption and anion exchange chromatography. Final purification was achieved by electrophoresis on glass fiber filter paper. The purification procedure of Thompson et al. (1979) from the high yield K-84 strain gives agrocin 84 of about 65% purity on a dry weight basis. Heip et al. (1975) have also reported on a partial purification of agrocin 84.

Physical and Chemical Properties

Part of the structure of agrocin 84 has been elucidated. The upper limit for the molecular weight of the triethylammonium salt is 1100 ± 100; ultraviolet spectral properties $\lambda_{max}^{H_2O\ pH7}$ 265 nm (ϵ = 19,860) (Roberts et al., 1977); the migration of agrocin 84 to the anode at pH 3.5 shows that it is a strong acidic substance (Heip et al., 1975). The ultraviolet properties of agrocin 84 in acid and alkali are characteristic of 6-*N*-acylated 9-substituted adenine nucleosides. The molar extinction coefficient is 22,675 at 264 nm (pH 7) and a minimum at 227 nm (264–227, ratio 6.00) (Thompson et al., 1979).

Structural Elucidation

Roberts et al. (1977) showed that agrocin 84 contains adenine, phosphorus, 3-deoxy-D-arabinose, and D-glucose in a molar ratio of 1:2:1:1. An unidentified, phosphorylated carbohydrate is also present. The infrared spectrum shows the presence of a P—O stretching frequency at 1225 cm^{-1}. Acid hydrolysis of agrocin 84 produces glucose and 3-deoxy-D-arabinose. Authentic 3-deoxy-D-arabinose, prepared by way of the methyl 2,5-di-O-acetyl-3-deoxy-β-threopentofuranoside (Casini and Goodman, 1964), is identical with 3-deoxy-D-arabinose from agrocin 84.

Alkaline hydrolysis of agrocin 84 produces crystalline 3′-deoxyarabinofuranosyladenine; ultraviolet spectral properties: $\lambda_{max}^{pH\,12.55}$ 260.5 nm (ϵ = 14,300); $\lambda_{max}^{pH\,6.92}$ 260 (ϵ = 14,400); $\lambda_{max}^{pH\,1.06}$ 258 (ϵ = 14,200) (Roberts et al., 1977). This is expected for a 9-substituted adenine nucleoside. The isolated 3′-deoxyarabinofuranosyladenine, 9-(3-deoxy-β-D-threopentofuranosyl)-adenine, is the same as that synthesized by Goodman and coworkers (Martinez et al., 1966). The presence of a 6-N-phosphoramidate group in agrocin 84 is based on the bathochromic shifts for 6-N-acylated 9-substituted adenine nucleoside derivatives and the acid lability of phosphoramidate (Ralph and Khorana, 1961; Schweizer et al., 1970). Furthermore, 6-N-(cyanoethylphosphoryl)adenosine and agrocin 84 have the same ultraviolet spectral characteristics: $\lambda_{max}^{pH\,1}$ 264 nm (ϵ = 19,500); $\lambda_{max}^{pH\,13}$ 267.5 nm (ϵ = 19,800). These data strongly suggest that agrocin 84 contains a 6-N-phosphoramidate linkage to adenine as shown in Fig. 1.29.

Inhibition of Growth

Agrocin 84 is isolated from the nonpathogenic strain *A. radiobacter* var. *radiobacter* strain 84. It is a general inhibitor of pathogenic strains of *Agrobacterium* of the same species or related species. Nearly all nonpathogens are unaffected (Kerr and Roberts, 1976). This type of inhibition is similar to that of the colicins. However, it is possible that agrocin 84 should be classed as a new type of bacteriocin (Roberts et al., 1977).

Biochemical Properties

Agrocin 84 is specifically taken up by sensitive strains of *Agrobacterium*, but is not taken up by nonpathogenic strains. Some pathogenic strains are nonsensitive to agrocin 84 (Kerr and Roberts, 1976). Purified agrocin 84 is very effective in preventing tumor induction by the *A. tumefaciens* C-58 strain and might be an effective treatment of crown gall (Thompson et al., 1979). Until the studies of McCardell and Pootjes (1976), it was not known how agrocin 84 affects the pathogenic organism. McCardell and Pootjes showed that agrocin 84 inhibits RNA, DNA, and protein synthesis (Fig. 1.30). Protein synthesis is inhibited more than DNA synthesis. RNA synthesis is inhibited after 20 min and the uptake of amino acids and cell motility are stopped immediately. These data indicate that agrocin 84 may affect energy-generating reactions. Fifty percent of *Agrobacterium* cells

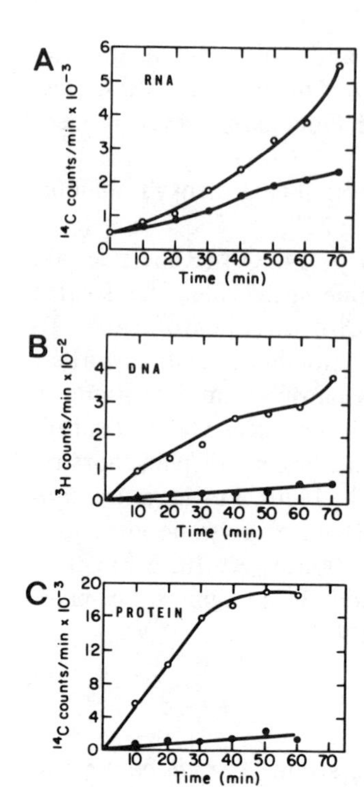

Figure 1.30 Effect of agrocin 84 on the synthesis of *A. tumefaciens* H-38-9. Agrocin added at zero time. (*A*) Ribonucleic acid synthesis; (*B*) deoxyribonucleic acid synthesis; (*C*) protein synthesis. Agocin 84 (●); control (○). Reprinted with permission from McCardell and Pootjes, 1976.

were killed in 15 min following contact with agrocin 84. The remaining surviving cells were also inhibited.

***Plasmid Required for Virulence of* A. tumefaciens.** Crown gall is a malignancy affecting dicotyledonous plants that occurs following inoculation of a wound site with *A. tumefaciens*. Evidence presented by Zaenen et al. (1974) showed that 11 virulent strains of *A. tumefaciens* contain a large plasmid. Eight avirulent strains lack such plasmids. Hamilton and Fall (1971) showed that virulent strain C-58 could be converted to a stable avirulent strain by passage at 37°C. Van Larebeke et al. (1974) suggested that the genes controlling crown gall are located on a large plasmid called the Ti plasmid. Unequivocal evidence that the large plasmid is important in determining the virulence of strain C-58 was provided by Nester and coworkers (Watson et al., 1975) (Fig. 1.31). They detected a small, higher-density peak of circular DNA in virulent strain C-58, but not in the avirulent NTl strain. There are at least three different Ti plasmids; all carry virulence genes but otherwise show very low homology to one another. Use of the Ti plasmids also codes for the metabolism of the unusual amino acid derivative nopaline and for sensitivity to agrocin 84. On exposure of a sensitive strain to agrocin 84, mutant strains that are resistant to agrocin 84 may appear. These are usually avirulent because they have lost the Ti plasmid or have deletions in it.

In terms of new concepts in the biology of host–parasite relations, the study of the crown-gall phenomenon has been very rewarding. The crown-gall formation by *A. tumefaciens* is the first example where parasitic and symbiotic situations between nonrelated organisms show a natural transfer of genetic information from the prokaryote into the genome of the eukaryote (Schell et al., 1979). In this case, the parasite forces its host to synthesize products (opines) that only the parasite can use. Schell et al. (1979) have called this type of interaction "genetic colonization."

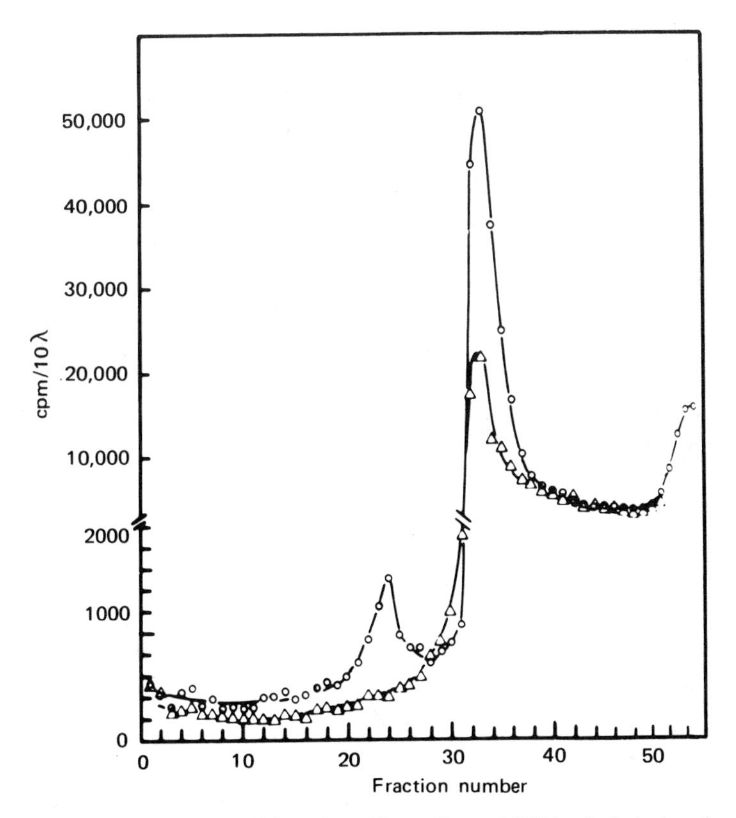

Figure 1.31 Cesium chloride–ethidium bromide gradient of DNA of virulent and avirulent strains: Strain C-58 (O); strain NT1 (Δ). Reprinted with permission from Watson et al., 1975.

1.11 SHOWDOMYCIN

Showdomycin, 2-(β-D-ribofuranosyl)maleimide (SHM) (Fig. 1.32), is a naturally occurring maleimide C-nucleoside analog elaborated by *S. showdoensis* (Nishimura et al., 1964). The structure was established by Darnall et al. (1967) and by Nakagawa et al. (1967). The X-ray data reported by Tsukuda and Koyama (1970) show that SHM, in the crystalline state, exists in the *syn* conformation; however, Townsend and coworkers (Chenon et al., 1973) could not draw a conclusion about the conformation of showdomycin in solution based on the ^{13}C nmr spectra. Elstner et al. (1973a) have reported on the isolation, structural elucidation, properties, and biosynthesis of the bicyclic maleimide analog, maleimycin (Fig. 1.32), from the culture filtrates of *S. showdoensis*. Ozaki et al. (1972) reported on the microbial transformation of showdomycin to isoshowdomycin (isoSHM) (Fig. 1.32). Although SHM is structurally similar to pseudouridine and uridine, its biological properties are markedly different.

The biological activities of SHM include radiosensitization of *E. coli* (Titani and Katsube, 1969), induction of volume changes in mitochondria (Hadler et al., 1968), inhibition of protein synthesis, nucleic acid synthesis, and growth of *E. coli* (Komatsu and Tanaka, 1968; Nishimura and Komatsu, 1968; Bermek et al., 1970), selective inhibition of certain enzymes (Roy-Burman et al., 1968; Kalman, 1972; Komatsu and Tanaka, 1971), and inhibition of uptake of sugars and amino acids in *E. coli* B (Roy-Burman et al., 1971).

The discovery, isolation, production, physical and chemical properties, structural elucidation, synthesis, and inhibition of growth of showdomycin have been reviewed (Suhadolnik, 1970; Townsend, 1975). This section reviews publications since 1970.

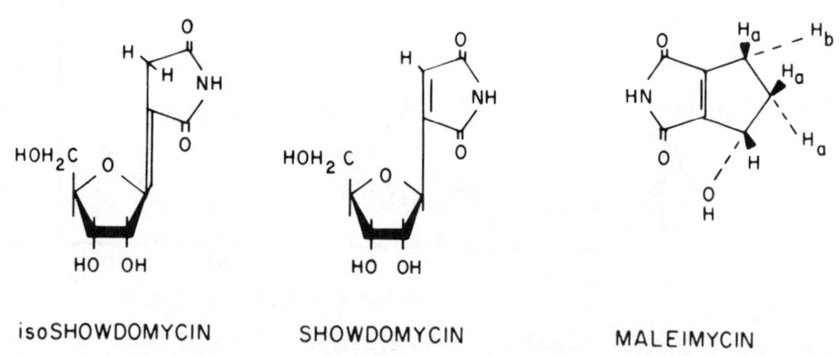

Figure 1.32 Structures of showdomycin, isoshowdomycin, and maleimycin.

Chemical Synthesis of Showdomycin and Showdomycin Analogs

The chemical synthesis of showdomycin and 3-methylshowdomycin has been described by Kalvoda (1976), Kalvoda et al. (1970), and Trummlitz and Moffatt (Trummlitz and Moffatt, 1973; Trummlitz et al., 1975).

Just and coworkers reported on the synthesis of the 2′-epimer of showdomycin, the arabinosyl analog, D,L-carbocyclic showdomycin, and *N*-carbomethoxy showdomycin, 2-(2′α3′α-dihydroxy-4′β-hydroxymethyl-*N*-carbomethoxypyrrolidin-1β-yl)maleimide (**6**), and D,L-2′-deoxyshowdomycin (**4**) (Just and Lim, 1977; Lim, 1976; Just and Donnini, 1977). The key step in the synthesis of 2′-deoxyshowdomycin is the reaction of furan (**1**) with methyl-β-nitroacrylate (**2**) to give the Diels-Alder adduct, **3**, which is used to synthesize D,L-2′-deoxyshowdomycin (**4**) (Scheme 1.6). Whereas the Diels-Alder approach was successful for the synthesis of the carbocyclic analog of showdomycin (Just and Lim, 1977), it failed for the *N*-substituted analog (**6**). Therefore, the teloidinone (**5**) was used as the starting substance for the synthesis of **6**. The ribofuranosyl ring oxygens of showdomycin and pyrazofurin play an important role because their replacement by a methylene group or a *N*-carbomethoxy group leads to the complete loss of antiviral, antibacterial, and antifungal activity (Just and Kim, 1976).

Enzymatic Transformation of Showdomycin to Isoshowdomycin

Microbial transformation of antibiotics has been reported by several laboratories (Perlman et al., 1966; (Argoudelis and Mason, 1969; Čapek et al., 1969). Ozaki et al. (1972) reported on the enzymatic isomerization of SHM to isoshowdomycin (Fig. 1.32). Whereas the λ_{max} of SHM is 222 nm, isoSHM has λ_{max} of 264 nm (log ϵ = 4.3); mp 215–216°C; $[\alpha]_D^{225}$ −34.0° (*c* 1, H_2O). The physical and chemical properties of isoSHM are identical with those of chemically synthesized isoSHM (Nakagawa et al., unpublished data). IsoSHM does not inhibit gram-negative or gram-positive bacteria. Therefore, the location of the double bond in the maleimide ring of SHM is crucial for antibacterial activity.

Phosphorylation of Showdomycin to Showdomycin 5′-Monophosphate

Shirato et al. (1968) reported the chemical synthesis of SHM 5′-phosphate using *Serratia marcescens* extracts and *p*-nitrophenylphosphate. Kalman (1972) used $POCl_3$ for the direct phosphorylation of SHM according to the method of Yoshikawa et al. (1969).

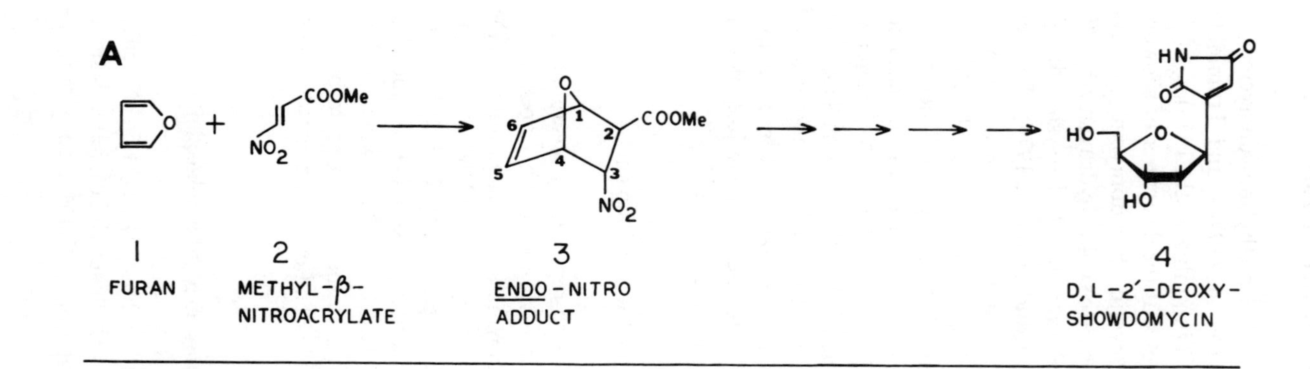

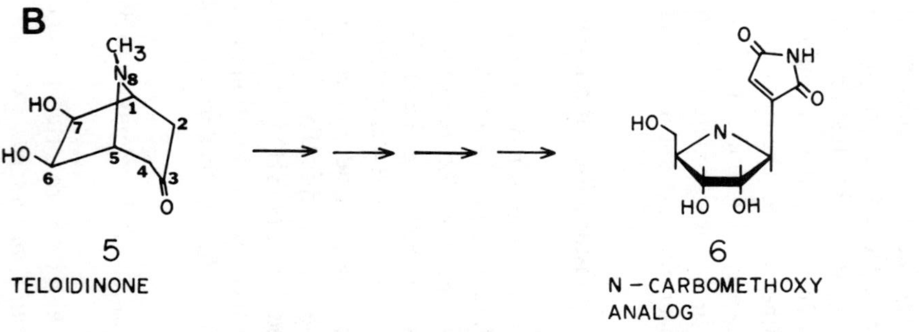

Scheme 1.6 Synthesis of (*A*) D,L-2′-deoxyshowdomycin and the (*B*) *N*-carbomethoxy analog of showdomycin. Reproduced (in modified form) by permission of the National Research Council of Canada from the *Canadian Journal of Chemistry*, Volume 55, pp. 2293–2297 and 2998–3006, 1977.

Biosynthesis of Showdomycin

Two mechanisms have been considered for the biosynthesis of showdomycin. The first involves the elimination of N-1 of pseudouridine followed by ring closure to form the maleimide ring. The second involves the utilization of four carbon atoms of α-ketoglutarate as either a symmetrical or asymmetrical four-carbon donor for the formation of the maleimide ring. The incorporation of ^{13}C and ^{14}C from labeled acetate and glutamate clearly shows that carbons 2–5 of glutamate serve as the four-carbon donor for the formation of the maleimide ring (Elstner and Suhadolnik, 1972; Elstner et al., 1973b). The enrichment of C-4 of the maleimide ring by ^{13}C from [1-^{13}C]acetate and the enrichment of C-2, C-3 and C-4 from [2-^{13}C]acetate are shown in Fig. 1.33. Similarly, ^{14}C from [5-^{14}C]glutamate and [5-^{14}C]α-ketoglutarate is incorporated exclusively into C-4, whereas ^{14}C from [1-^{14}C]α-ketoglutarate is not incorporated into SHM (Fig. 1.34). These data further substantiate that acetate, once it enters the Krebs cycle, is converted

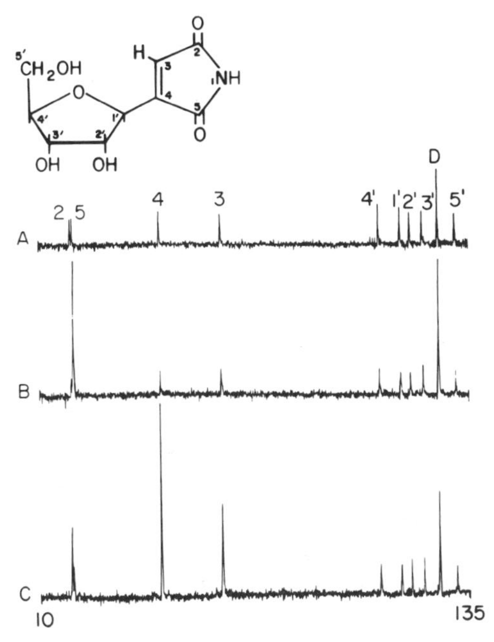

Figure 1.33 Proton-decoupled ^{13}C Fourier transform NMR spectra of aqueous showdomycin (plus dioxane). (*A*) Natural-abundance of ^{13}C spectrum of 0.24 *M* showdomycin; (*D*) Resonance of dioxane; (*B*) 0.11 *M* showdomycin after [1-^{13}C]acetate incorporation; (*C*) 0.13 *M* showdomycin after [2-^{13}C]acetate (1800 μ*M*) incorporation. Reprinted with permission from Elstner et al., 1973b.

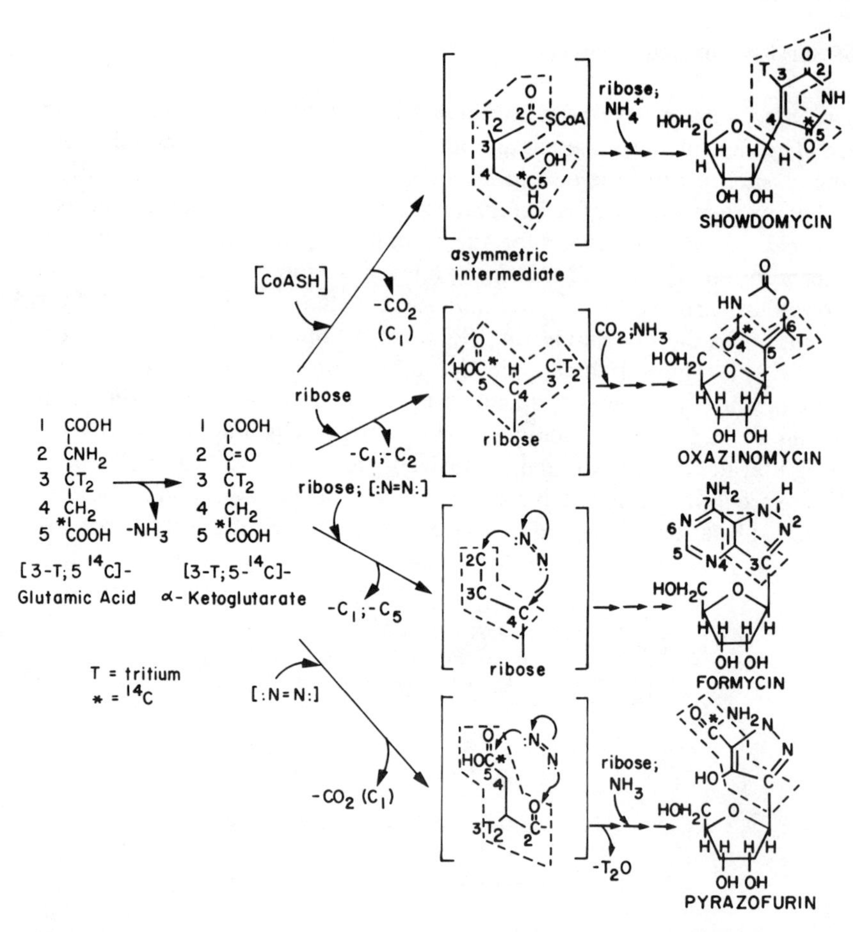

Figure 1.34 Glutamic acid, the common precursor for the C_3 or C_4 carbon skeleton of the aglycon of showdomycin, oxazinomycin, formycin, and pyrazofurin.

to α-ketoglutarate, which is decarboxylated to a four-carbon precursor for the maleimide ring. Proof that a C_4 asymmetric intermediate is formed following the decarboxylation of α-ketoglutarate was shown by the loss of tritium from C-4 of glutamate, but the asymmetric incorporation of ^{14}C into C-4 of SHM. These data were obtained by the addition of [4-^{3}H;5-^{14}C]glutamate to SHM-producing cultures of *S. showdoensis* (Fig. 1.34).

Biochemical Properties

Inhibitory Effects of Showdomycin: Permeability and Alkylation Properties. SHM is not a substrate for either nucleoside kinase or nucleoside phosphorylase (Roy-Burman et al., 1968). SHM probably exerts its inhibitory effect on bacteria and tumors without conversion to the nucleotide (Nishimura et al., 1964; Matsuura et al., 1964). The cytotoxic action of SHM results from two properties. First, as a nucleoside analog, SHM can easily enter the cell. Second, once inside the cell, the maleimide ring of SHM can act as an alkylating agent for sulfhydryl, amino acid, and imidazole groups (Gregory, 1955; Leslie, 1965; Morell et al., 1964; Sharpless and Flavin, 1966; Watanabe, 1970). The inhibitory effects of SHM are reversed by preincubation of *E. coli* with cysteine, nucleosides, or nucleoside analogs (Roy-Burman et al., 1971; Doskočil and Holý, 1974a). SHM also blocks the transport of glucose and other nutrients by reacting with thiol groups of the respective transporting systems (Doskočil and Holý, 1974a). SHM is not limited in its action as an alkylating agent.

The entry of showdomycin labeled at C-4 into *E. coli* is competitively inhibited by nucleosides (Komatsu, 1971a). *E. coli* mutants resistant to SHM take up very little [^{14}C]SHM and show an altered ability to take up nucleosides; these SHM-resistant mutants were found to be very sensitive to NEM (Komatsu, 1971b). Roy-Burman et al. (1971) noted that the difference between NEM and SHM is that the nucleosides cannot reverse the inhibitory effects of NEM on the uptake of glucose, whereas most nucleosides overcome the inhibitory effect of SHM on glucose uptake. This led to the speculation that resistance to SHM by bacteria is due to an inhibition of transport of the drug into the cell.

Because some nucleosides protect *E. coli* against SHM, there must be competition between SHM and nucleosides for a common transport system. SHM-resistant *E. coli* have marked reduction in uridine, cytidine, and SHM transport, but show normal guanosine and adenosine transport; exogenous SHM, which inhibits transport of glucose, guanosine, uridine, cytidine, and adenosine, has no effect on these processes in SHM-resistant *E. coli* (Roy-Burman and Visser, 1972). SHM-resistant *E. coli* mutants also have an impaired ability to take up deoxynucleosides (Komatsu and Tanaka, 1972). Transport systems for deoxyadenosine and adenosine were found to be defective in SHM-resistant *E. coli* B (Roy-Burman and Visser, 1975). Leung and Visser (1977) solved the problem involved in the transport of uracil, uridine, and the ribose moiety of uridine by the use of *E. coli* mutants. They demonstrated that SHM-resistant *E. coli* had a defective uridine transport system (Table 1.4). The data in Table 1.4 compare the uptake of uracil, uridine, and cytidine in *E. coli* with the three mutants.

Mutant U^- is similar to the parent *E. coli* strain in that it lacks the uracil transport system, but is able to transport intact uridine and cytidine. Mutant U^-UR^- cannot transport either uracil or uridine and cytidine transport is decreased 60%. The transport of adenosine, guanosine, deoxy-

Table 1.4 Uptake of Uracil, Uridine, and Cytidine in *E. coli* B, Mutant U⁻ (Lacks Uracil Transport), Mutant U⁻UR⁻ (Lacks Uracil and Uridine Transport), and Mutant NUC⁻ (Showdomycin Resistant)

Additions	Initial Concentration (μM)	Uptake (%)[a]			
		E. coli B	Mutant U^-	Mutant U^-UR^-	Mutant NUC^-
[2-^{14}C] Uracil	0.5	100 (106)[b]	0.3	0	117
	10	100 (186)	3.8	2.4	107
[2-^{14}] Uridine	5	100 (343)	75.5	1.4	3.9
	100	100 (520)	74.8	1.3	5.5
[U-^{14}C] Uridine	5	100 (–)	96.7	42.3	1.4
[2-^{14}C] Cytidine	5	100 (257)	97.1	44.5	3.8

Reprinted with permission from Leung and Visser, 1977.
[a] Uptake was measured under standard assay conditions in the presence of 5 m*M* glucose. Assays for cytidine uptake were carried out in the presence of 205 μM tetrahydrouridine as described in the text. Values are expressed as percentage of uptake in *E. coli* B (100%).
[b] Values in parentheses indicate uptake in pmol/10 s.

adenosine, deoxyguanosine, adenine, and guanine with mutants U^- and U^-UR^- was the same as with the parent strain. With showdomycin mutant NUC^-, the uracil transport is normal, but the capacity to transport intact uridine, cytidine, or the ribosyl moiety of uridine is lost.

Because binding sites are important for nucleoside transport in bacteria, the type of binding sites and how SHM affected these binding sites was studied. Two types of binding sites are known: (i) the constitutive site with high substrate affinity and (ii) the inducible sites, which have lower substrate affinity. Doskočil and Holý (1974b) found that *E. coli* mutants resistant to SHM affected the constitutive nucleoside binding sites, but not the inducible nucleoside binding sites. This was demonstrated experimentally by Komatsu and Tanaka (1973), who reported that *E. coli* K12 has two deoxycytidine transport systems. SHM is transported by a high affinity deoxycytidine transport system that is not affected by guanosine or xanthosine. In SHM-resistant mutants, this high affinity deoxycytidine transport system is no longer detectable, while the low affinity deoxycytidine transport system is not changed. In *E. coli* SHM does not inhibit uridine phosphorylase in the periplasmic space, but in Ehrlich ascites cells, SHM inhibits uridine phosphorylase.

Inhibition of Thymidylate Synthetase and Pseudouridylate Synthetase. Kalman (1971) suggested that the functional nucleophile in thymidylate synthetase is an enzyme–SH group that participates in the formation of a transient thioether linkage between the enzyme and C-6 of dUMP. Kalman (1972; 1974) used SHM and SHM 5′-phosphate to elucidate the catalytic mechanism of thymidylate synthetase and pseudouridylate synthetase. SHM and its 5′-phosphate alkylate the sulfhydryl at the active site of thymidylate synthetase. This is consistent with a sulfhydryl addition–elimination mechanism of the enzyme catalyzed reaction (Fig. 1.35). There is a concerted nucleophilic attack of the functional sulfhydryl group in thymidylate synthetase across the 5,6 double bond of dUMP (Fig. 1.35*a*) followed by enzyme inactivation (Fig. 1.35*b*).

Radiosensitization of* E. coli *B/r by Showdomycin. Titani and Kasube (1969) reported a marked radiosensitizing effect of *E. coli* B/r with SHM. Under anoxic conditions, the radiosensitizing ability of SHM is derived mainly from the maleimide moiety. Under aerated conditions, the hydroxyl groups of ribose in showdomycin are essential to radiosensitizing ability.

Oxidative Phosphorylation and Carcinogenesis. One of the hypotheses postulated for carcinogenesis is that carcinogens damage mitochondria by

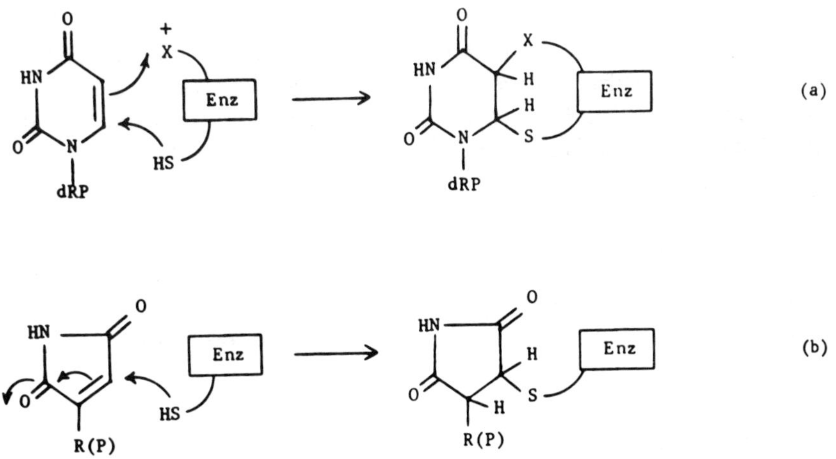

Figure 1.35 Analogy between the postulated catalytic role of an active site SH group and its interaction with showdomycin and showdomycin 5′-phosphate. X^+, electrophilic precursor of the methyl group of thymidylate; dR, deoxyribosyl; P, 5′-phosphate. Reprinted with permission from Kalman, 1972.

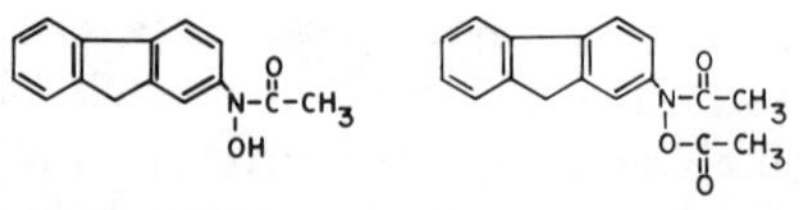

N-OH-AAF N-OAc-AAF

interfering with the flux of energy in the mitochondria. When this damage occurs, the membranes become leaky and there is a release of mitochondrial genetic material that behaves like an oncogenic virus and enters the genome of the cell (Hadler et al., 1971; Hadler, 1974). One of the well-known carcinogens is *N*-hydroxy-*N*-acetyl-2-aminofluorene (*N*-OH-AAF). *N*-OH-AAF induces an ATP-energized mitochondrial volume change when combined with SHM. *N*-OH-AAF or SHM alone cannot induce ATP-energized mitochondrial volume changes. *N*-OH-AAF exposes a pivotal mitochondrial sulfhydryl, which is alkylated by showdomycin.

There are three other positional isomers of *N*-OH-AAF in which the nitrogen is on the 1,3, and 4 positions of the fluorene nucleus. Each of these four isomers in combination with SHM induced ATP-energized mitochondrial changes whose magnitudes paralleled their carcinogenicity in rats (Hadler et al., 1971). SHM also stimulated the ATP-energized mitochondrial volume change in the presence of *N*-OAc-AAF. Of interest is the observation that *N*-OAc-AAF, which is favored by many investigators to be a model of the ultimate electrophilic carcinogenic agent derived from *N*-OH-AAF, is also enzymatically reconverted to *N*-OH-AAF by mitochondria (Hadler and Demetriou, 1975).

Therefore, carcinogenesis is the result of a disruption of the symbiotic relationship that exists between the mitochondria and cell. This symbiosis was established during the course of evolution. In effect, this hypothesis suggests that carcinogenesis represents a partial reversal of evolution.

1.12 CROTONOSIDE

Crotonoside (isoguanosine, 2-hydroxyadenosine) (Fig. 1.36) inhibits the inducible binding sites of *E. coli*. It has been isolated from *Croton tiglium* L

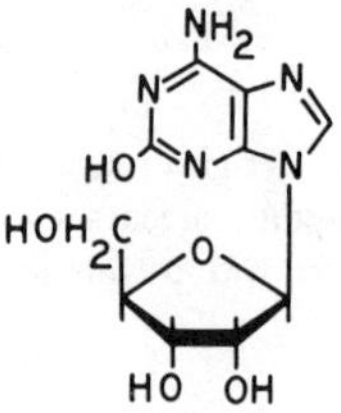

Figure 1.36 Structure of crotonoside (isoguanosine).

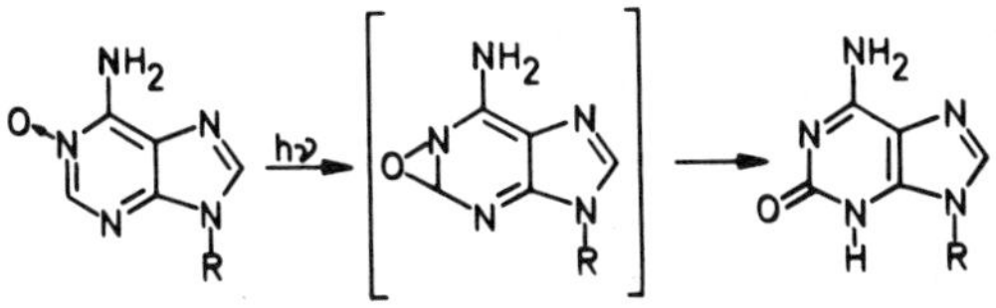

Figure 1.37 Photoisomerization of adenine 1-oxide nucleotides to the isoguanine nucleotides. R = ribosyl mono-, di-, or triphosphate. Reprinted with permission from Mantsch et al., *Biochemistry*, **14**, 5593–5601 (1975). Copyright by the American Chemical Society.

by Cherbuliez and Bernhard (1932). More recently Pettit et al. (1976) isolated isoguanine from the wings of the butterfly *Prioneris thestylis*. Crotonoside and 2′-aminoguanosine (see page 93) are the only two naturally occurring analogs of guanosine. The physical and chemical properties, chemical synthesis, and synthesis of crotonoside derivatives have been reviewed by Suhadolnik (1970). Although the isolation of the 5′-triphosphate of isoguanosine has not been reported from biological sources, Mantsch et al. (1975) have recently reported on the synthesis of the isoguanine nucleotides by way of photoisomerism of adenine 1-oxide nucleotides (Fig. 1.37).

Biochemical Properties

The biochemical properties of isoguanosine as reported prior to 1970 have been reviewed (Suhadolnik, 1970). Lowy et al. (1952) reported that [2-^{14}C]isoguanosine was incorporated into the nucleic acids following intraperitoneal administration to adult rats. With bacteria the labeled isoguanosine was not incorporated into the nucleic acids (Balis et al., 1952).

Because binding sites on membrane surfaces are important for nucleoside-transporting systems, Doskočil and Holý (1974b) studied the effect of several nucleoside analogs on the inducible binding sites and the constitutive sites of *E. coli*. They reported that isoguanosine preferentially inhibits the inducible binding sites, whereas the nucleoside analog, showdomycin, interferes more strongly with the constitutive function. Therefore, by using these nucleoside analogs, Doskočil and Holý showed that the inducible binding sites may indeed function independently of the constitutive sites.

Effect on Glutamine Dehydrogenase and cAMP. Isoguanosine 5′-di- and 5′-triphosphates bind strongly and inhibit glutamic acid dehydrogenase (Mantsch et al., 1975). Crotonoside has also been reported to cause a significant accumulation of cAMP (Huang et al., 1972).

Allosteric Modifier Activities with Isocitrate Dehydrogenase. Plaut et al. (1979) have studied the specificity of bovine heart NAD^+-linked isocitrate dehydrogenase for the configuration of cosubstrate (NAD^+) and allosteric effector (ADP). Isoguanosine 5′-diphosphate was neither an activator nor an inhibitor competitive with ADP.

SUMMARY

There are 12 naturally occurring nucleoside analogs or analog groups that inhibit cell wall synthesis, transport and/or are inhibitors of yeast and fungi.

Tunicamycin (and presumably the streptovirudins) is a structural analog of UDP-*N*-acetylglucosamine and dolichol. Tunicamycin inhibits the incorporation of *N*-acetylglucosamine and UDP-*N*-acetylmuramyl-pentapeptide into glycoproteins, glycosaminoglycan, peptidoglycans, and teichoic acid (Fig. 1.38). The glycosylation of interferon is not necessary for its secretion. In addition, nonglycosylated interferon, produced by exposure of viral infected cells to tunicamycin, has full antiviral activity.

Agrocin 84 completely inhibits amino acid uptake and stops cell motility immediately. RNA, DNA, and protein synthesis are inhibited completely in *A. radiobacter* var. *tumefaciens* by agrocin 84. Crotonoside preferentially inhibits the inducible nucleoside binding site, whereas showdomycin interferes with the constitutive function. Showdomycin has the addi-

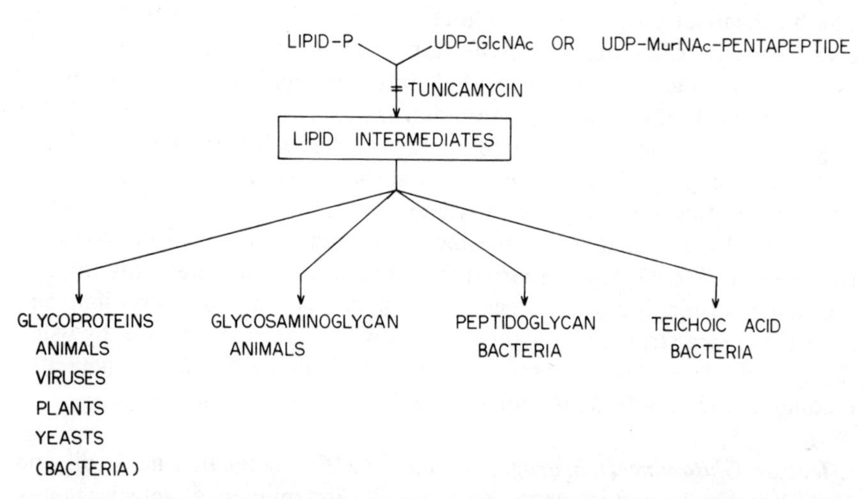

Figure 1.38 Inhibition of the lipid-linked sugar intermediates necessary for the biosynthesis of glycoproteins, glycosaminoglycan, peptidoglycan, and teichoic acid by tunicamycin. Reprinted with permission from Takatsuki, 1978.

tional effect of alkylating C-6 of dTMP, which inhibits thymidylate synthetase and also damages the mitochondrial membrane.

Amipurimycin, the ezomycins, nikkomycin, the polyoxins, septacidin, sinefungin and A9145C and thraustomycin are antifungal agents. Sinefungin (A9145) has the added property of inhibiting the methylases by acting as a competitive inhibitor of *S*-adenosylmethionine. Nikkomycin and the polyoxins inhibit chitin synthetase. Yeasts are inhibited by platenocidin and sinefungin.

REFERENCES

Argoudelis, A. D., and D. J. Mason, *J. Antibiot.*, **22,** 289 (1969).

Argoudelis, A. D., and S. A. Misak, *J. Antibiot.*, **29,** 818 (1976).

Azuma, T., K. Isono, and J. A. McCloskey, *Nucleic Acids Res.*, Spec. Publ. No. 2, s19 (1976).

Azuma, T., T. Saita, and K. Isono, *Chem. Pharm. Bull.*, **25,** 1740 (1977a).

Azuma, T., K. Isono, P. F. Crain, and J. A. McCloskey, *J. Chem. Soc. Chem. Commun.*, **1977b,** 159.

Balis, M. E., D. H. Levin, G. B. Brown, G. B. Elion, H. Vanderwerff and G. H. Hitchings, *J. Biol. Chem.*, **199,** 277 (1952).

Bartnicki-Garcia, S., and E. Lippman, *J. Gen. Microbiol.*, **71,** 301 (1972).

Benitez, T., T. G. Villa, and I. Garcia Acha, *Arch. Microbiol.*, **108,** 183 (1976).

Bermek, E., W. Kramer, H. Monkemeyer, and H. Matthaei, *Biochem. Biophys. Res. Commun.*, **40,** 1311 (1970).

Berry, D. R., and B. J. Abbott, *J. Antibiot.*, **31,** 185 (1978).

Bettinger, G. E., and F. E. Young, *Biochem. Biophys. Res. Commun.*, **67,** 16 (1975).

Bettinger, G. E., A. N. Chatterjee, and F. E. Young, *J. Biol. Chem.*, **252,** 4118 (1977a).

Bettinger, G. E., and F. E. Young, in *Microbiology*, D. Schlessinger, Ed., American Society for Microbiology, Washington, D.C., 1977b, p. 69.

Boeck, L. D., G. M. Clem, M. M. Wilson, and J. E. Westhead, *Fermentation Antimicrob. Agents Chemother.*, **3,** 49 (1973).

Bracha, R., and L. Glaser, *Biochem. Biophys. Res. Commun.*, **72,** 1091 (1976).

Brandner, G., and A. I. Virtanen, *Acta Chem. Scand.*, **17,** 2563 (1963).

Brimacombe, J. S., and Webber, J. M., *Mucopolysaccharides*, BBA Library, Vol. 6, Elsevier, Amsterdam, 1964.

Brink, A. J., J. Coetzer, O. G. DèVilliers, R. H. Hall, A. Jordaan, and G. J. Kruger, *Tetrahedron*, **32,** 965 (1976).

Cabib, E., and F. A. Keller, *J. Biol. Chem.*, **246,** 167 (1971).

Čapek, A., A. Šimek, E. Svátek, and M. Buděšínský, *Folia Microbiol. (Prague)*, **14,** 557 (1969).

Casini, G., and L. Goodman, *J. Am. Chem. Soc.*, **86,** 1427 (1964).

Chen, W. W., and W. J. Lennarz, *J. Biol. Chem.*, **251,** 7802 (1976).

Chenon, M.-T., R. J. Pugmire, D. M. Grant, R. P. Pazica, and L. B. Townsend, *J. Heterocyc. Chem.*, **10,** 427 (1973).

Cherbuliez, E., and K. Bernhard, *Helv. Chim. Acta*, **15,** 464 (1932).

Dähn, U., J. H. Hagenmaier, H. Höhne, W. A. König, G. Wolf, and H. Zähner, *Arch. Microbiol.*, **107,** 143 (1976).

Damodaran, N. P., G. H. Jones, and J. G. Moffatt, *J. Am. Chem. Soc.*, **93,** 3812 (1971).

Darnall, K. R., L. B. Townsend, and R. K. Robins, *Proc. Natl. Acad. Sci. U.S.*, **57,** 548 (1967).

Deutcher, J. D., M. H. von Saltza, and F. E. Pansy, *Antimicrob. Agents Chemother.*, 83 (1963).

de Vries, O. M. H., and J. G. H. Wessels, *Arch. Microbiol.*, **102,** 209 (1975).

Doskočil, J., and A. Holý, *Nucleic Acids Res.*, **1,** 491 (1974a).

Doskočil, J., and A. Holý, *Nucleic Acids Res.*, **1,** 645 (1974b).

Duda, E., and M. Schlesinger, *J. Virol.*, **15,** 416 (1975).

Duksin, D., and P. Bornstein, *Proc. Natl. Acad. Sci. U.S.*, **74,** 3433 (1977a).

Duksin, D., and P. Bornstein, *J. Biol. Chem.*, **252,** 955 (1977b).

Eckardt, K., H. Thrum, G. Bradler, R. Fugner, E. Tonew, M. Tonew, and H. Tkocz, *Z. Allg. Mikrobiol.*, **13,** 625 (1973).

Eckardt, K., H. Thrum, G. Bradler, T. Tonew, and M. Tonew, *J. Antibiot.*, **28,** 274 (1975).

Eguchi, J., S. Sasaki, N. Ohta, T. Akashiba, T. Tsuchiyama, and S. Suzuki, *Am. Phytopathol. Soc.*, **34,** 280 (1968).

Elstner, E. F., and R. J. Suhadolnik, *Biochemistry*, **11,** 2578 (1972).

Elstner, E. F., D. M. Carnes, R. J. Suhadolnik, G. P. Kreishman, M. P. Schweizer, and R. K. Robins, *Biochemistry*, **12,** 4992 (1973a).

Elstner, E. F., R. J. Suhadolnik, and A. Allerhand, *J. Biol. Chem.*, **248,** 5385 (1973b).

Endo, A., and T. Misato, *Biochem. Biophys. Res. Commun.*, **37,** 718 (1969).

Endo, A., K. Kakiki, and J. Misato, *J. Bacteriol.*, **104,** 189 (1970).

Ericson, M. C., J. T. Gafford, and A. D. Elbein, *J. Biol. Chem.*, **252,** 7431 (1977).

Forsee, W. T., and A. D. Elbein, *J. Biol. Chem.*, **250,** 9283 (1975).

Forsee, W. T., G. Valkovich, and A. D. Elbein, *Arch. Biochem. Biophys.*, **174,** 469 (1976).

Fujisawa, J., Y. Iwakura, and Y. Kawade, *J. Biol. Chem.*, **253,** 8677 (1978).

Fuller, R. W., and R. Nagarajan, *Biochem. Pharmacol.*, **27,** 1981 (1978).

Funayama, S., and K. Isono, *Biochemistry*, **14,** 5568 (1975).

Funayama, S., and K. Isono, *Agric. Biol. Chem.* (*Tokyo*), **40,** 1039 (1976).

Funayama, S., and K. Isono, *Biochemistry*, **16,** 3121 (1977).

Garoff, H., and R. T. Schwarz, *Nature*, **274,** 487 (1978).

Gibson, R., Schlesinger, S., and Kornfeld, S., *J. Biol. Chem.*, in press (1979).

Gooday, G. W., A. de Rousset-Hall, and D. Hunsley, *Trans. Br. Mycol. Soc.*, **67,** 193 (1976).

Gordee, R. S., and T. F. Butler, *J. Antibiot.*, **26,** 466 (1973).

Gregory, J. D., *J. Am. Chem. Soc.*, **77,** 3922 (1955).

Hadler, H. I., *Medikon*, **3,** 22 (1974).

Hadler, H. I., and J. M. Demetriou, *Biochemistry*, **14,** 5374 (1975).

Hadler, H. I., B. E. Claybourn, and T. P. Tschang, *Biochem. Biophys. Res. Commun.*, **31,** 25 (1968).

Hadler, H. I., B. G. Daniel, and R. D. Pratt, *J. Antibiot.*, **24,** 405 (1971).

Hamill, R. L., and M. M. Hoehn, *J. Antibiot.*, **26,** 463 (1973).

Hamill, R. L., C. B. Carrell, S. M. Nash, and R. Nagarajan, 17th Annual ICAAC Meeting, New York, 1977, Abstr. 48.

Hamilton, R. H., and M. Z. Fall, *Experientia*, **27,** 229 (1971).

Hancock, I. C., G. Wiseman, and J. Baddiley, *FEBS Lett.*, **69,** 75 (1976).

Harada, S., and T. Kishi, *J. Antibiot.*, **30,** 11 (1977).

Hart, G. W., and W. J. Lennarz, *J. Biol. Chem.*, **253,** 5795 (1978).

Hashimoto, T., M. Kito, T. Takeuchi, M. Hamada, K. Maeda, Y. Okami, and H. Umezawa, *J. Antibiot.*, **21A,** 37 (1968).

Hasilik, A., and W. Tanner, *Antimicrob. Agents Chemother.*, **10,** 402 (1976).

Hatanaka, S., *Acta Chem. Scand.*, **16,** 513 (1962).

Heip, J., G. C. Chatterjee, J. Vandekerckhove, M. van Montagu, and J. Schell, *Arch. Int. Physiol. Biochem.*, **83,** 974 (1975).

Hickman, S., A. Kulczycki, Jr., R. G. Lynch, and S. Kornfeld, *J. Biol. Chem.*, **252,** 4402 (1977).

Hickman, S., and S. Kornfeld, *J. Immunol.*, **121,** 990 (1978).

Honke, T., M. Tanaka, and S. Nakamura, *J. Antibiot.*, **30,** 439 (1977).

Hopwood, J. J., and A. Dorfman, *Biochem. Biophys. Res. Commun.*, **75,** 472 (1977).

Hori, M., K. Kakiki, S. Suzuki, and T. Misato, *Agr. Biol. Chem.*, **35,** 1280 (1971).

Hori, M., K. Kakiki, and T. Misato, *Agr. Biol. Chem.*, **38,** 691 (1947a).

Hori, M., K. Kakiki, and T. Misato, *Agr. Biol. Chem.*, **38,** 699 (1974b).

Hori, M., K. Kakiki, and T. Misato, *J. Antibiot.*, **28,** 237 (1975).

Hori, M., K. Kakiki, and T. Misato, *J. Pestic. Sci.*, **1,** 31 (1976).

Hori, M., K. Kakiki, and T. Misato, *J. Pestic. Sci.*, **2,** 139 (1977).

Huang, M., H. Shimizu, and J. W. Daly, *J. Med. Chem.*, **15,** 462 (1972).

Ishizaki, H., K. Mitsuoka, and H. Unoh, *Ann. Phytopathol. Soc. Jap.*, **40,** 433 (1974).

Ishizaki, H., K. Mitsuoka, M. Kohno, and H. Kunoh, *Ann. Phytopathol. Soc. Jap.*, **42,** 35 (1976).

Isono, K., and R. J. Suhadolnik, *Arch. Biochem. Biophys.*, **173,** 141 (1976).

Isono, K., and S. Suzuki, *Agr. Biol. Chem.*, **30,** 813 (1966).

Isono, K., and S. Suzuki, 156th Meeting, American Chemical Society, Atlantic City, N.J., Abstracts Medi, 1968, p. 35.

Isono, K., J. Nagatsu, Y. Kawashima, and S. Suzuki, *Agr. Biol. Chem.*, **29,** 848 (1965).

Isono, K., J. Nagatsu, K. Kobinata, K. Sasaki, and S. Suzuki, *Agr. Biol. Chem.*, **31,** 190 (1967).

Isono, K., K. Asahi, and S. Suzuki, *J. Am. Chem. Soc.*, **91,** 7490 (1969).

Isono, K., T. Azuma, and S. Suzuki, *Chem. Pharm. Bull.*, **19,** 505 (1971).

Isono, K., S. Suzuki, M. Tanaka, T. Nanbata, and K. Shibuya, *Agr. Biol. Chem.*, **36,** 1571 (1972).

Isono, K., P. F. Crain, T. J. Odiorne, J. A. McCloskey, and R. J. Suhadolnik, *J. Am. Chem.* Soc., **95,** 5788 (1973).

Isono, K., P. F. Crain, and J. A. McCloskey, *J. Am. Chem. Soc.*, **97,** 943 (1975a).

Isono, K., S. Funayama, and R. J. Suhadolnik, *Biochemistry*, **14,** 2992 (1975b).

Isono, K., T. Sato, K. Hirasawa, S. Funayama, and S. Suzuki, *J. Am. Chem. Soc.*, **100,** 3937 (1978).

Ito, T., Y. Kodama, K. Kawamura, K. Suzuki, A. Takatsuki, and G. Tamura, *Agr. Biol. Chem.*, **41,** 2303 (1977).

Ito, T., Y. Kodama, K. Kawamura, K. Suzuki, A. Takatsuki, and G. Tamura, *Agr. Biol. Chem.*, **42,** (1978).

Iwasa, T., T. Kishi, K. Matsuura, and O. Wakae, *J. Antibiot.*, **30,** 1 (1977).

Just, G., and S. Kim, *Tetrahedron Lett.*, **1976,** 1063.

Just, G., and G. P. Donnini, *Can. J. Chem.*, **55,** 2998 (1977).

Just, G., and M.-I. Lim, *Can. J. Chem.*, **55,** 2993 (1977).

Kalman, T. I., *Biochemistry*, **10,** 2567 (1971).

Kalman, T. I., *Biochem. Biophys. Res. Commun.*, **49,** 1007 (1972).

Kalman, T. I., Abstracts, 168th Meeting of the American Chemical Society, 1974, Biol. #128.

Kaluza, G., C. Scholtissek, and R. Rott, *J. Gen. Virol.*, **14,** 251 (1972).

Kaluza, G., M. F. G. Schmidt, and C. Scholtissek, *Virology*, **54,** 179 (1973).

Kalvoda, L., *J. Carbohydr. Nucleosides Nucleotides*, **3,** 47 (1976).

Kalvoda, L., J. Farkas, and F. Šorm, *Tetrahedron Lett.*, **1970,** 2297.

Katoh, Y., A. Kuninaka, H. Yoshino, A. Takatsuki, M. Yamasaki, and G. Tamura, *J. Gen. Appl. Microbiol.*, **22,** 247 (1976).

Katoh, Y., A. Kuninaka, H. Yoshino, A. Takatsuki, M. Yamasaki, and G. Tamura, *J. Gen. Appl. Microbiol.*, **24,** 233 (1978).

Katz, F. N., J. E. Rothman, V. R. Lingappa, G. Blobel, and H. F. Lodish, *Proc. Natl. Acad. Sci. U.S.*, **74,** 3278 (1977).

Keiley, M. L., G. S. McKnight, and R. T. Schimke, *J. Biol. Chem.*, **251,** 5490 (1976).

Keller, F. A., and E. Cabib, *J. Biol. Chem.*, **246,** 160 (1971).

Keller, R. K., and G. Swank, *Biochem. Biophys. Res. Commun.*, **85,** 762 (1978).

Kerr, A., and W. P. Roberts, *Physiol. Plant Pathol.* **9,** 205 (1976).

Klenk, H.-D., C. Scholtissek, and R. Rott, *Virology*, **49,** 723 (1972).

Kneifel, H., W. A. König, G. Wolf, and H. Zähner, *J. Antibiot.* **27,** 20 (1974).

Komatsu, Y., *Agr. Biol. Chem.*, **35,** 1328 (1971a).

Komatsu, Y., *J. Antibiot.*, **24,** 876 (1971b).

Komatsu, Y., and K. Tanaka, *Agr. Biol. Chem.*, **32,** 1021 (1968).

Komatsu, Y., and K. Tanaka, *Agr. Biol. Chem.*, **35,** 526 (1971).

Komatsu, Y., and K. Tanaka, *Biochim. Biophys. Acta.* **288,** 390 (1972).

Komatsu, Y., and K. Tanaka, *Biochim. Biophys. Acta*, **311,** 496 (1973).

Kuo, S.-C., and J. O. Lampen, *J. Bacteriol.*, **111,** 419 (1972).

Kuo, S.-C., and J. O. Lampen, *Biochem. Biophys. Res. Commun.*, **58,** 287 (1974).

Kuo, S.-C., and J. O. Lampen, *Arch. Biochem. Biophys.*, **172,** 574 (1976).

Kuzuhara, H., H. Ohrui, and S. Emoto, *Tetrahedron Lett.*, **1973,** 5055.

Lambert, P. A., I. C. Hancock, and J. Baddiley, *Biochim. Biophys. Acta*, **472,** 1 (1977).

Larsen, P. O., *Acta Chem. Scand.*, **21,** 1592 (1967).

Leavitt, R., S. Schlessinger, and S. Kornfeld, *J. Virol.*, **21,** 375 (1977).

Lehle, L., and W. B. Tanner, *FEBS Lett.*, **71,** 167 (1976).

Leslie, J., *Anal. Biochem.*, **10,** 162 (1965).

Leung, K.-K., and D. W. Visser, *J. Biol. Chem.*, **252,** 2492 (1977).

Lim, M., Ph.D. Thesis, McGill University, Montreal, 1976.

Lindahl, U., *Methods Enzymol.*, **28,** 676 (1972).

Lopez-Romero, E., and J. Ruiz-Herrera, *Antonie van Leeuwenhoek, J. Microbiol. Serol.*, **42,** 261 (1976).

Lowy, B. A., J. Davoll, and G. B. Brown, *J. Biol. Chem.*, **197,** 591 (1952).

Lucas, J. J., C. J. Waechter, and W. J. Lennarz, *J. Biol. Chem.*, **250,** 1992 (1975).

McArthur, H.A.I., F. M. Roberts, I. C. Hancock, and J. Baddiley, *FEBS Lett.*, **86,** 193 (1978).

McCardell, B. A., and C. F. Pootjes, *Antimicrob. Agents Chemother.*, **10,** 498 (1976).

Mantsch, H. H., I. Goia, M. Kezdi, O. Bârzu, M. Dânşoreanu, G. Jebeleanu, and N. G. Ty, *Biochemistry*, **14,** 5593 (1975).

Martinez, A. P., W. W. Lee, and L. Goodman, *J. Org. Chem.*, **31,** 3263 (1966).

Matsuura, S., O. Shiratori, and K. Katagiri, *J. Antibiot.* **17,** 234 (1964).

Mizrahi, A., J. A. O'Malley, W. A. Carter, S. Takatsuki, G. Tamura, and E. Sulkowski, *Biol. Chem.*, **253,** 7612 (1978).

Mizuno, M., Y. Shimojima, T. Sugawara, and I. Takeda, J., *Antibiot.* **24A,** 896 (1971).

Morell, S. A., V. E. Ayers, T. J. Greenwalt, and P. Hoffman, *J. Biol. Chem.*, **239,** 2696 (1964).

Nagarajan, R., B. Chao, D. E. Dorman, S. M. Nash, J. L. Occolowitz, and A. Schabel, 17th Annual ICAAC Meeting, New York, 1977, Abstr. 50.

Nakagawa, Y., H. Kano, Y. Tsukuda, and H. Koyama, *Tetrahedron Lett.*, 4105 (1967).

Nakamura, K., and R. W. Compans, *Virology*, **84,** 303 (1978).

Nakamura, S., M. Arai, K. Karasawa, and H. Yonehara, *J. Antibiot.*, **10A,** 248 (1957).

Nishimura, H., and Y. Komatsu, *J. Antibiot.*, **21,** 250 (1968).

Nishimura, H., M. Moyama, Y. Komatsu, H. Kato, N. Shimaoka, and Y. Tanaka, *J. Antibiot.*, **17A,** 148 (1964).

Ohta, N., K. Kakiki, and T. Misato, *Agr. Biol. Chem.*, **34,** 1224 (1970).

Olden, K., R. M. Pratt, and K. M. Yamada, *J. Cell Biol.*, **75,** Abstr. CJ885 (1977).

Ozaki, M., T. Kariya, H. Kato, and T. Kimura, *Agr. Biol. Chem.*, **36,** 451 (1972).

Palmiter, R. D., T. Oka, and R. T. Schimke, *J. Biol. Chem.*, **250,** 724 (1971).

Perlman, D., A. B. Mauger, and H. Weissbach, *Antimicrob. Agents Chemother.*, **1966,** 581.

Pettit, G. R., R. H. Ode, R. M. Coomes, and S. L. Ode, *Lloydia*, **39,** 363 (1976).

Plaut, G. W. E., C. P. Cheung, R. J. Suhadolnik, and T. Aogaichi, *Biochemistry*, (1979).

Pugh, C. S. G., R. T. Borchardt, and H. O. Stone, *J. Biol. Chem.*, **253,** 4075 (1978).

Ralph, R. K., and H. G. Khorana, *J. Am. Chem. Soc.*, **83,** 2926 (1961).

Richmond, M. E., S. DeLuca, and J. E. Silbert, *Biochemistry*, **12,** 3898 (1973).

Roberts, W. P., M. E. Tate, and A. Kerr, *Nature*, **265,** 379 (1977).

Roberts, F. M., H. A. I. McArthur, I. C. Hancock, and J. Baddiley, *FEBS Lett.*, **79,** 211 (1979).

Rodén, L., J. R. Baker, T. Helting, N. B. Schwartz, A. C. Stoolmiller, S. Yamagata, and T. Yamagata, *Methods Enzymol.*, **28,** 638 (1972).

Rothman, E. J., and H. F. Lodish, *Nature*, **269,** 775 (1977).

Roy-Burman, S., and D. W. Visser, *Biochim. Biophys. Acta*, **282,** 383 (1972).

Roy-Burman, S., and D. W. Visser, *J. Biol. Chem.*, **250,** 9270 (1975).

Roy-Burman, S., P. Roy-Burman, and D. W. Visser, *Cancer Res.*, **28,** 1605 (1968).

Roy-Burman, S., Y. H. Huang, and D. W. Visser, *Biochem. Biophys. Res. Commun.*, **42,** 445 (1971).

Sakata, K., and J. Uzawa, *Agr. Biol. Chem.*, **41,** 413 (1977).

Sakata, K., A. Sakurai, and S. Tamura, *Agr. Biol. Chem.*, **37,** 697 (1973).

Sakata, K., A. Sakurai, and S. Tamura, *Tetrahedron Lett.*, **1974a,** 4327.

Sakata, K., A. Sakurai, and S. Tamura, *Tetrahedron Lett.*, **1974b,** 1533.

Sakata, K., A. Sakurai, and S. Tamura, *Agr. Biol. Chem.*, **38,** 1883 (1974c).

Sakata, K., A. Sakurai, and S. Tamura, *Agr. Biol. Chem.*, **39,** 885 (1975a).

Sakata, K., A. Sakurai, and S. Tamura, *Tetrahedron Lett.*, **1975b,** 3191.

Sakata, K., A. Sakurai, and S. Tamura, *Agr. Biol. Chem.*, **40,** 1993 (1976).

Sakata, K., A. Sakurai, and S. Tamura, *Agr. Biol. Chem.*, **41,** 2027 (1977a).

Sakata, K., A. Sakurai, and S. Tamura, *Agr. Biol. Chem.*, **41,** 2033 (1977b).

Sakata, K., J. Uzawa, and A. Sakurai, *Org. Magn. Resonance*, **10,** 230 (1977c).

Schell, J., M. van Montagu, M. De Beuckeleer, M. De Block, A. Depicker, M. De Wilde, G. Engler, C. Genetello, J. P. Hernalsteens, M. Holsters, J. Seurinck, B. Silva, F. van Vliet, and R. Villarroel, in Proceedings Royal Society of London, in press (1979).

Scholtissek, C., R. Rott, G. Hau, and G. Kaluza, *J. Virol.*, **13,** 1186 (1974).

Scholtissek, C., R. Rott, and H. D. Klenk, *Virology*, **63,** 191 (1975).

Schwarz, R. T., and H. D. Klenk, *J. Virol.*, **14,** 1023 (1974).

Schwarz, R. J., J. M. Rohrschneider, and M. F. G. Schmidt, *J. Virology*, **19,** 782 (1976).

Schwiezer, M. P., K. McGrath, and L. Baczynski, *Biochem. Biophys. Res. Commun.*, **40,** 1046, (1970).

Sharpless, N. E., and M. Flavin, *Biochemistry*, **5,** 2963 (1966).

Shibuya, K., M. Tanaka, T. Nanbata, K. Isono, and S. Suzuki, *Agr. Biol. Chem.*, **36,** 1229 (1972).

Shirato, S., K. Yoshide, and Y. Miyazaki, *J. Ferment. Technol.*, **46,** 233 (1968).

Shoji, J., R. Sakazaki, M. Mayama, Y. Kawamura, and Y. Yasuda, *J. Antibiot.* (*Tokyo*), **23,** 295 (1970).

Shomura, T., K. Hamamoto, T. Ohashi, S. Amano, J. Yoshida, C. Moriyama, and T. Niida, *Sci. Rep. Meiji Seika Kaisha*, **9,** 5 (1967).

Silbert, J. E., and A. C. Repucci, Jr., *J. Biol. Chem.*, **251,** 3948 (1976).

Struck, D. K., and W. J. Lennarz, *J. Biol. Chem.*, **252,** 1007 (1977).

Struck, D. K., P. B. Siuta, M. D. Lane, and W. J. Lennarz, *J. Biol. Chem.*, **253,** 5332 (1978).

Suhadolnik, R. J., *Nucleoside Antibiotics*, Wiley, New York, 1970.

Takaoka, K., T. Kuwayama, and A. Aoki, Japanese Patent 615, 332 (1971).

Takatsuki, A., *J. Agr. Chem. Soc. Japan*, **52,** R167 (1978).

Takatsuki, A., and G. Tamura, *J. Antibiot.*, **24,** 224 (1971a).

Takatsuki, A., and G. Tamura, *J. Antibiot.*, **24,** 785 (1971b).

Takatsuki, A., and G. Tamura, *J. Antibiot.*, **25,** 362 (1972).

Takatsuki, A., and G. Tamura, *Agr. Biol. Chem.*, **42,** 275 (1978).

Takatsuki, A., K. Arima, and G. Tamura, *J. Antibiot.*, **24,** 215 (1971).

Takatsuki, A., K. Shimizu, and G. Tamura, *J. Antibiot.*, **25,** 75 (1972).

Takatsuki, A., K. Kohno, and G. Tamura, *Agr. Biol. Chem.*, **39,** 2089 (1975).

Takatsuki, A., K. Kohno, M. Nishimura, and G. Tamura, Abstracts, Annual Meeting of the Agricultural Chemical Society of Japan, April 1976, p. 131.

Takatsuki, A., K. Kawamura, M. Okina, Y. Kodama, T. Ito, and G. Tamura, *Agr. Biol. Chem.*, **41,** 2307 (1977a).

Takatsuki, A., Y. Fukui, and G. Tamura, *Agr. Biol. Chem.*, **41,** 425 (1977b).

Takatsuki, A., M. Munekata, M. Nishimura, K. Kohno, K. Onodera, and G. Tamura, *Agr. Biol. Chem.*, **41,** 1831 (1977c).

Takatsuki, A., T. Kobayashi, T. Sasaki, and G. Tamura, Abstracts of Papers, Annual Meeting of the Agricultural Chemical Society of Japan, p. 352 (1978a).

Takatsuki, A., Y. Fukui, and G. Tamura, *Agr. Biol. Chem.*, **42,** 1621 (1978b).

Takatsuki, A., K. Kawamura, Y. Kodama, T. Ito, and G. Tamura, *Agr. Biol. Chem.*, **43,** 761 (1979).

Tamura, G., T. Sasaki, M. Matsuhashi, A. Takatsuki, and M. Yamasaki, *Agr. Biol. Chem.*, **40,** 447 (1976).

Thompson, T. J., R. H. Hamilton, and C. F. Pootjes, *Plant Physiol.* (*Suppl.*), **59,** 110 (1977).

Thompson, R. J., R. H. Hamilton, and C. F. Pootjes, *Antimicrob. Agents Chemotherapy*, in press.

Thrum, H., K. Eckardt, G. Bradler, R. Fugner, E. Tonew, and M. Tonew, *J. Antibiot.*, **28,** 514 (1975).

Titani, Y., and Y. Katsube, *Biochim. Biophys. Acta*, **192,** 367 (1969).

Tkacz, J. S., and J. O. Lampen, *Biochem. Biophys. Res. Commun.*, **65,** 248 (1975).

Tkacz, J. S., S.-C. Kuo, and J. O. Lampen in *Alcohol, Industry and Research*, O. Forsander, K. Eriksson, E. Oura, and P. Jounela-Eriksson, Eds., Research Laboratories of the Finnish State Alcohol Monopoly, Helsinki, 1977, p. 147.

Tonew, E., M. Tonew, K. Eckardt, H. Thrum, and B. Gumpert, *Acta Virol.*, **19,** 311 (1975).

Townsend, L. B., in *Handbook of Biochemistry and Molecular Biology, Nucleic Acids, Vol. I*, 3rd ed, G. D. Fasman, Ed., CRC Press, Cleveland, 1975, p. 271.

Trummlitz, G., and J. G. Moffatt, *J. Org. Chem.*, **38,** 1841 (1973).

Trummlitz, G., D. B. Repke, and J. G. Moffatt, *J. Org. Chem.*, **40,** 3352 (1975).

Tsukuda, Y., and H. Koyama, *J. Chem. Soc.* (*B*), **1970,** 1209.

Tsuruoka, T., H. Yumoto, N. Ezaki, and T. Niida, *Sci. Rep. Meiji Seika Kaisha*, **9,** 1 (1967).

Turco, S. J., and E. C. Heath, *J. Biol. Chem.*, **252,** 2918 (1977).

Turner, J. R., T. F. Butler, R. W. Fuller, and N. V. Owen, Abstracts, 17th Annual ICAAC Meeting, New York, 1977, Abstr. 49.

Uramoto, M., J. Uzawa, S. Suzuki, K. Isono, J. G. Liehr, and J. A. McCloskey, *Nucleic Acids Res.*, Spec. Publ. No. 5, *s327* (1978).

van der Valk, P., and J. G. H. Wessels, *Protoplasma*, **90,** 65 (1976).

van Larebeke, N., G. Engler, M. Holsters, S. van den Elsacker, I. Zaenen, R. A. Schilperoort, and J. Schell, *Nature*, **252,** 169 (1974).

van Larebeke, N., C. Genetillo, J. Schell, R. A. Schilperoort, A. K. Hermans, J. P. Hernalsteens, and M. van Montagu, *Nature*, **255,** 742 (1975).

Vardanis, A., *Experientia*, **34,** 228 (1977).

Virtanen, A. I., and T. Ettala, *Acta Chem. Scand.*, **11,** 182 (1957).

Waechter, C. J., and W. J. Lennarz, *Ann. Rev. Biochem.*, **45,** 95 (1976).

Ward, J. B., *FEBS Lett.*, **78,** 151 (1977).

Watanabe, S., *J. Antibiot.*, **23,** 313 (1970).

Watson, B., T. C. Currier, M. P. Gordon, M. D. Chilton, and E. W. Nester, *J. Bacteriol.*, **123,** 255 (1975).

Weston, A., and H. R. Perkins, *FEBS Lett.*, **76,** 195 (1977).

Wood, W. A., *The Enzymes*, **7,** 281 (1972).

Wyke, A. W., and J. B. Ward, *J. Bacteriol.*, **130,** 1055 (1977).

Yoshikawa, M., T. Kato, and T. Takenishi, *Bull. Chem. Soc. Jap.*, **42,** 3505 (1969).

Zaenen, I., N. van Larabeke, H. Teuchy, M. van Montagu, and J. Schell, *J. Mol. Biol.*, **86,** 109 (1974).

Chapter 2 Inhibition of Protein Synthesis

2.1 THE AMINOACYLAMINOHEXOSYLCYTOSINE ANALOGS 73

Blasticidin H 73
- Discovery, Production, and Isolation 73
- Physical and Chemical Properties 74
- Structural Elucidation 74
- Inhibition of Growth 75
- Biosynthesis of Blasticidin S and H 75

Hikizimycin 77
- Discovery, Production, and Isolation 77
- Physical and Chemical Properties 77
- Structural Elucidation 78
- Inhibition of Growth 78

Mildiomycin 79
- Discovery, Production, and Isolation 79
- Physical and Chemical Properties 80
- Structural Elucidation 80
- Inhibition of Growth 80

Norplicacetin 80
- Discovery, Production, and Isolation 80
- Physical and Chemical Properties 81
- Structural Elucidation 81
- Inhibition of Growth 81

Oxamicetin 81
- Discovery, Production, and Isolation 81
- Physical and Chemical Properties 82
- Structural Elucidation 82
- Inhibition of Growth 84

"Aspiculamycin": Its Identity with Gougerotin 84
- Chemical Synthesis 84

Biochemical Properties of the Aminoacylaminohexosylcytosine Analogs 86

The three groups of naturally occurring analogs that inhibit protein synthesis are (i) the aminoacylaminohexosylcytosine analogs,* (ii) the purine nucleoside analogs, and (iii) the clindamycin ribonucleotides.

2.1 THE AMINOACYLAMINOHEXOSYLCYTOSINE ANALOGS

Ten naturally occurring pyrimidine nucleoside analogs (aminoacylaminohexosylcytosine analogs) have been isolated from the *Streptomyces*. They are amicetin, bamicetin, blasticidin S, blasticidin H, gougerotin, hikizimycin, mildiomycin, norplicacetin, oxamicetin, and plicacetin.

These 10 nucleoside analogs have similar structures and similar inhibitory actions (Černá et al., 1973; Lichtenthaler et al., 1975a; Menzel and Lichtenthaler, 1975). Because of the common structure of the aminoacylaminohexosylcytosine analogs (Lichtenthaler and Trummlitz, 1974), they all inhibit peptidyl transferase, which subsequently blocks the transfer of amino acids from aminoacyl tRNA to polypeptide (Clark and Chang, 1965). Amicetin (amicetin A), bamicetin (amicetin C), blasticidin S, gougerotin (aspiculamycin, asteromycin, moroyamycin), and plicacetin (Fig. 2.1) have been reviewed in detail (Suhadolnik, 1970; Townsend, 1975; Bloch, 1978; Nakamura and Kondo, 1977). Their structures are shown here and their biochemical properties are described on page 86 with those of all the aminoacylaminohexosyl cytosine analogs because their mode of action is so closely related. The discovery, production, isolation, structural elucidation, physical and chemical properties, and chemical synthesis of the new aminoacylaminohexosyl cytosine analogs, blasticidin H, hikizimycin, mildiomycin, oxamicetin, and norplicacetin, are treated in separate sections. The structural studies on gougerotin and "aspiculamycin" are included here because, contrary to earlier reports, gougerotin and "aspiculamycin" have the same structure.

Blasticidin H

While screening for compounds with ultraviolet adsorption characteristics similar to those of 1-substituted cytosine from the culture filtrates of *S. griseochromogenes*, Seto and Yonehara (1977) isolated blasticidin H (Fig. 2.2).

Discovery, Production, and Isolation. Blasticidin H is isolated and purified by ion exchange chromatography, adsorption onto active carbon, elution with 60% aqueous acetone, and passage through a column of Sephadex G-

* Also referred to as the 4-aminohexose pyrimidine nucleosides.

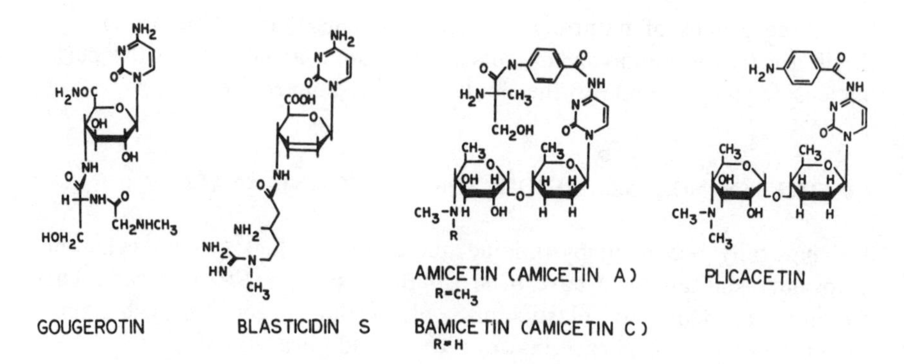

Figure 2.1 Structures of the aminoacylaminohexosylcytosine nucleosides.

15. It is crystallized from water and recrystallized from dilute HCl–ethanol (Seto and Yonehara, 1977).

Physical and Chemical Properties. The molecular formula of blasticidin H is $C_{17}H_{28}N_8O_6 \cdot 2\ HCl \cdot H_2O$; mol wt 530; mp 230–235°C (decomp.); pK_a 2.8 (carboxylic acid), 4.2 (amino group of cytosine), 8.2 (amine), and 12.5 (guanidine); ultraviolet spectral properties: λ_{max}^{H+} 277 nm (ϵ = 13,500) and λ_{max}^{OH-} 270 (ϵ = 9600) (Seto and Yonehara, 1977).

Structural Elucidation. The comparison of the nmr spectra of blasticidins H and S is shown in Fig. 2.3. The amino acid moiety (blastidic acid) and the unsubstituted amino function are present in blasticidin H. The olefinic protons of blasticidin S disappear in blasticidin H. Acid hydrolysis of blasti-

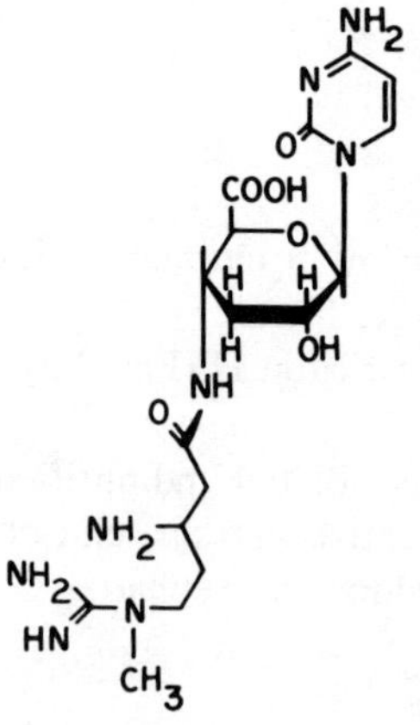

Figure 2.2 Structure of blasticidin H (Seto and Yonehara, 1977).

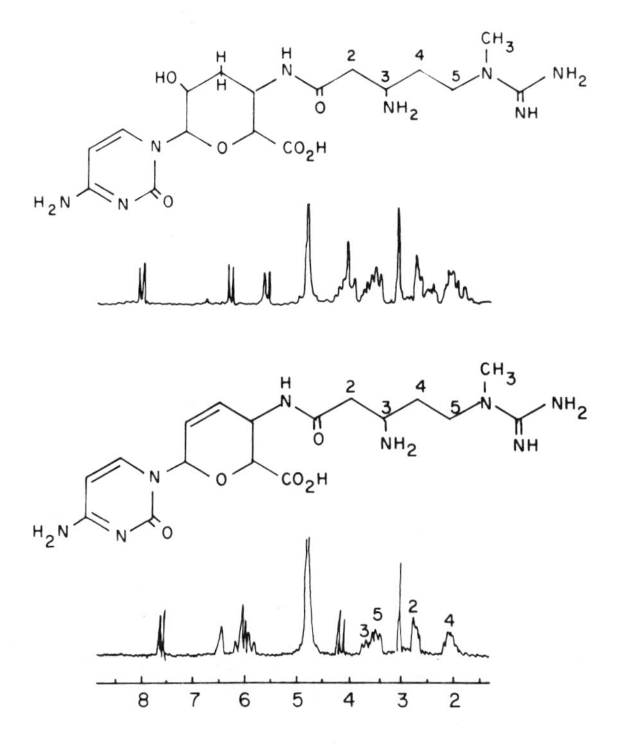

Figure 2.3 Blasticidin S (lower trace) and blasticidin H (upper trace) ¹HNMR spectra. Modified from Seto and Yonehara, 1977, reprinted with permission.

cidin H gives rise to pentopyranamine D (see page 155). Acetylation of the product following alkaline treatment of blasticidin H, cytomycin, gave the *N,O*-diacetate whose ir, uv, and nmr spectra show that the hydroxy function is at C-2 and not C-4. Because the amino group at C-4 of the sugar is not free, it must be connected through an amide bond. Therefore, the structure assigned to blasticidin H is 1-(4-blastidylamino-3,4-dideoxy-β-D-ribohexopyranosyluronic acid) cytosine (Seto and Yonehara, 1977).

Inhibition of Growth. Blasticidin H is slightly active against *P. oryzae* (Seto and Yonehara, 1977).

Biosynthesis of Blasticidin S and H. The isolation of the 3′-deoxynucleoside, pentopyranine C (see page 155 for structure), suggests that the double bond in the sugar moiety of cytosinine can be introduced by dehydration of the 2′-hydroxyl and 3′-hydrogen and not its 2′-hydrogen and 3′-hydroxyl isomer (Seto et al., 1973).

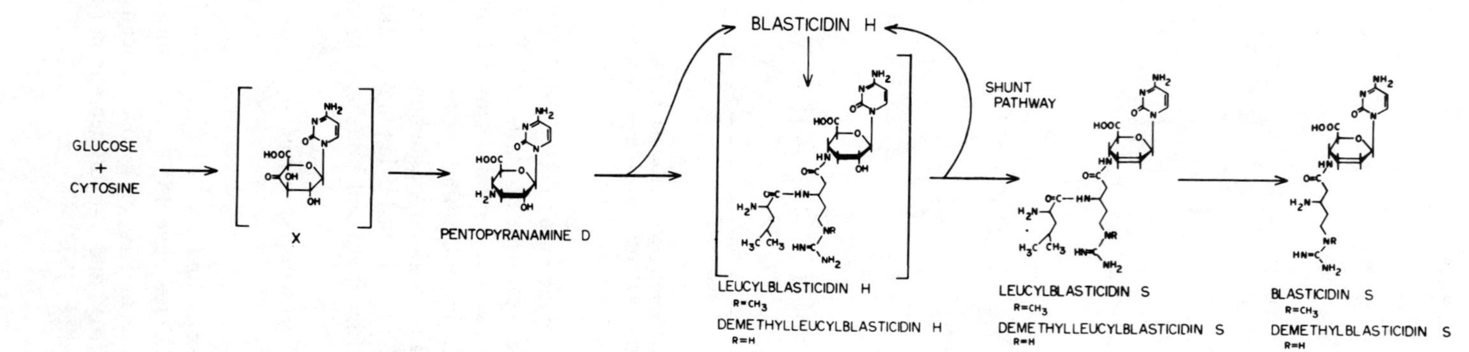

Figure 2.4 Biosynthesis of blasticidin S (Seto, personal communication).

Yonehara and Ōtake (1965) reported that cytosinine and blastidic acid were taken up by washed cultures of *Streptomyces griseochromogenes*, which resulted in increased formation of blasticidin S. The uptake of [U-^{14}C]cytidine resulted in the incorporation of ^{14}C into the cytosine of blasticidin S. Seto et al. (1976) suggest that the isolation of pentopyranic acid (see page 155 for structure) from the culture medium of *S. griseochromogenes* is evidence that this acid is a precursor of the hypothetical intermediate (Fig. 2.4). The oxidation of the hydroxymethyl group to the carboxyl group of the hexose moiety occurs prior to the oxidation at C-4′. The isolation of blasticidin H and demethylblasticidin S (Seto and Yonehara, 1977) suggests that these two compounds may be intermediates for the biosynthesis of blasticidin S (Fig. 2.4).

Hikizimycin

This naturally occurring pyrimidine nucleoside disaccharide analog was first isolated by Hamill and Hoehn (1964) from *Streptomyces longissmus* and subsequently by Uchida et al. (1971) from *Streptomyces* A-5. They assigned the names anthelmycin and hikizimycin, respectively (Fig. 2.5). The importance of this nucleoside is its anthelmintic activity.

Discovery, Production, and Isolation. The culture medium, growth conditions, and isolation procedure for hikizimycin (anthelmycin) have been described (Hamill and Hoehn, 1964). Hikizimycin is crystallized as colorless plates by the addition of acetone to water containing the nucleoside; yield 150 g/800 liters culture medium.

Physical and Chemical Properties. The molecular formula for hikizimycin is $C_{21}H_{37}N_5O_{14}$; mol wt 583; mp >200°C (decomp.); $[\alpha]_{10}^{25}$ 17.5° (*c* 1.58,

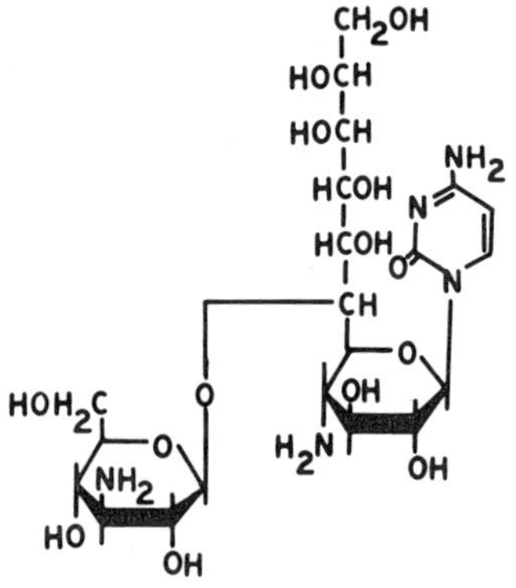

CH2OH
HOCH
HOCH
HCOH
HCOH
CH
NH2
N
O
N
O
OH
H2N
OH
HOH2C
O
O
NH2
HO
OH

Figure 2.5 Structure of hikizimycin (anthelmycin).

water); pK_a 3.3, 6.8, and 7.9 (in H_2O); ultraviolet spectral properties: $\lambda_{max}^{pH\,7,\,H_2O}$ 234 nm ($E_{1\,cm}^{1\,percent}$ 127), λ_{max}^{H+} 274 nm ($E_{1\,cm}^{1\,percent}$ 198) (ϵ = 8000), λ_{max}^{OH-} 267 nm ($E_{1\,cm}^{1\,percent}$ 135.5); infrared absorption spectral properties in mineral oil: 2.99, 6.04, 6.23, 6.69, 7.1 (sh), 7.77, 8.27, 8.65 (sh), 9.32, 9.7 (broad sh), 10.5 (broad sh), 11.3 and 12.8 μ (Hamill and Hoehn, 1964; Ennifar et al., 1977; Vuilhorgne et al., 1977). Hikizimycin is soluble in water, acid, and methanol (7 mg/ml) at room temperature, but insoluble in ethanol, acetone, chloroform, butanol, and ether (Hamill and Hoehn, 1964). The ^{13}C nmr spectral analysis has been described by Vuilhorgne et al. (1977) and Ennifar et al. (1977). Decoupling experiments in the ^1H nmr spectrum of hikizimycin trihydrobromide indicate that the N-anomeric proton at 5.70 is coupled to the adjacent 2′-proton at 3.88 in the C_{11} sugar (Ennifar et al., 1977). The X-ray powder patterns and the mass spectrum of hikizimycin have been described by Nagarajan and Nash (1974) and Ennifar et al. (1977). Hikizimycin is composed of 3-amino-3-deoxy-D-glucopyranose, the aminoundecose sugar, hikosamine, and cytosine.

Structural Elucidation. The first structural studies suggested a 1 → 2 or 1 → 3 glycosidic linkage between kanosamine (3-amino-3-deoxy-D-glucopyranose) and 4-amino-4-deoxyundecose which has been named hikosamine (Uchida and Das, 1972; Uchida, 1976; Nagarajan and Nash, 1974). However, mass spectrometry of derivatives of hikizimycin and the ^{13}C nmr rules out the 1 → 2 and 1 → 3 glycosidic linkage between 3-amino-3-deoxy-D-glycopyranose and hikosamine (Ennifar et al., 1977). That the sugar linkage is a β-1 → 6 glycosidic linkage was established by Vuilhorgne et al. (1977). The linkage of hikosamine is a β-nucleosidic linkage with cytosine (Das et al., 1972). The configuration of the polyol side chain of hikosamine resulted from the isolation and characterization of D-glycero-D-galactoheptose obtained by periodate oxidation of the C-4 and C-5 bond of hikosamine (Uchida, 1976). The X-ray powder patterns, the ^{13}C nmr spectra of the N,N'-diacetylhikizimycin and N,N'-diacetylanthelmycin established that the two antibiotics are identical (Ennifar et al., 1977; Nagarajan and Nash, 1974).

By hydrogenation, methylation, acid hydrolysis, and mass spectral, ^1H nmr, and ^{13}C nmr studies, the data obtained unequivocally establish that

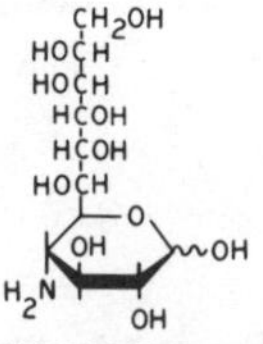

HIKOSAMINE

hikizimycin (anthelmycin) is a nucleoside analog consisting of 3-amino-3-deoxy-D-glucopyranose, linked β-1 $\rightarrow$ 6 to the C_{11} undecosamine, hikosamine, which has a β-link to cytosine.

Inhibition of Growth. Although hikizimycin exhibits weak broad antimicrobial activity, it is very effective in removing parasites such as pinworms, roundworms, whipworms, and strongyles from infected host animals (Hamill and Hoehn, 1964). In addition to its anthelmintic properties, the inhibitor also has housefly larvicidal activity. Concerning its mode of action, hikizimycin appears to be an inhibitor of transpeptidation with an inhibitory pattern similar to that of gougerotin (Uchida and Wolf, 1974).

Mildiomycin

Harada and Kishi (1978) described a new nucleoside analog, mildiomycin (Fig. 2.6), that shows strong activity against microorganisms that cause plant diseases (Iwasa et al., 1978). An interesting structural feature of mildiomycin is that the carboxyguanidino butyl group is covalently bound to the unsaturated pyranoside by a carbon–carbon bond (Harada et al., 1978).

Discovery, Production, and Isolation. Mildiomycin is isolated from the culture filtrates of *Streptoverticillium rimofaciens* B98891 found in a soil sample from Papua, New Guinea. The test assays for isolation and purification are based on *in vivo* tests using *Erysiphe graminis* on barley or inhibitory activity against *Rhodotorula rubra;* the fermentation medium is described by Iwasa et al. (1978). The isolation and purification are accomplished by ion exchange and adsorption chromatography. Mildiomycin is precipitated from methanol as a hygroscopic basic substance; yield 15 g/60 liters.

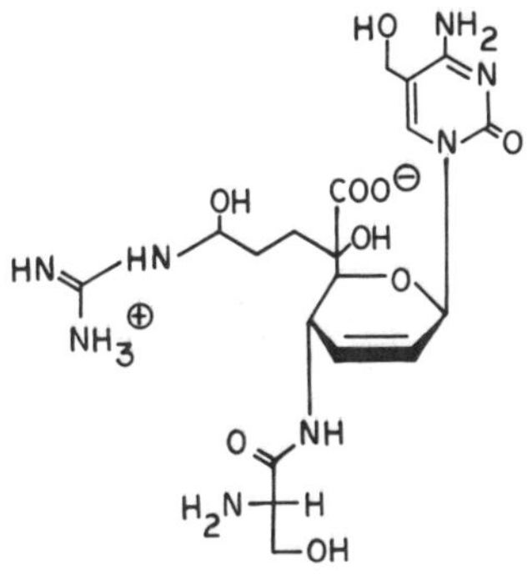

Figure 2.6 Structure of mildiomycin (Harada et al., 1978).

Physical and Chemical Properties. The molecular formula of mildiomycin is $C_{19}H_{30}N_8O_9 \cdot H_2O$; mol wt 532.53; mp > 300°C (decomp.), $[\alpha]_D^{23}$ + 100° (*c* 0.5, H_2O); p*K*a′ values: 2.8 (COO^-), 4.2 (3-NH^+=), 7.2 (2″-NH_3^+), and > 12 (guanidine); ultraviolet spectral properties: $\lambda_{max}^{pH\ 7.0,\ 0.1\ N\ NaOH}$ 271 nm (ϵ = 8720); $\lambda_{max}^{0.1 N HCl}$ 280 nm (ϵ = 13,100); mildiomycin is a water-soluble, basic nucleoside that is sparingly soluble in pyridine, dimethyl sulfoxide, dioxane or tetrahydrofuran; high pressure liquid chromatography tests for purity have been reported; 5-hydroxymethyl cytosine and L-serine are isolated following acid hydrolysis (2 *N* HCl, 100°C, 1 h) (Kishi et al., 1977; Harada and Kishi, 1978; Harada et al., 1978).

Structural Elucidation. The nmr and ^{13}C nmr spectra showed the presence of the six-membered ring; the guanidino butyl group was established by periodic oxidation; the CD spectra of mildiomycin follow blasticidin S and gougerotin; and ^{13}C nmr spectral evidence of mildiomycin and degradation products establish the structure (Harada et al., 1978).

Inhibition of Growth. Mildiomycin is strongly active against the powdery mildew fungus of barley and other plants; it shows weak activity against most fungi and bacteria. Of 1400 strains of yeast tested, 21 strains of *Rhodotorula* were highly susceptible; *Mycobacterium phlei* was strongly inhibited. Several phytopathogenic fungi are moderately susceptible to mildiomycin (Iwasa et al., 1978). It is also effective against certain species of plant-parasitic mites (*Tetranychus urticae*) in practical crop cultivation (Morimoto, personal communication).

Norplicacetin

Of the pyrimidine disaccharide nucleoside analogs this is the most recently isolated (Evans and Weare, 1977). Norplicacetin (Fig. 2.7) is the C-4″ mono-*N*-demethylated form of plicacetin and thus comprises the major structural features of bamicetin (Fig. 2.1), except for the α-methylseryl portion. Consequently, it may be considered a metabolic degradation product of bamicetin or the biosynthetic precursor thereof, a relationship that similarly applies to the pair plicacetin/amicetin.

Discovery, Production, and Isolation. Evans and Weare (1977) isolated norplicacetin from the culture filtrates of a *Streptomyces* isolated from a soil sample from Ghana. The fermentation medium, conditions for maximal production of norplicacetin, and the isolation have been described (Evans and Weare, 1977). Norplicacetin is readily crystallized from methanol as

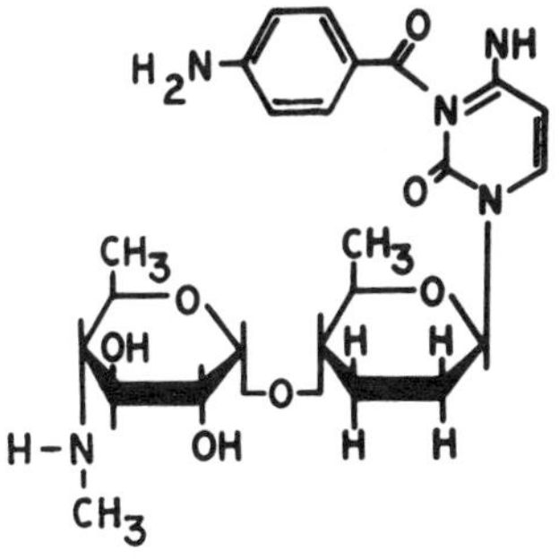

Figure 2.7 Structure of norplicacetin (Evans and Weare, 1977).

white crystals. Norplicacetin, amicetin, bamicetin, and plicacetin are isolated from the fermentation broth of *S. plicacetus* (Haskell et al., 1958), as well as from the *Streptomyces* isolated from the soil sample from Ghana.

Physical and Chemical Properties. Molecular formula: norplicaetin is $C_{24}H_{35}N_5O_7$; mol wt 507 (determined by mass spectrometry); mp 168–171°C; $[\alpha]_D^{23}$ +125° (*c* 1, methanol); ultraviolet spectral properties: $\lambda_{max}^{methanol}$ 253 nm ($E_{1cm}^{1 percent}$ 230), 325 nm ($E_{1cm}^{1 percent}$ 500) (Evans and Weare, 1977).

Structural Elucidation. The structure of norplicacetin was determined primarily by mass and nmr spectroscopy. The nmr spectrum is similar to that of plicacetin. The difference is that the singlet at τ 7.16 (NCH_3 protons) integrated for three protons. These data suggest that norplicacetin is a new member of the amicetin group of nucleoside antibiotics (Evans and Weare, 1977).

Inhibition of Growth. The antimicrobial effect of norplicacetin resembles that of plicacetin in that it is moderately inhibitory to gram-positive bacteria and mycobacteria (Evans and Weare, 1977).

Oxamicetin

Oxamicetin (Fig. 2.8) was isolated by Konishi et al. (1973) from *Streptomyces oxamicetus*. It is structurally similar to amicetin (Fig. 2.1). Its biochemical properties are probably similar to the aminoacylaminohexosyl cytosine analogs.

Discovery, Production, and Isolation. Oxamicetin is produced by *Arthrobacter oxamicetus* and is isolated by the shake culture method of Konishi et

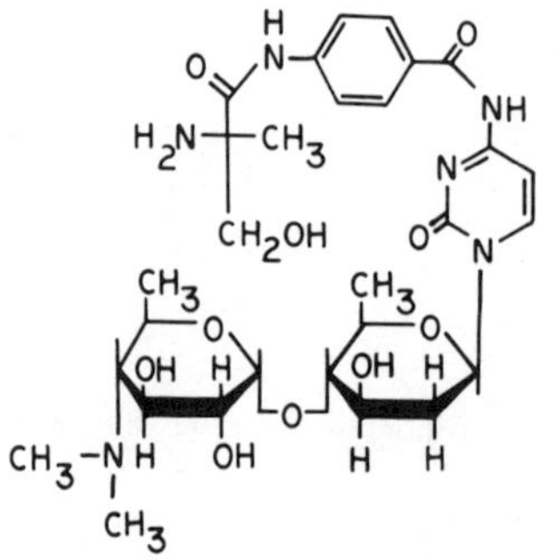

Figure 2.8 Structure of oxamicetin.

al. (1973) and Tomita et al. (1973). Colorless needlelike crystals form when crystallized overnight from water–HCl. Crystallization has also been accomplished from aqueous methanol at room temperature.

Physical and Chemical Properties. The molecular formula of oxamicetin is $C_{29}H_{42}N_6O_{10}$; mol wt 634.7 (mol. wt. by vapor pressure osmometry 639); mp (free base) 176–179°C; (hydrochloride) 205–216°C; $[\alpha]_D^{25}$ +66° (c 0.4, water); ultraviolet spectral properties: $\lambda_{max}^{(water)}$ 305 nm (ϵ = 31,700), $\lambda_{max}^{0.1\,N\,HCl}$ 316 nm (ϵ = 26,600), $\lambda_{max}^{0.1\,N\,NaOH}$ 322 nm (ϵ = 21,900); the ir spectrum and nmr spectrum have been reported (Konishi et al., 1973). Oxamicetin is a basic antibiotic. It is soluble in water, methanol, ethanol, and *n*-butanol, but insoluble in acetone, ethyl acetate, and other organic solvents (Konishi et al., 1973).

Structural Elucidation. On the basis of the physicochemical properties of oxamicetin described above, Konishi et al. (1973) suggest that the structure of oxamicetin is closely related to that of amicetin. Hydrolysis of oxamicetin (**1**) in methanolic HCl resulted in the isolation of cytimidine (**2**), and the *N*,*N*-dimethyl-4-amino sugar methyl glycoside, amosaminide (**3**) (Fig. 2.9). Further purification of the acid-hydrolyzed oxamicetin resulted in the isolation of nucleoside **4**, 1-(2,6-dideoxy-β-D-arabinohexopyranosyl)-cytosine. Compound **3** is stable to acid hydrolysis under conditions that cleave the amicetose–cytosine linkage in amicetin. Hydroxy cytosamine (**5**) is isolated by alkaline hydrolysis of **1**. The structure of **1** was elucidated by nmr and mass spectrometry (Konishi et al., 1973). The hexose attached to the cytosine of oxamicetin (**1**) has been determined to be 2,6-dideoxy-D-arabinohexose, 3-hydroxyamicetose. This sugar is also a constitutent of chromomycins and olivomycin (Miyamoto et al., 1964; Berlin et al., 1964).

The unequivocal synthesis of the nucleoside portion of oxamicetin, 1-(2,6-dideoxy-β-D-arabinohexopyranosyl)cytosine (**4**), was achieved by

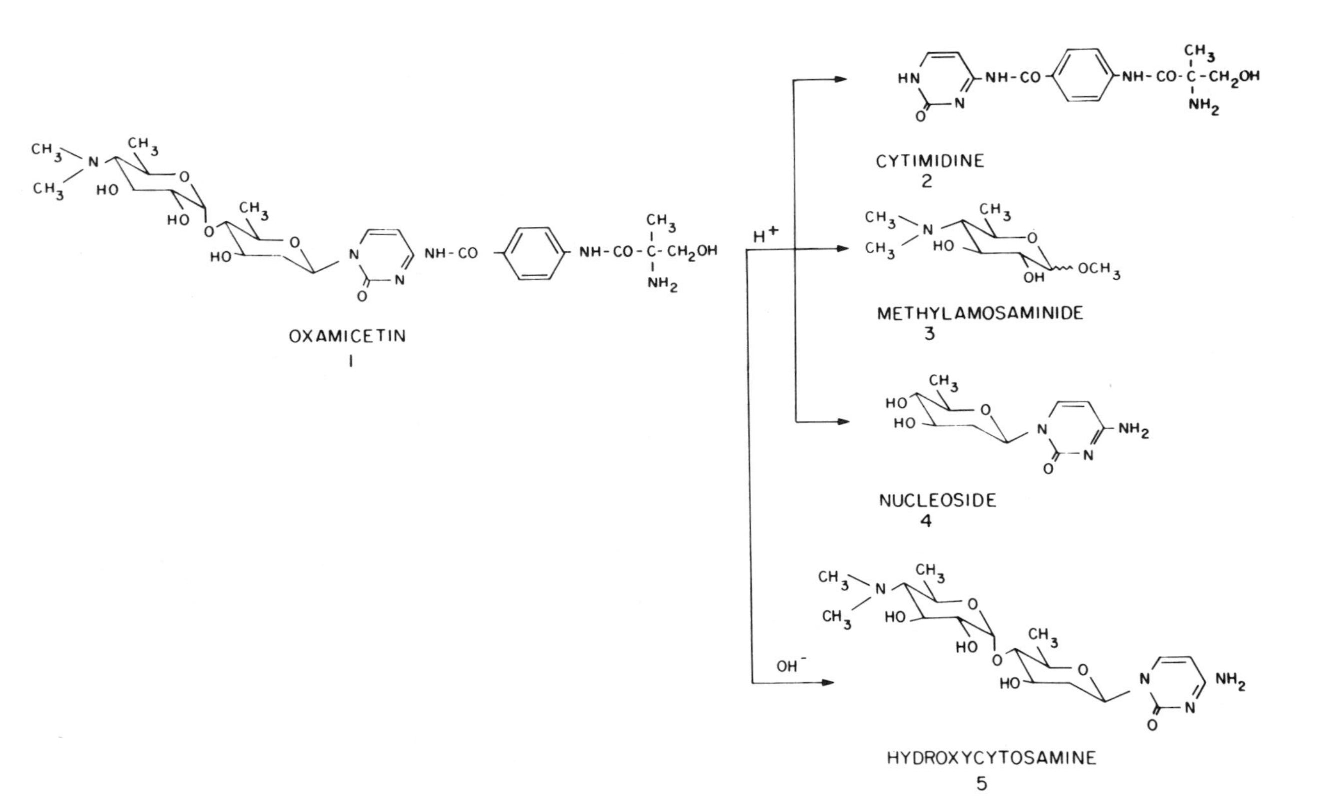

Figure 2.9 Degradation of oxamicetin. Reprinted with permission from Konishi et al., 1973.

Lichtenthaler and Kulikowski (1976) and its chemical and physical properties were found to be identical with those of the product isolated from acid methanolysis of oxamicetin, confirming the structure of oxamicetin.

Inhibition of Growth. Oxamicetin protects mice from *Staphylococcus aureus* and *E. coli* infections. The LD_{50} (i.v.) is 200 mg/kg. It is nontoxic at 400 mg/kg (subcutaneously) (Konishi et al., 1973). The effect of oxamicetin on ribosomal peptidyl transferase parallels that of amicetin, indicating that biological activity is due to interference with protein synthesis at the peptide chain elongation stage (Lichtenthaler et al., 1975a).

"Aspiculamycin": Its Identity with Gougerotin

Chemical Synthesis. Although gougerotin has been reviewed (Suhadolnik, 1970; Townsend, 1975; Fox et al., 1966), it is briefly discussed here because aspiculamycin has been given erroneous structural assignments and is the same as gougerotin (**12**) (Fig. 2.10). The elegant studies of Lichtenthaler and coworkers (Lichtenthaler et al., 1975b,c,d) have shown that aspiculamycin (**9**) and gougerotin (**12**) are identical by the chemical synthesis of gougerotin and "seryl gougerotin" (aspiculamycin) (Fig. 2.10).

The synthetic aspiculamycin proved to be different from aspiculamycin as isolated from *Streptomyces toyocaensis* (Arai et al., 1974; Haneishi et al., 1974). The discrepancy in the structure proposed for aspiculamycin was resolved by Lichtenthaler et al. (1975b) when they reinvestigated the structure of the product isolated from *S. toyocaensis* var. *aspiculamyceticus*. They showed that aspiculamycin (**9**) is identical with gougerotin (**12**), 1-[4-deoxy-4-(sarcosylseryl)amino-β-D-glucopyranuronamide]cytosine (Lichtenthaler et al., 1975b). Because aspiculamycin is identical with gougerotin, Lichtenthaler suggested that the description *S. toyocaensis* var. *gougerotii* is more appropriate (Lichtenthaler et al., 1975c). When the review on gougerotin was written in 1970 (Suhadolnik, 1970), the synthesis of gougerotin was completed only to substance C. Since then, Watanabe et al. (1972) and Lichtenthaler et al. (1975c) have described 17-step and 13-step syntheses of gougerotin from methyl α-D-galactoside. An additional contribution by Lichtenthaler and coworkers (Lichtenthaler et al., 1975d) is that in synthesizing "seryl gougerotin" they have opened the way for the preparation of gougerotin analogs that can be used to better understand the biochemical properties of these nucleosides as inhibitors of peptidyl transferase.

Asteromycin and moroyamycin are also identical with gougerotin (Ikeuchi et al., 1972; Sakagami et al., 1972; Sakagami, personal communication).

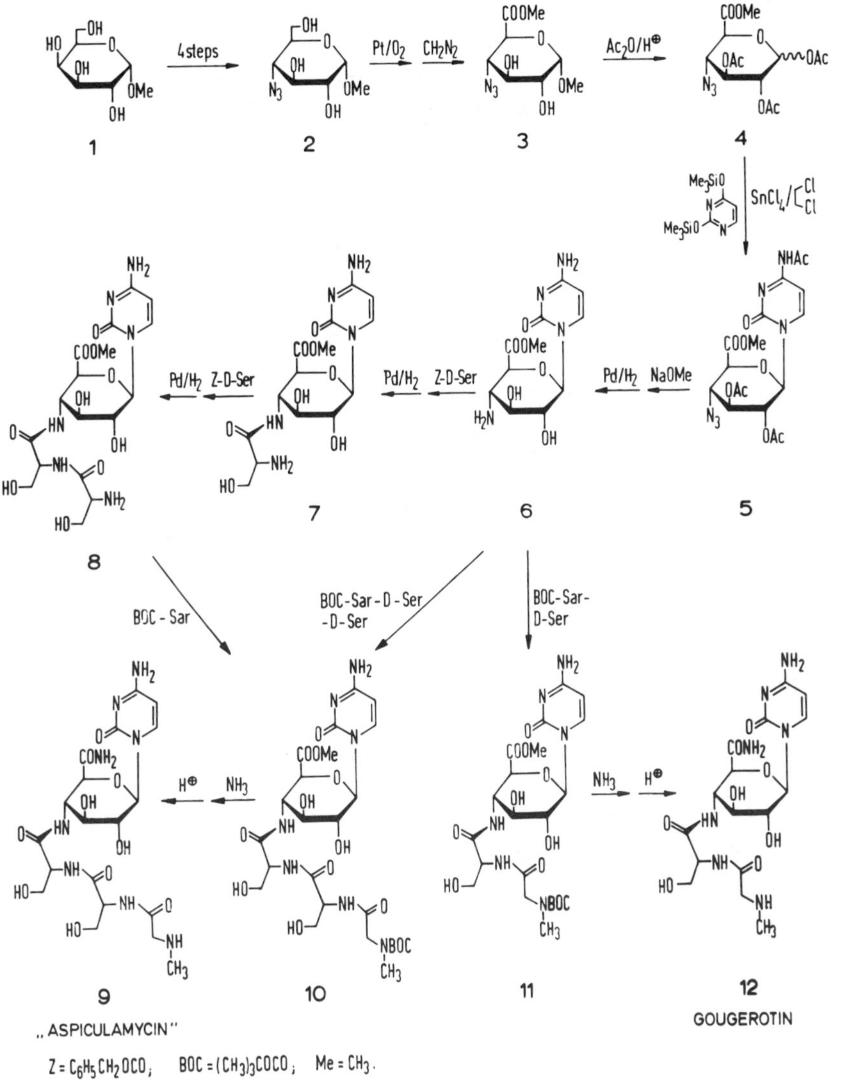

,, ASPICULAMYCIN''

$Z = C_6H_5CH_2OCO$; $BOC = (CH_3)_3COCO$; $Me = CH_3$.

Figure 2.10 Total synthesis of "Aspiculamycin" and gougerotin. Reprinted with permission from Lichtenthaler et al., 1975c.

Biochemical Properties of the Aminoacylaminohexosylcytosine Analogs

There are numerous reports on the mechanism by which the aminoacylaminohexosylcytosine nucleoside analogs, as nonfunctional analogs of aminoacyl-tRNA, block protein synthesis. The mode of action of these analogs involves the inhibition of ribosomal peptidyl transferase, which blocks the 70 S promoted transfer of *N*-Ac-aminoacyl CACCA to puromycin (Černá et al., 1973; Lichtenthaler et al., 1975a; Menzel and Lichtenthaler, 1975). The fragment reaction (CCA-fmet) has been extensively used to study the effect of these pyrimidine nucleoside analogs on the formation of a peptide bond. Černá et al. (1973) and Sikorski et al. (1977) reported on the role of gougerotin, blasticidin S, amicetin, bamicetin, and plicacetin in peptide formation. They studied 80 S ribosomes from wheat germ and *E. coli* ribosomal peptidyl transferase by using puromycin with either $tRNA^{AcPhe}$ or $(lys)_n$-tRNA (Fig. 2.11). Blasticidin S is the most inhibitory in either the transfer of lysine peptides from $(lys)_n$-tRNA to puromycin or the transfer of N-AcPhe from $tRNA^{AcPhe}$ to puromycin. The structural requirements essential for the inhibition of the transpeptidation step have been reported by Černá et al. (1973). They studied the increase in the binding of the donor substrate to the acceptor site. Plicacetin was the least inhibitory. Other laboratories reported similar findings (Monro et al., 1970; Celma et al., 1970; Černá et al; 1971; Hishizawa and Pestka, 1971). With amicetin and bamicetin, the terminal methylseryl group appears to be essential for the inhibition of transpeptidation. Plicacetin, which has no seryl group, is less inhibitory than amicetin and bamicetin. Blasticidin S and gougerotin are less inhibitory with the CACCA-Phe. The nucleoside antibiotics tested increased the binding of acceptor substrate to the acceptor site (Table 2.1) (Monro et al., 1970; Celma et al., 1970; Černá et al., 1971, 1973; Hishizawa and Pestka, 1971). Apparently, gougerotin, blasticidin S, amicetin, bamicetin, and oxamicetin have characteristic features that represent the

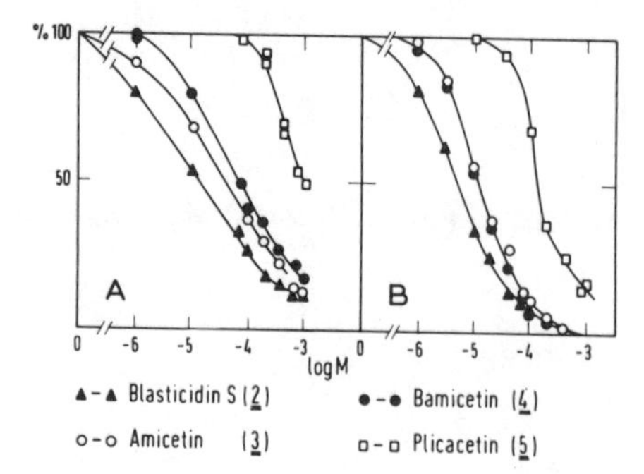

Figure 2.11 The effect of blasticidin S (**2**), amicetin (**3**), bamicetin (**4**), and plicacetin (**5**) on the transfer (*A*) of lysine peptides from $(Lys)_n$-tRNA and (*B*) of the AcPhe residue from tRNA to puromycin. Log *M*, concentration of 4-aminohexose pyrimidine nucleosides. Percent AcPhe-puromycin or $(Lys)_n$-puromycin formed relative to control without inhibitor. Reprinted with permission from Černá et al., 1973.

Table 2.1 The Effect of Antibiotics on the CACCA-[^{3}H]Phe and CACCA-(Ac[^{14}C]Leu) Binding to Ribosomes[a]

		CACCA-[^{3}H]Phe		CACCA-(Ac[^{14}C]Leu)	
		cpm	%	cpm	%
Control	—	702	100	530	100
Amicetin	10^{-4}	405	58	578	109
	10^{-3}	321	46	620	117
Bamicetin	10^{-4}	492	70	680	128
	10^{-3}	303	43	690	130
Plicacetin	10^{-4}	747	106	584	110
	10^{-3}	609	87	604	114
Blasticidin S	10^{-4}	588	84	1216	229
	10^{-3}	486	69	1300	245
Gougerotin	10^{-4}	611	87	753	142
	10^{-3}	386	55	870	164

Reprinted with permission from Černá et al., 1973.

[a] Reaction mixtures containing CACCA-[^{3}H]Phe (3520 cpm), 70 S ribosomes (six A_{260} units) and 20% (v/v) ethanol were incubated at 24°C for 20 min.

minimal requirements for the inhibition of protein synthesis (Lichtenthaler and Trummlitz, 1974). These detailed studies by Lichtenthaler and Trummlitz (1974) were further substantiated following the synthesis of pyrimidine analogs and their effect on protein synthesis. A very thorough study of the structure–activity relationship of 17 analogs of gougerotin has been described by Coutsogeorgopoulos et al. (1975). They reported that replacement of the sarcosyl-D-seryl side chain with various aminoacyl residues did not lead to acceptor (puromycin-like) activity with *N*-Ac-Phe-tRNA as the donor.

Pestka and coworkers (Pestka, 1972a, b; Pestka et al., 1972) observed that many compounds considered to be inhibitors of peptide bond synthesis do not inhibit peptidylpuromycin synthesis when native polyribosomes are used from *E. coli* (Pestka, 1972a). This conclusion is based on studies by Pestka and coworkers (Pestka, 1972b; Pestka et al., 1972) with systems such as the formation of fMet-, *N*-AcPhe-, and polylysylpuromycin from synthetic donors with ammonium chloride-washed ribosomes. Therefore, Pestka cautions that the data obtained with model systems need not predict the behavior of the antibiotics in the intact cell. Pestka and coworkers (Pestka, 1972b; Pestka et al., 1972) reported that amicetin, gougerotin, and blasticidin S are both competitive and noncompetitive inhibitors of the peptidyl puromycin synthesis with native polyribosomes from *E. coli*. The two types of inhibition observed with the nucleoside analogs could mean that either two sites exist for interaction with peptidyltransferase or there are two classes of ribosomal states that are amenable to inhibition.

To determine the quantitative interaction of gougerotin with *E. coli* and *S. cerevisiae* ribosomes, Barbacid and Vazquez (1974) showed that there is one homogeneous binding site for [^{3}H-G]gougerotin per *E. coli* ribosome, whereas the binding with *S. cerevisiae* ribosomes to gougerotin is heterogeneous and shows two different kinds of conformation that can be detected by binding of [^{3}H-G]gougerotin. Gougerotin has a much stronger affinity for washed *E. coli* ribosomes than for ribosomes that are reconstituted from subunits (Fig. 2.12). The binding of [^{3}H-G]gougerotin at the peptidyl transferase binding site of prokaryotic and eukaryotic ribosomes is completely inhibited by blasticidin S. Studies with ribosomes from rat liver or brain also showed two types of inhibition of peptidyl puromycin synthesis (Carrasco and Vazquez, 1972). Vazquez and coworkers reported that the

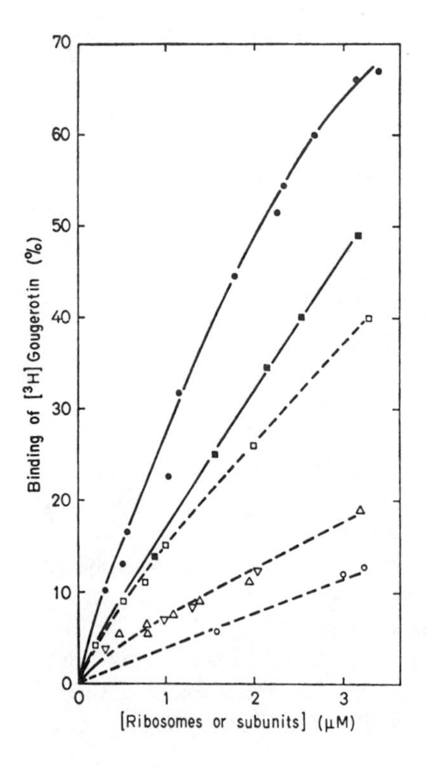

Figure 2.12 Binding of [^{3}H-G]gougerotin to *E. coli* ribosomes. Data are taken from assays following the sedimentation method. The ribosomes or ribosomal complexes were preincubated in 100 μl volumes at 30°C for 15 min prior to the addition of [G-^{3}H]gougerotin. Poly(U), when required, was used at a constant ratio of 100 μg/mmol ribosomes, whereas tRNA^{Phe}, when indicated, was added at the same concentration as that of the ribosomes. Either untreated control ribosomes or ribosomes kept for 6 h at 0°C under nondissociating conditions (●); "dissociated and reconstituted ribosomes" (■); ribosomal complex reconstituted from preparations of 30 and 50 S subunits, poly(U), and tRNA^{Phe} (□); ribosomal complex reconstituted from preparations of 30 and 50 S subunits and poly(U) (▽); ribosomes reconstituted from preparations of 30 and 50 S subunits (△); 50 S ribosomal subunits (○). Reprinted with permission from Barbacid and Vazquez, 1974.

binding of gougerotin to the ribosomes from human tonsil and yeast is inhibited by blasticidin S and amicetin (Carrasco and Vazquez, 1972; Vazquez, 1974; Battaner and Vazquez, 1971).

The amino group of cytosine is essential for biological activity of blasticidin S. Yamaguchi et al. (1975) reported on the isolation and purification of blasticidin S deaminase, the enzyme catalyzing the deamination of the cytosine moiety of blasticidin S. Caskey and Beaudet (1972) also reported on the effect of amicetin and gougerotin on total peptide chain termination.

2.2 THE PURINE NUCLEOSIDE ANALOGS

The six naturally occurring purine nucleoside analogs that inhibit protein synthesis described in this section are puromycin, nucleocidin, homocitrullylaminoadenosine, lysylaminoadenosine, antibiotic A201A, and 2′-aminoguanosine. These purine analogs have been extremely useful biochemical probes in the illustration of the complex steps involved in peptide bond formation. The discovery, production, and isolation, physical and chemical properties, chemical synthesis, and inhibition of growth of puromycin, nucleocidin, homocitrullylaminoadenosine, and lysylaminoadenosine have been reported in detail (Suhadolnik, 1970; Townsend, 1975) and are not repeated here. Antibiotic A201A and 2′-aminoguanosine are described in detail because they were discovered after 1970.

Antibiotic A201A

Hamill and Hoehn (1976) reported the isolation of the adenine nucleoside, analog A201A (Fig. 2.13). This new nucleoside antibiotic contains N^6-dimethyladenine, an aromatic acid, and three different monosaccharides. 3-Amino-3-deoxy-D-ribose is attached to adenine and 3,4-di-*O*-methylrhamnose is the terminal sugar. The third sugar is an unsaturated hexose. The structure of antibiotic A201A is comparable to that of puromycin (Fig. 2.13). Antibiotic A201A inhibits protein synthesis at the formation of the first peptide bond.

Discovery, Production, and Isolation. Antibiotic A201A is produced by growing *Streptomyces capreolus* on a vegetative medium at 30°C for 72 h in shake flasks followed by inoculation into fermentation media containing glucose, soybean grits, blackstrap molasses, calcium carbonate, and distilled water at 30°C for 96 h. The isolation of antibiotic A201A is accomplished by extraction with chloroform, concentration to dryness, dissolution in

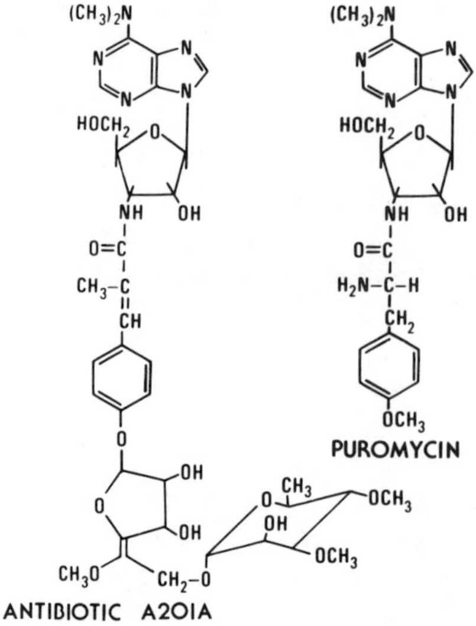

Figure 2.13 Structure of antibiotic A201A and comparison with puromycin (Hamill and Hoehn, 1976; Kirst et al., 1976).

methanol, filtration, evaporation to dryness, solubilization in chloroform, and addition of petroleum ether to precipitate the antibiotic. The crude antibiotic is dissolved in chloroform, chromatographed on silica gel, and eluted with acetone–methanol. The eluant is concentrated and the antibiotic is crystallized from acetone (Hamill and Hoehn, 1976).

Physical and Chemical Properties of Antibiotic A201A. Antibiotic A201A is a neutral, white crystalline compound; the molecular formula is $C_{37}H_{50}N_6O_{14}$; mol wt 802; mp 170–172°C; $[\alpha]_D^{25}$ −129.4° (*c* 1, methanol); ultraviolet spectral properties: λ_{max}^{pH7} 212 nm (ϵ = 41,500) and 278 nm (ϵ = 37,000); the infrared spectrum indicates hydroxyl groups, an amide carbonyl, and an aromatic moiety; the nmr spectrum indicates the presence of —C—CH_3, —N—CH_3, —O—CH_3, aromatic protons, and sugars. Antibiotic A201A is soluble in alcohols, acetone, chloroform, and ethyl acetate and insoluble in water, hydrocarbons, and ether (Hamill and Hoehn, 1976; Kirst et al., 1976).

Structural Elucidation. Kirst et al. (1976) reported that antibiotic A201A is structurally similar to puromycin. However, it is chemically distinct from puromycin and the other aminoacyl nucleosides which accounts for differences in its biological activity. The structure of antibiotic A201A was elucidated by a detailed analysis of spectral data of the parent compound and derivatives from degradation.

Inhibition of Growth. Antibiotic A201A is very active against gram-positive bacteria, mycobacteria, mycoplasma species, and fungi. It is effective against infections of *Mycoplasma gallisepticum*, *Entamoeba histolytica*, and *Borrelia novyi* (Ensminger and Wright, 1976). Antibiotic A201A is used to improve weight gain in poultry and swine. The LD_{50} of A201A (administered i.p.) is greater than 400 mg/kg. Antibiotic A201A inhibits all the *Staphylococcus aureus* strains sensitive and resistant to penicillin and methicillin. It also inhibits 91% of the *Streptococcus* Group A strains and 96% of *Streptococcus pneumoniae* strains. It is somewhat less active against Group D *Streptococcus*, *Hemophilus influenzae*, and *Niesseria gonorrhoeae*. When administered subcutaneously into bacterial infections in mice, antibiotic A201A is very effective against infections of *S. aureus* and *S. pyogenes*, but not against infections of *S. pneumoniae*. Guinea pigs treated intramuscularly with antibiotic A201A are cured of an infection of *Clostridium chauvoei*.

Seventy-five percent of a subcutaneous dose of antibiotic A201A is excreted by way of the intestine and only 15–20% is eliminated in the urine. Blood levels of antibiotic A201A decrease 60 min after administration (Ensminger and Wright, 1976).

Biochemical Properties. Antibiotic A201A inhibits the incorporation of leucine into protein, but has no effect on RNA or DNA synthesis (Fig. 2.14). However, antibiotic A201A (unlike puromycin) does not accept polypeptides and fails to stimulate the release of nascent polypeptides. Because

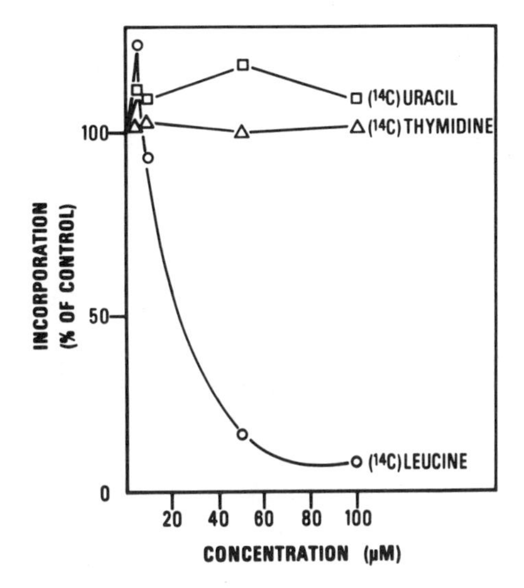

Figure 2.14 Inhibition of RNA, DNA, and protein synthesis in *Staphylococcus aureus* by antibiotic A201A. Reprinted with permission from Epp and Allen, 1976.

puromycin is structurally similar to aminoacyl-tRNA, puromycin accepts the growing polypeptide chain. Since antibiotic A201A does not have a reactive amino group like puromycin, the possibility of acceptor activity for antibiotic A201A is eliminated. Although antibiotic A201A does not inhibit peptidyl transferase activity, it inhibits poly(Phe) synthesis by 50% at 2 μM. When antibiotic A201A is tested against cellfree poly(U)-directed synthesis of poly(Phe) on *E. coli* ribosomes, it does not interfere with the binding of Phe-tRNA to the ribosomal A site, nor does it inhibit the binding of *N*-AcPhe-tRNA to the salt-washed ribosomes in the presence of initiation factors. The reaction between bound *N*-AcPhe-tRNA and puromycin is very sensitive to antibiotic A201A (Epp and Allen, 1976), which selectively inhibits dipeptide synthesis by interfering with the formation of a puromycin reactive 70 S initiation complex. Antibiotic A201A inhibits the formation of the first dipeptide bond. It appears that antibiotic A201A interferes with the joining of an initiation complex to the 50 S subunit. If antibiotic A201A inhibits the "joining reaction," the result should be a "runoff" of polyribosomes, and growing cells exposed to A201A should finish a round of synthesis but fail to start to translate another message. Sucrose density gradients after centrifugation of lysates from exponentially growing *E. coli* treated with puromycin or antibiotic A201A show that antibiotic A201A causes a runoff of ribosomes just as occurs with puromycin (Fig. 2.15). These data agree with the suggestion that antibiotic A201A interferes with

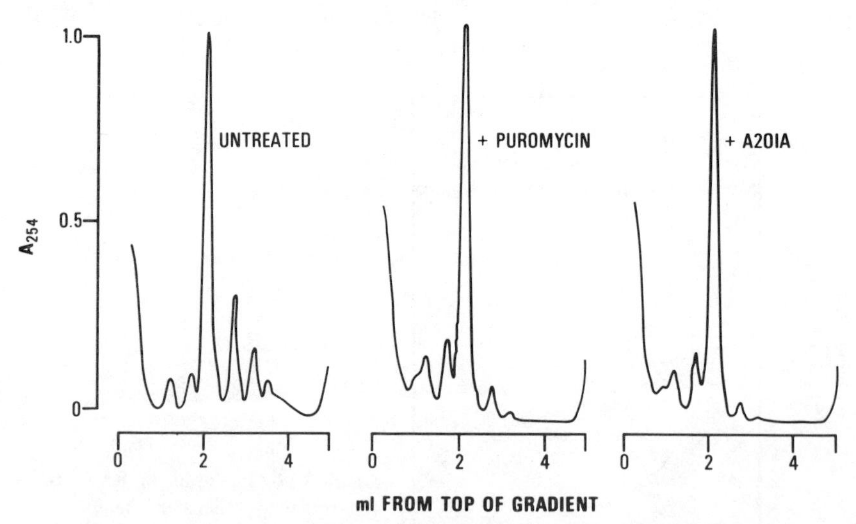

Figure 2.15 Sucrose density gradients after centrifugation showing polysome runoff in lysates of exponentially growing *E. coli* cells treated with either 100 μM antibiotic A201A or puromycin. Reprinted with permission from Epp and Allen, 1976.

the formation of a puromycin-reactive ribosome, but does not inhibit the peptidyltransferase enzyme.

Physical and Chemical Properties of Antibiotic A201B. Antibiotic A201B is a minor factor isolated as an oil from the culture filtrates of *S. capreolus;* mol wt 417 (osmotic method); ultraviolet spectral properties: λ_{max} 222 nm ($E_{1\,cm}^{1\,percent}$ 515), 242 nm ($E_{1\,cm}^{1\,percent}$ 405), 327 nm ($E_{1\,cm}^{1\,percent}$ 150), and 340 nm ($E_{1\,cm}^{1\,percent}$ 225). Antibiotic A201B inhibits phytopathogenic fungi and gram-positive bacteria (Hamill and Hoehn, 1976).

2′-Aminoguanosine

2′-Aminoguanosine (2AG, 2′-amino-2′-deoxyguanosine) (Fig. 2.16) and isoguanosine (crotonoside) (Fig. 1.36) are the first two guanosine analogs isolated from natural sources. 2AG inhibits *E. coli* strain KY3591 and has antitumor activity against HeLa cells and Sarcoma 180. 2AG primarily inhibits protein synthesis; it has a limited effect on RNA synthesis and no effect on DNA synthesis (Nakanishi et al., 1977). Guanosine reverses the inhibition of 2AG noncompetitively.

Discovery, Production, and Isolation. 2AG is isolated from a strain of *Enterobacter cloacae* KY3071. The fermentation conditions, isolation, purification, and assay are described by Nakanishi et al. (1974, 1977). 2AG crystallizes as white plates from water; yield 400 mg/liter after 24 h incubation.

Physical and Chemical Properties. The molecular formula of 2AG is $C_{10}H_{14}N_6O_4$; mol wt 294; crystallization (H_2O) $C_{10}H_{14}N_6O_4 \cdot \frac{1}{2}\ H_2O$; mp 252–254°C; $[\alpha]_D^{26}$-56.6° (*c* 0.54, H_2O) (Nakanishi et al., 1974); ultraviolet spectral properties $\lambda_{max}^{H_2O}$ 252 nm, 275 nm; (sh) $\lambda_{max}^{pH\,12}$ 256 nm, 268 nm (Ikehara et al., 1976); infrared spectral properties: bands at 3520, 3200 cm^{-1} ($-NH_2$, $-OH$), and 1725, 1488 cm^{-1} (typical of guanine C=C, C=N, $-OH$, and $-NH_2$); the nmr spectrum is similar to that of guanosine except

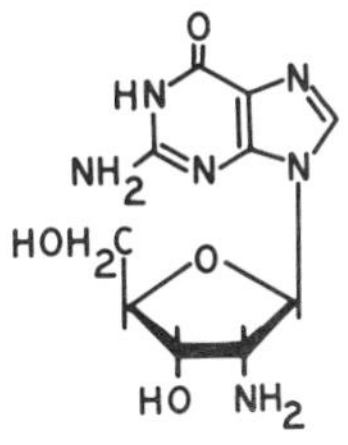

Figure 2.16 Structure of 2′-aminoguanosine.

for a doublet signal for the anomeric proton at 55 ppm (J = 5.5 Hz). With guanosine this signal shifted to 6.15 ppm (J = 4.0 Hz).

Structural Elucidation. 2AG is hydrolyzed by HCl to guanine and 2-amino-2-deoxy-D-ribose. The identification of the 2-amino sugar was established by the Tsuji reaction; proof of the furanose ring was determined by periodate oxidation (Nakanishi et al., 1974). Because the occurrence of 2-amino-2-deoxy-D-ribose has not been reported in nature, Nakanishi et al. (1976) reported a detailed study of the ^{13}C nmr of this nucleoside analog, guanosine, and deoxyguanosine. The chemical shift of C-2′ for the analog (δ 58.15) differs from that of guanosine (δ 71.02), which suggests that the amino group of the aminopentose is located at C-2′. Nakanishi et al. (1976) also described the nmr spectra of N-diacetyl-2′-aminoguanosine and tetraacetyl-2′-aminoguanosine ($CDCl_3$) . The downfield shift of the anomeric hydrogen and the 2′-hydrogen in the N-diacetyl derivative confirmed the amino group on C-2′. These data firmly establish the structure of 2′-aminoguanosine.

Chemical Synthesis. 2AG has been synthesized chemically by Ikehara et al. (1976). The starting compound is 8,2′-anhydro-8-oxy-9-β-D-arabinofuranosylguanine (R = H) (Fig. 2.17). Simultaneous with the report of Ikehara, Hobbs and Eckstein (1977) described their method for the synthesis of 2′-azido-2′-deoxyribofuranosyl and 2′-amino-2′-deoxyribofuranosylpurines. The method involves the condensation of the triacetate of 2′-azido-2′-deoxyribose with N^6-octanoyladenine. Deacylation affords the α- and β-anomers of 2′-azido-2′-deoxyadenosine. For the synthesis of the 2′-azidoguanosine and 2AG nucleosides, the N^6-octanoyladenine is replaced by N^2-palmitoylguanine.

Inhibition of Growth. Of the bacteria tested, only strain KY3591 of *E. coli* is inhibited by 2AG (Nakanishi et al., 1974). The antibacterial activity is readily reversed by guanosine and other purine nucleosides (Nakanishi et al., 1977). The growth of HeLa S-3 cells was suppressed at 10 μg/ml. Sarcoma 180 was inhibited *in vivo*.

Biochemical Properties. Resistant *E. coli* KY3591 colonies appear after 2 h exposure to 2AG. Adenosine and guanosine reverse the inhibition; guanosine reverses the inhibition noncompetitively. Guanine, adenine, hypoxanthine, xanthine, and xanthosine do not reverse inhibition. Guanosine prevents uptake of 2AG (Nakanishi et al., 1977). Purine nucleosides are transported into *E. coli* after hydrolysis by purine nucleoside phosphorylase or purine phosphoribosyltransferase (Hochstadt-Ozer, 1972). The

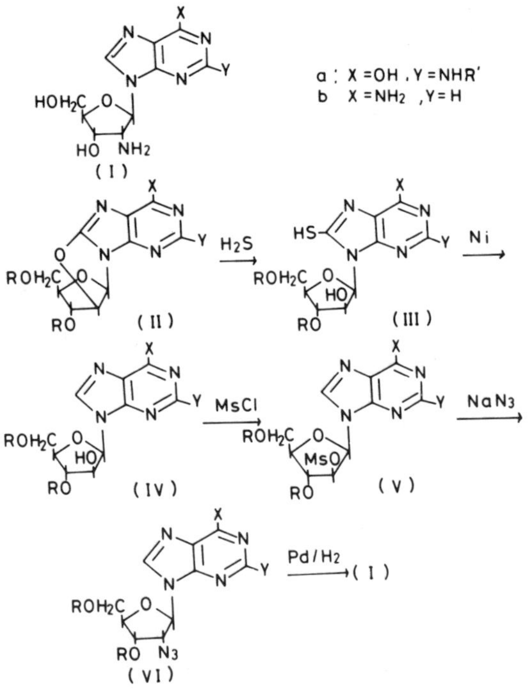

Figure 2.17 Synthesis of 2′-aminoguanosine. Modified from Ikehara et al., 1976, reprinted with permission.

bases are released into the medium. 2AG prevents the release of guanine into the medium. 2AG is taken up by strain KY3591 and converted to its 5′-mono-, 5′-di-, and 5′-triphosphates. Protein synthesis in KY3591 is inhibited after 15 min; RNA synthesis is inhibited after 15–30 min; DNA synthesis is not inhibited (Fig. 2.18). 2AG primarily inhibits protein synthesis, but also has some effect on RNA synthesis. Eighty-six percent of 2AG is in the acid-soluble fraction as the nucleotides. The limited amount located in the acid-insoluble fraction is in the RNA and not in the DNA.

Nakanishi et al. (1977) proposed three mechanisms for the inhibition of 2AG on protein synthesis. First, the 5′-triphosphate acts as an analog of GTP, which is required for the initiation step of protein synthesis. Initiation is blocked and the elongation reaction cannot continue for several minutes. Second, the 5′-triphosphate acts as an analog of GTP, which is required for the elongation step of protein synthesis. Third, the 5′-triphosphate is incorporated into messenger, transfer, and ribosomal RNA in place of GMP; subsequently, these RNAs do not function normally, resulting in an inhibition of protein synthesis. The analog is incorporated into RNA, which

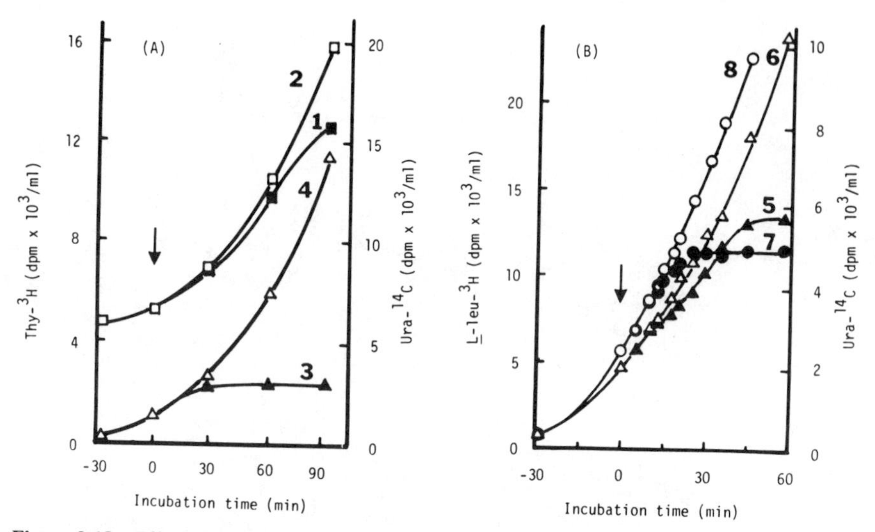

Figure 2.18 Effects of 2AG on RNA, DNA, and protein synthesis in *E. coli* KY3591. (*A*) Incorporation of [^{3}H]thymine and [^{14}C]uracil into the acid-insoluble fraction. Incorporation of [^{3}H]thymine in the presence (*1*) or absence (*2*) of 2AG. Incorporation of [^{14}C]uracil in the presence (*3*) or absence (*4*) of 2AG. (*B*) Incorporation of [^{14}C]uracil and L-[^{3}H]leucine into acid-insoluble fraction. Incorporation of [^{14}C]uracil in the presence (*5*) or absence (*6*) of 2AG. Incorporation of L-[^{3}H]leucine in the presence (*7*) or absence (*8*) of 2AG. The arrows indicate the time of 2AG addition. Reprinted with permission from Nakanishi et al., 1977.

results in the formation of nonfunctional RNA and the subsequent inhibition of protein synthesis.

Puromycin

Biosynthesis. The role of demethylated puromycin as a precursor in the biosynthesis of puromycin has been reviewed (Suhadolnik, 1970).

Biochemical Properties. Puromycin is an aminoacyl nucleoside analog elaborated by *S. alboniger* that is structurally similar to aminoacyl tRNA (Fig. 2.19). It is a broad-spectrum antibiotic with antitumor activity that inhibits protein synthesis *in vivo* and *in vitro* (Suhadolnik, 1970; Nathans, 1967). The synthesis of the "reversed" nucleoside has been described by Leonard and coworkers (Leonard and Carraway, 1966; Leonard et al., 1968). Nair and Emmanuel (1977) reported the synthesis of "reversed" puromycin. The analog incorporates all the features of puromycin, but is devoid of the structural components that are toxic to animals.

Puromycin inhibits protein synthesis by the premature release of nascent polypeptide chains as peptidylpuromycin (Benne and Voorma, 1972; Aharonowitz and Ron, 1975; Dubnoff et al., 1972). Puromycin substitutes for the incoming coded aminoacyl-tRNA. It is likely that there is a binding site on peptidyltransferase for the adenine ring (Crystal et al., 1974). The liquid and solid-state conformation of puromycin has been determined by X-ray crystallography (Sundaralingam and Arova, 1972). The synthesis of two affinity analogs of puromycin that label the ribosome has been reported: 5′-*O*-(*N*-bromoacetyl-*p*-aminophenylphosphoryl)-3′-*N*-L-phenylalanylpuromycin (Monro and Vazquez, 1967; Jayaraman and Goldberg, 1968) and α-*N*-iodoacetylpuromycin (Goldberg and Mitsugi, 1967). Pestka and Moldave and coworkers have written experimental procedures for the peptidylpuromycin synthesis in *E. coli* and rat liver or brain (Pestka, 1974a, 1974b; Edens et al., 1974).

The Puromycin Reaction. Puromycin has been used to study (i) the mechanism of peptide bond formation on the ribosome by transfer of the nascent peptide chain from peptidyl tRNA to the α-amino group of tRNA with the acylated amino acid (Harris and Pestka, 1975; Edens et al., 1975; Weissbach and Pestka, 1977), (ii) the mode of action of elongation (Pestka, 1972a), (iii) the inhibition of protein synthesis (Nair and Emmanuel, 1977), and (iv) the movement of the protein synthesis initiator (fMet-tRNA^{fmet}) (Benne and Voorma, 1972).

The 70 S initiation complex, which is sensitive to puromycin, is used as the model to study the mechanism of polypeptide chain initiation. Figure

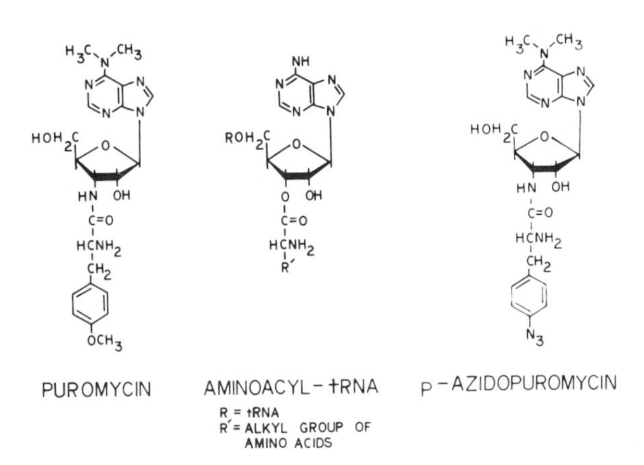

Figure 2.19 Structures of puromycin, aminoacyl tRNA, and *p*-azidopuromycin.

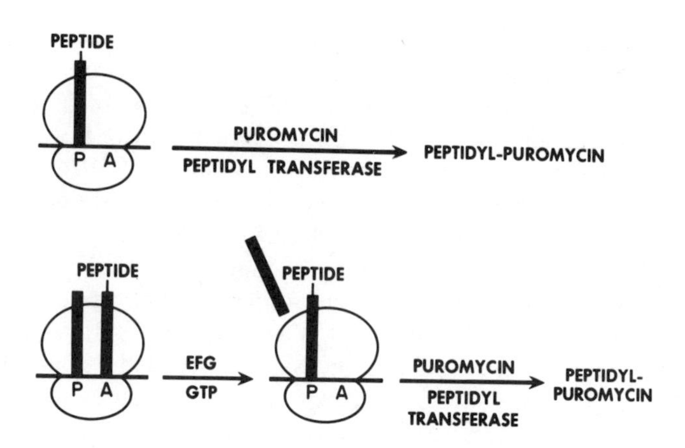

Figure 2.20 The puromycin reaction with peptidyl-tRNA. Reprinted with permission from Weissbach and Pestka, 1977.

2.20 shows the puromycin reaction with peptidyl-tRNA. It is well established that methionylvaline is the first peptide bond formed in lysed rabbit reticulocytes (Crystal et al., 1974; Fresno et al., 1976). Cheung et al. (1973) have questioned the validity of the methionylpuromycin as a model. Sparsomycin, which inhibits peptidyltransferases (Monro and Vazquez, 1967; Jayaraman and Goldberg, 1968; Goldberg and Mitsugi, 1967), does not inhibit either the first dipeptide or the pactamycin-induced dipeptide accumulation (Cheung et al., 1973). Pactamycin, which inhibits methionylpuromycin formation (Kappen et al., 1973; Seal and Marcus, 1972), causes an accumulation of methionylvaline with globin mRNA (Cheung et al., 1973; Wu et al., 1977). These findings suggest that the formation of the first peptide in reticulocytes is a unique process. Either the ribosomes undergo a special conformation when the first peptide bond is formed or a different peptidyltransferase is used to synthesize the dipeptide.

The Fate of Ribosomes Upon Release of Growing Protein Chains by Puromycin. In bacterial cultures, puromycin causes the breakdown of polyribosomes, which results in an accumulation of 70 S monomers and increases the exchange of ribosomal subunits (Kohler et al., 1968; Subramanian et al., 1969; Azzam and Algranati, 1973; Subramanian and Davis, 1971) (Fig. 2.21). The ribosome dissociation factor IF3 dissociates the 70 S ribosome (Sabol et al., 1970; Subramanian and Davis, 1970; Dubnoff and Maitra, 1971; Stringer et al., 1977). Puromycin causes the ribosome to be detected as the 70 S particle. The 70 S particle then dissociates and equilibrates with the pool of subunits (Azzam and Algranati, 1973; Kaempfer, 1970; Subramanian and Davis, 1973). Azzam and Algranati (1973)

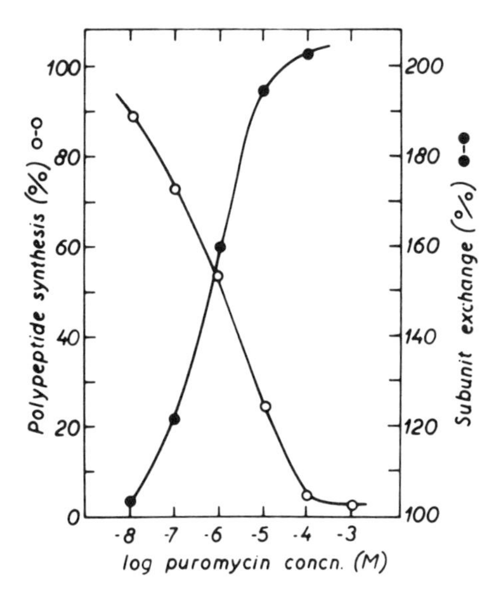

Figure 2.21 Effect of puromycin concentrations on polypeptide synthesis (○) and subunit exchange (●) with increasing concentrations of puromycin. Reprinted with permission from Azzam and Algranati, 1973.

demonstrated that detachment of ribosome occurred as 70 S particles after the addition of puromycin followed by dissociation and equilibration with the pool of subunits. Subramanian and Davis (1973) studied the form in which ribosomes are released from messenger by puromycin. The puromycin causes premature release of nascent polypeptide at internal codons located within the gene. Addition of puromycin to a mixture of heavy (^{32}P) and light (^{3}H) polysomes immediately stops amino acid incorporation. The ribosomes that are released are indistinguishable from those ribosomes obtained without puromycin, that is, the labeled heavy ribosomes do not dissociate into subunits and then reassociate.

Photoaffinity Labeling of Ribosomes. Because affinity labeling is such a powerful tool for exploring ligand–receptor interactions, Cooperman and coworkers (Jaynes et al., 1978; Nicholson and Cooperman, 1978) synthesized *p*-azidopuromycin, 6-dimethylamino-9-[3′-deoxy-3′-(*p*-azido-L-phenylalanylamino)-β-D-ribofuranosyl]purine, to better understand the binding of puromycin to ribosomes.

The *E. coli* ribosome is a complex organelle. It is composed of 54 different proteins and 3 RNA chains. Two of the RNA chains have 1600 and 3300 bases, respectively (Wittmann, 1976). A principal goal of the research on the intact ribosome is the construction of a structure–function map in which the RNA areas and the protein would be located and eventually assigned roles in the protein synthesis process. The affinity labeling by which a covalent bond forms between a ligand and its receptor site for several nucleoside analogs has already provided evidence as to the locations of the peptidyltransferase center and the mRNA codon site (Cooperman, 1976; Cooperman et al., 1979; Bayley and Knowles, 1977; Johnson and Cantor, 1977; Zamir, 1977). It appears that the photoincorporation of puromycin occurs at ribosomal protein L23 (50 S subunit) and proceeds by affinity labeling (Fig. 2.22). In addition, proteins S14 and S18 (30 S subunit) and an unidentified site in the RNA fraction of the 50 S subunit are also affinity labeled.

Cooperman and coworkers concluded that the photolabeling of protein L23 by *p*-azidopuromycin is added proof that protein L23 is a true component of the puromycin binding site on the 50 S subunit. *N*-Ethyl-2-diazomalonylpuromycin differs in that it is directed more toward the 30 S site (Cooperman et al., 1979).

Thompson and Moldave (1974) compared directly the reaction catalyzed on 60 S particles with the reaction on 80 S ribosomes or on 80 S–mRNA complexes. To do so, it was necessary to measure the phenylalanylpuromycin formation with combinations of poly(U), 40 S, and 60S subunits. To make this comparison, it was essential to determine the effect of *N*-acetylphenylalanyl-tRNA and 60 S subunit concentrations on peptidyltransferase. This was accomplished by the addition of puromycin to the assay mixture. At 54 pmol of 60 S subunits and 121 pmol of 60 S subunits, the initial rate was proportional to the concentration of substrate up to 25 and 50 pmol of *N*-acetylphenylalanyl-tRNA (Fig. 2.23).

Effect of Puromycin on the Nervous System. Puromycin has been studied in another role in the brain related to protein synthesis. Mice resistant to audiogenic seizures can be rendered susceptible to sound-induced convulsions following an intense acoustic stimulus at a critical period of neural development. This is referred to as "acoustic priming" (Henry, 1967). Puromycin or purine amino nucleoside (PAN) (Fig. 3.37) blocks the process, which suggests that puromycin affects protein synthesis per se. However, Lieff et al. (1976) suggest that puromycin inhibits peptide synthesis which then interferes with normal neurohumoral transmission. The memory process in mice and goldfish and the role of protein synthesis in this process has been studied by using puromycin (Barondes and Cohen, 1966; Flexner and Flexner, 1966; Agranoff et al., 1966; Barondes, 1970; Paggi and Toschi, 1971). Puromycin and PAN are reversible mixed inhibitors of acetylcholine esterase (Wulff, 1973). Moss and Fahrney (1976) reported that puromycin blocked the postjunctional response to acetylcholine. The interference with the acetylcholine receptor by puromycin and the inhibition of acetylcholine esterase may be important to the interpretation of experiments in which puromycin interferes with memory and amnesia for learned responses (Wulff, 1973; Moss et al., 1974).

Structural Requirements of Puromycin for Biological Activity. The structural requirements for maximum inhibition of protein synthesis are the diaminonucleoside and an aromatic amino acid of puromycin (Suhadolnik, 1970). However, Vince and Symons and their coworkers have synthesized analogs of puromycin to show that the dimethyl groups, the methoxyl group, the furanosyl oxygen, and the 5′-hydroxyl group of puromycin are

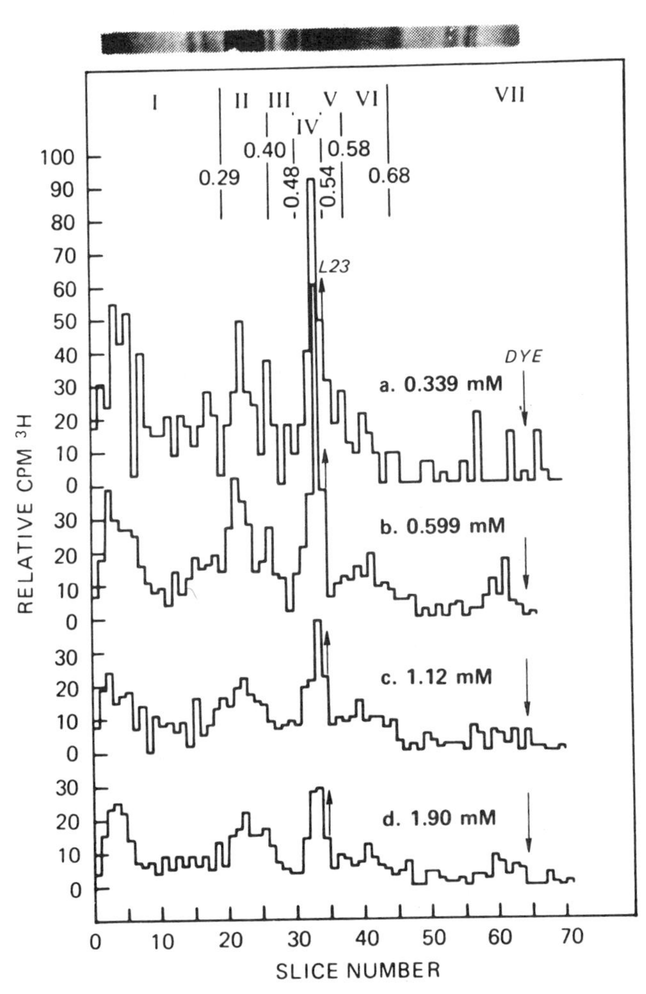

Figure 2.22 Polyacrylamide gel pattern of labeled proteins from 50 S particles as a function of puromycin concentration. A stained gel is shown above the graph. The arrow pointing up marks the position of authentic L23. Experimental conditions: 111 A_{260} units/ml ribosomes; photolysis was for 8 min at 2737 Å. The specific activities of puromycin were: (*a*) 890 Ci/mol; (*b*) 504 Ci/mol; (*c*) 269 Ci/mol; (*d*) 159 Ci/mol. Reported counts per minute are for protein from 3.5 A_{260} units of 50 S particles. Reprinted with permission from Jaynes et al., *Biochemistry*, **17**, 561 (1978). Copyright by the American Chemical Society.

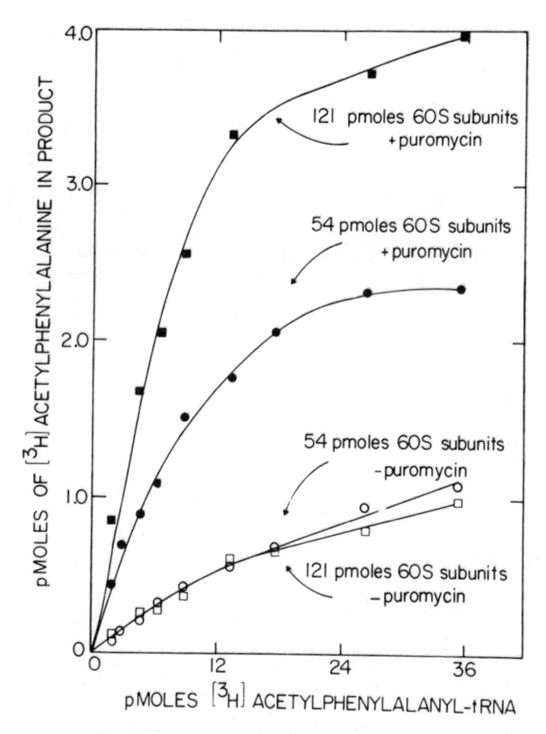

Figure 2.23 The effect of acetylphenylalanyl-tRNA and 60 S subunit concentrations on peptidyltransferase. Reprinted with permission from Thompson and Moldave, *Biochemistry*, **13**, 1348 (1974). Copyright by the American Chemical Society.

not necessary to inhibit protein synthesis (Daluge and Vince, 1972; Vince and Isakson, 1973; Vince et al., 1976; Symons et al., 1969). Vince and coworkers showed that a carbocyclic puromycin analog, in which the furanosyl oxygen is replaced by a methylene moiety, has antimicrobial antitumor activity (Daluge and Vince, 1972; Vince et al., 1972). Four cyclohexyl puromycin derivatives have been synthesized. All four inhibit protein synthesis (Vince and Daluge, 1977).

Nucleocidin

Biochemical Properties. Nucleocidin (4′-fluoro-5′-*O*-sulfamoyladenosine, antibiotic T-3018) (Fig. 2.24) is synthesized by *S. clavus*. The structural elucidation of nucleocidin is based on nmr, mass spectral studies, and chemical synthesis (Morton et al., 1969; Jenkins et al., 1971, 1976).

Florini et al. (1966) reported that nucleocidin inhibits the incorporation of leucine into rat liver protein *in vivo* and *in vitro*. Nucleocidin is a more

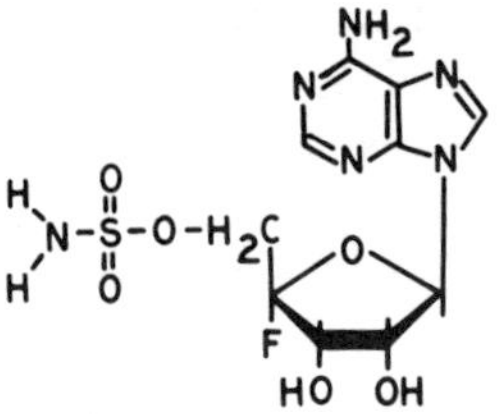

Figure 2.24 Structure of nucleocidin.

potent inhibitor than puromycin *in vivo*. However, the *in vitro* inhibition of protein synthesis by these two nucleosides is essentially the same. The differences between the *in vivo* and *in vitro* inhibition of protein synthesis by nucleocidin and puromycin are apparently due to the slower metabolism and excretion of nucleocidin compared to puromycin. Nucleocidin inhibits protein synthesis by binding to the ribosomes, which subsequently inhibits peptide bond formation. There is no effect on the binding of tRNA to ribosomes nor is there an inhibition of RNA synthesis by nucleocidin (Florini et al., 1966).

Homocitrullylaminoadenosine and Lysylaminoadenosine

Biochemical Properties. Homocitrullylaminoadenosine and lysylaminoadenosine (Fig. 2.25) are naturally occurring adenosine analogs found in the culture filtrates of *Cordyceps militaris* (Guarino and Kredich, 1964). Guarino et al. (1963) reported that homocitrullylaminoadenosine inhibits protein synthesis by either blocking the activation of amino acids or the transfer of tRNA. There is an inhibition of the incorporation of amino acids from aminoacyl tRNA into protein. Apparently, lysylaminoadenosine inhibits much the same way.

The chemical syntheses of N^6-desmethylpuromycin, 3′-L-homocitrullylamino-3′-deoxyadenosine, and 3′-L-lysylamino-3′-deoxyadenosine have been described (Lichtenthaler et al., 1979). Homocitrullyl-3′-deoxyadenosine and

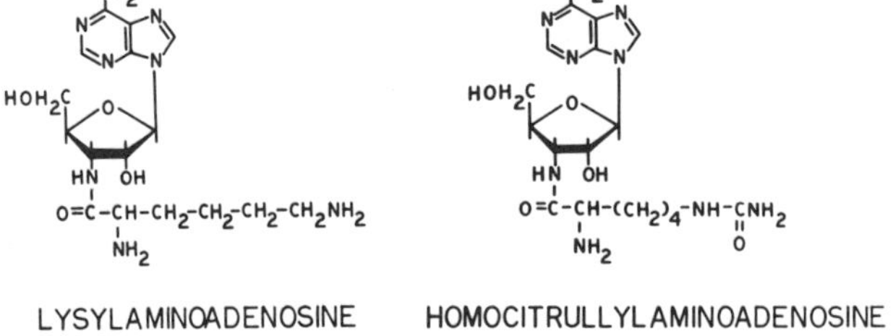

LYSYLAMINOADENOSINE HOMOCITRULLYLAMINOADENOSINE

Figure 2.25 Structures of homocitrullylaminoadenosine and lysylaminoadenosine.

lysylamino-3′-deoxyadenosine inhibit poly(U)-directed polyphenylalanine synthesis analogous to puromycin. The N^6-desmethylpuromycin is as inhibitory as puromycin; homocitrullyl 3′-deoxyadenosine and lysylamino-3′-deoxyadenosine are less inhibitory by factors of twenty- and forty-fold, respectively, relative to puromycin. This latter finding was not unexpected, since the dual hydrophobic–hydrophilic nature of the positively charged lysyl side chain does not have an optimum fit into the hydrophobic pocket of the ribosome (Lichtenthaler et al., 1979).

2.3 THE CLINDAMYCIN RIBONUCLEOTIDES

Bacteria on continued exposure to biological antagonists develop resistant strains. Therefore, structural modifications are necessary to overcome the resistant strains. The formation of clindamycin ribonucleotides by means of biomodifications is an example of this type of structural change. Argoudelis and coworkers have reported on the biomodifications of analogs elaborated by the *Actinomycetes*. They described the acylation of chloramphenicol by *Streptomyces coelicolor*, the phosphorylation of lincomycin by *S. rochei*, the conversion of clindamycin (**1**) to 1-demethyclindamycin and clindamycin sulfoxide by *S. punipalus* and *S. armentosus*, and the phosphorylation of clindamycin by whole cells and lysates of *S. coelicolor* (Argoudelis and Coats, 1969, 1971; Coats and Argoudelis, 1971; Argoudelis et al., 1969, 1977). Clindamycin is a clinically useful nucleoside analog that is produced by the chlorination of lincomycin (Argoudelis and Coats, 1971). Coats and Argoudelis (1971) reported that the addition of clindamycin to growing cultures of *S. coelicolor* resulted in an *in vitro* inactive clindamycin 3-phosphate (**2**) that could be regenerated to clindamycin by treatment with alkaline phosphatase. During the isolation procedure, four additional compounds were observed. Argoudelis et al. (1977) established the structures of the new nucleotides as clindamycin 3′-(5′-cytidylate) (**3**), clindamycin 3′-(5′-adenylate) (**4**), clindamycin 3′-(5′-uridylate) (**5**), and clindamycin 3′-(5′-guanylate) (**6**) (Fig. 2.26).

Discovery, Production, and Isolation

Fermentation conditions have been described (Argoudelis and Coats, 1971). Clindamycin was added to whole cell cultures of *S. coelicolor* 24 h after inoculation. The cultures were harvested after 48 h. The clindamycin biosynthetic products were isolated by ion exchange chromatography, counter-double current distribution, DEAE-Sephadex, and Amberlite XAD-

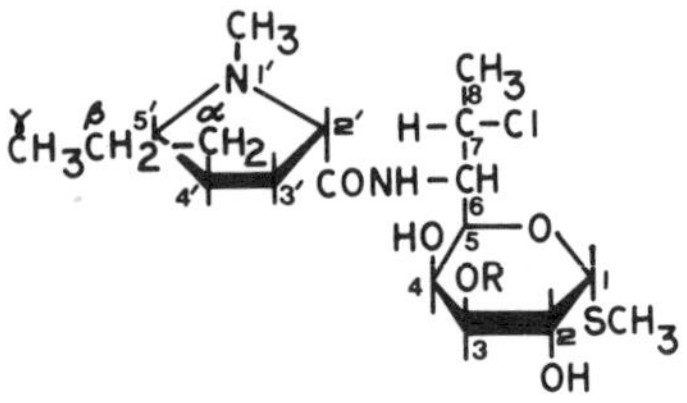

CLINDAMYCIN (1)
R=H

CLINDAMYCIN 3-PHOSPHATE (2)
R= $-P(=O)(OH)-OH$

R = CYTIDYL-5′-YL (3)
R = ADENYL-5′-YL (4)
R = URIDYL-5′-YL (5)
R = GUANYL-5′-YL (6)

Figure 2.26 Structures of clindamycin (**1**), clindamycin 3-phosphate (**2**), and the clindamycin ribonucleotides (**3–6**) (Argoudelis et al., 1977).

2. The clindamycin 3-ribonucleotides and clindamycin 3-phosphate were isolated as amorphous, colorless substances. They are soluble in water and alcohol, but insoluble in acetone and hydrocarbon solvents (Argoudelis et al., 1977).

Physical and Chemical Properties

The chemical and physical properties of clindamycin 3-phosphate and the ribonucleotides are shown in Table 2.2. The assignment of the nucleoside 5′-phosphate clindamycin linkage was established by hydrolysis with snake venom phosphodiesterase.

Structural Elucidation

The assignment of the phosphate to the C-3 position is based on periodate oxidation studies; 1 *N* HCl treatment of the adenosine and guanosine clindamycin nucleotides afforded adenine, guanine, and ribose, while the pyrimidine nucleotides gave cytidine, uridine, and ribose; 0.1 *N* NaOH treatment gave the corresponding nucleosides. The nmr spectra of the clindamycin ribonucleotides are in agreement with the assigned structures.

Table 2.2 Physical and Chemical Properties of the Clindamycin Ribonucleotides

Compound	Molecular Formula	Mol wt		$[\alpha]_D^{25}$ (°)[b]	$[M]_D$(°)	Uv $[\lambda_{max}(\epsilon \times 10^{-3})]$		
		Calcd.	Found[a]			pH 2.0	pH 7.0	pH 11.0
2	$C_{18}H_{34}N_2O_8ClPS$	504	530	+91.3	+458	No Uv absorption	269 (6.80)	271 (6.60)
3	$C_{27}H_{45}N_5O_{12}ClPS$	729	742	+61	+445	279 (9.60)	269 (6.80)	271 (6.60)
4	$C_{28}H_{43}N_7O_{11}ClPS$	753	726	+62.9	+473	257 (12.60)	261 (12.50)	261 (12.70)
5	$C_{27}N_{44}N_4O_{13}ClPS$	732	764	+79.5	+578	261 (8.20)	262 (8.40)	262 (6.50)
6	$C_{28}H_{45}N_7O_{12}ClPS$	769	750	+69	+530	256 (11.10)	254 (12.50)	259 (10.70)
						273 (8.00)(sh)	273 (8.00)(sh)	266 (10.60)

Reprinted with permission from Argoudelis et al., 1977.

[a] Molecular weights were determined by vapor pressure osmometry in methanol.

[b] Specific rotation was determined in water (*c* 1).

Inhibition of Growth

The clindamycin ribonucleotides are inactive against *S. aureus in vitro*. However, they protect *S. aureus* infected mice.

Biochemical Properties

The formation of these nucleotides resembles the adenylation of streptomycin and spectinomycin by an R factor carrying *E. coli* (Umezawa et al., 1968; Beneviste et al., 1969). Weisblum and Davies (1968) reported that streptomycin and spectinomycin inhibit protein synthesis in bacteria.

Coats and Argoudelis (1971) and Argoudelis et al. (1977) reported that clindamycin 3-ribonucleotides and clindamycin 3-phosphate did not inhibit cultures of *S. aureus*. However, these nucleotides protect *S. aureus*-infected mice when administered subcutaneously. The *in vivo* activity of clindamycin 3-ribonucleotides and clindamycin 3-phosphate is presumably due to the conversion to clindamycin following hydrolysis by alkaline phosphatase or phosphodiesterase.

SUMMARY

There are three groups of naturally occurring nucleoside analogs that inhibit protein synthesis. One group, the 4-aminohexose pyrimidine nucleosides (aminoacylaminohexosylcytosine analogs) is comprised of amicetin, bamicetin, blasticidin H, blasticidin S, gougerotin, hikizimycin, mildiomycin, norplicacetin, oxamicetin, and plicacetin. They are broad spectrum antibiotics that inhibit human carcinoma cells and viruses. These cytidine analogs are structurally similar to the aminoacyl adenylyl-terminus of tRNA. They act as inhibitors of protein synthesis by blocking peptide chain elongation by binding to the ribosome–mRNA peptidyl complex. Unlike puromycin, these nucleoside analogs do not form covalent bonds with nascent peptides.

The second group of inhibitors of protein synthesis is the adenine nucleoside analogs. These are puromycin, nucleocidin, homocitrullylaminoadenosine, lysylaminoadenosine, antibiotic A201A, and 2′-aminoguanosine. Puromycin acts as a codon-independent analog of aminoacyl-tRNA by catalyzing the release of incomplete peptide chains from the peptidyl–tRNA–ribosome complex. Polypeptide synthesis terminates with the formation of a covalent peptidyl–puromycin bond. The fluoro-containing nucleoside antibiotic, nucleocidin, inhibits protein synthesis by binding to the ribosomes and inhibiting peptide formation. RNA synthesis is not inhibited. Neither the binding of tRNA to ribosomes nor the puromycin reaction is inhibited by nucleocidin. Some site other than the puromycin binding site is affected by nucleocidin. Homocitrullylaminoadenosine and lysylaminoadenosine may inhibit protein synthesis by affecting the overall conversion of amino acids from tRNA charged with amino acids. The 3′-amino group of the *p*-methoxytyrosine group in puromycin, the homocitrulline group in homocitrullylaminoadenosine, and the lysine group in lysylaminoadenosine may play a common role in the inhibition of protein synthesis. Antibiotics A201A and A201B inhibit protein synthesis. However, they do not have the free amino group by which they can accept incomplete polypeptide chains as does puromycin. Antibiotic A201A interferes with the formation of the first peptide bond. 2′-Aminoguanosine has select antibacterial activity in that few *E. coli* strains are inhibited. Protein synthesis is probably inhibited by its incorporation into mRNA, which produces a nonfunctional mRNA.

The third group of inhibitors of protein synthesis is the clindamycin ribonucleotides. They are formed by biomodification of clindamycin. Although clindamycin 3-ribonucleotides cannot penetrate the bacterial cell wall, they do offer protection to mice infected with *S. aureus*. The protection is probably through the enzymatic hydrolysis of the 3-phosphate.

REFERENCES

Agranoff, B. W., Davis, R. E., and J. J. Brink, *Brain Res.*, **1,** 303 (1966).

Aharonowitz, Y., and E. Z. Ron, *FEBS Lett.*, **52,** 25 (1975).

Arai, M., T. Haneishi, R. Enokita, and H. Kayamori, *J. Antibiot.* **27,** 329 (1974).

Argoudelis, A. D., and J. H. Coats, *J. Antibiot.*, **22,** 341 (1969).

Argoudelis, A. D., and J. H. Coats, *J. Antibiot.*, **24,** 206 (1971).

Argoudelis, A. D., J. H. Coats, D. J. Mason, and O. K. Sebek, *J. Antibiot.*, **22,** 309, (1969).

Argoudelis, A. D., J. H. Coats, and S. A. Mizsak, *J. Antibiot.*, **30,** 474 (1977).

Azzam, M. E., and I. D. Algranati, *Proc. Natl. Acad. Sci. U.S.*, **70,** 3866 (1973).

Barbacid, M., and D. Vazquez, *Eur. J. Biochem.*, **44,** 445 (1974).

Barondes, S. H., *Int. Rev. Neurobiol.*, **12,** 177 (1970).

Barondes, S. H., and H. D. Cohen, *Science*, **151,** 594 (1966).

Battaner, E., and D. Vazquez, *Biochim. Biophys. Acta*, **254,** 316 (1971).

Bayley, H., and J. R. Knowles, *Methods Enzymol.*, **46,** (1977).

Beneviste, R. E., B. W. Ozanne, and J. Davies, *Bacteriol. Proc.*, **1969,** 48.

Benne, R., and H. O. Voorma, *FEBS Lett.*, **20,** 347 (1972).

Berlin, Y. A., S. E. Esipov. M. N. Kolosov. M. M. Shemyakin, and M. G. Brazhnikoya, *Tetrahedron Lett.*, **1964,** (3513).

Bloch, A., in *Encyclopedia of Chemical Technology*, Vol. 2, Wiley, New York, p. 962, 1978.

Carrasco, L., and D. Vazquez, *J. Antibiot.*, **25,** 732 (1972).

Caskey, C. T., and A. L. Beaudet, in *Molecular Mechanism of Antibiotic Action on Protein Biosynthesis and Membranes*, E. Munoz, F. Garcia-Ferrandiz, and D. Vazquez, Eds., Elsevier, New York, 1972, p. 326.

Celma, M. L., R. E. Monro, and D. Vazquez, *FEBS Lett.*, **6,** 273 (1970).

Černá, J., F. W. Lichtenthaler, and I. Rychlík, *FEBS Lett.*, **14,** 45 (1971).

Černá, J., I. Rychlík, and F. W. Lichenthaler, *FEBS Lett.*, **30,** 147 (1973).

Cheung, C. P., M. L. Stewart, and N. K. Gupta, *Biochem. Biophys. Res. Commun.*, **54,** 1092 (1973).

Clark, J. M., and A. Y. Chang, *J. Biol. Chem.*, **240,** 4734 (1965).

Coats, J. H., and A. D. Argoudelis, *J. Bacteriol.*, **108,** 459 (1971).

Cooperman, B. S. in *Aging, Carcinogenesis, and Radiation Biology: The Role of Nucleic Acid Addition Reactions*, K. C. Smith, Ed., Plenum, New York, 1976, p. 315.

Cooperman, B. S., P. G. Grant, R. A. Goldman, M. A. Luddy, A. Minella, A. W. Nicholson, and W. A. Strycharz, *Methods Enzymol.*, **59,** 796 (1979).

Coutsgeorgopoulos, C., A. Bloch, K. A. Watanabe, and J. J. Fox, *J. Med. Chem.*, **18,** 771 (1975).

Crystal, R. G., N. A. Elson, and W. F. Anderson, *Methods Enzymol.*, **30,** Part F, 113 (1974).

Daluge, S., and R. Vince, *J. Med. Chem.*, **15,** 171 (1972).

Das, B. C., J. Defaye, and K. Uchida, *Carbohydr. Res.*, **22,** 293 (1972).

Dubnoff, J. S., and U. Maitra, *Proc. Natl. Acad. Sci. U.S.*, **68,** 313 (1971).

Dubnoff, J. S., A. H. Lockwood, and U. Maitra, *J. Biol. Chem.*, **247,** 2884 (1972).

Edens, B., H. A. Thompson, and K. Moldave, in *Lippmann Symposium: Energy, Biosynthesis and Regulation in Molecular Biology*, Walter de Gruyter, New York, 1974, p. 179.

Edens, B., H. A. Thompson, and K. Moldave, *Biochemistry*, **14,** 54 (1975).

Ennifar, S., B. C. Das, S. M. Nash, and R. Nagarajan, *J. Chem. Soc. Chem. Commun.* **1977,** 41.

Ensminger, P. W., and W. E. Wright, Abstracts, 16th Annual ICAAC Meeting, Chicago, 1976, Abstr. 62.

Epp, J. K., and N. E. Allen, Abstracts, 16th Annual ICAAC Meeting Chicago, 1976, Abstr. 63.

Evans, J. R., and G. Weare, *J. Antibiot.*, **30,** 604 (1977).

Flexner, L. B., and J. B. Flexner, *Proc. Natl. Acad. Sci. U.S.*, **55,** 369 (1966).

Florini, J. R., H. H. Bird, and P. H. Bell, *J. Biol. Chem.*, **241,** 1091 (1966).

Fox, J. J., K. A. Watanabe, and A. Bloch, *Prog. Nucleic Acids Res. Mol. Biol.*, **5,** 251 (1966).

Fresno, M., L. Carrasco, and D. Vazquez, *Eur. J. Biochem.*, **68,** 355 (1976).

Goldberg, J. H., and K. Mitsugi, *Biochemistry*, **6,** 383 (1967).

Guarino, A. J., and N. M. Kredich, *Fed. Proc.*, **23,** 371 (1964).

Guarino, A. J., M. L. Ibershof, and R. Swain, *Biochim. Biophys. Acta*, **72,** 62 (1963).

Hamill, R. L., and M. M. Hoehn, *J. Antibiot.*, **17,** 100 (1964).

Hamill, R. L., and M. M. Hoehn, Abstracts, 16th Annual ICAAC Meeting, Chicago, 1976, Abstr. 60.

Haneishi, T., A. Tetrahara, and M. Arai, *J. Antibiot.*, **27,** 334 (1974).

Harada, S., and T. Kishi, *J. Antibiot.*, **31,** 519 (1978).

Harada, S., E. Mizuta, and T. Kishi, *J. Am. Chem. Soc.*, **100,** 4895 (1978).

Harris, R. J., and S. Pestka, in *Molecular Mechanisms of Protein Biosynthesis*, H. Weissbach, and S. Pestka, Eds., Academic Press, New York, 1975, p. 246.

Haskell, T. H., A. Ryder, R. P. Frohardt, S. A. Fusari, Z. L. Jakubowski, and Q. R. Bartz, *J. Am. Chem. Soc.*, **80,** 743 (1958).

Henry, K. B., *Science*, **158,** 938 (1967).

Hishizawa, T., and S. Pestka, *Arch. Biochem. Biophys.*, **147,** 624 (1971).

Hobbs, J. B., and F. Eckstein, *J. Org. Chem.*, **42,** 714 (1977).

Hochstadt-Ozer, J., *J. Biol. Chem.*, **247,** 2419 (1972).

Ikehara, M., T. Maruyama, and H. Miki, *Tetrahedron Lett.*, **1976,** 4485.

Ikeuchi, T., F. Kitame, M. Kikuchi, and N. Ishida, *J. Antibiot.*, **25,** 548 (1972).

Iwasa, T., K. Suetomi, and T. Kusaka, *J. Antibiot.*, **31,** 511 (1978).

Jayaraman, J., and J. H. Goldberg, *Biochemistry*, **7,** 418 (1968).

Jaynes, E. N., Jr., P. G. Grant, G. Giangrande, R. Wieder, and B. S. Cooperman, *Biochemistry*, **17,** 561 (1978).

Jenkins, I. D., and J. P. H. Verheyden, *J. Am. Chem. Soc.*, **93,** 4323 (1971).

Jenkins, I. D., J. P. H. Verheyden, and J. G. Moffatt, *J. Am. Chem. Soc.*, **98,** 3346 (1976).

Johnson, A. E., and C. R. Cantor, *Methods Enzymol.*, **46,** 180 (1977).

Kaempfer, R., *Nature*, **228,** 534 (1970).

Kappen, L. S., H. Suzuki, and J. H. Goldberg, *Proc. Natl. Acad. Sci. U.S.*, **70,** 22 (1973).

Kirst, H. A., D. E. Dorman, J. L. Occolowitz, E. F. Szymanski, and J. W. Paschal, Abstracts, 16th Annual ICAAC Meeting, Chicago, 1976, Abstr. 61.

Kishi, T., T. Iwasa, T. Kusaka, and S. Harada, U.S. Patent 4,007,267 (1977).

Kohler, R. E., E. Z. Ron, and B. D. Davis, *J. Mol. Biol.*, **36,** 71 (1968).

Konishi, M., M. Kimeda, H. Tsukiura, H. Yamamoto, T. Hoshiya, T. Mujaki, K.-I. Fujisawa, H. Koshiyama, and H. Kawaguchi, *J. Antibiot.*, **26,** 752 (1973).

Leonard, N. J., and K. L. Carraway, *J. Heterocycylic Chem.*, **3,** 485 (1966).

Leonard, N. J., F. C. Sciavolino, and V. Nair, *J. Org. Chem.*, **33,** 3169 (1968).

Lichtenthaler, F. W., and T. Kulikowski, *J. Org. Chem.*, **41,** 600 (1976).

Lichtenthaler, F. W., and G. Trummlitz, *FEBS Lett.*, **38,** 237 (1974).

Lichtenthaler, F. W., J. Černá, and I. Rychlík, *FEBS Lett.*, **53,** 184 (1975a).

Lichtenthaler, F. W., T. Morino, and H. M. Menzel, *Tetrahedron Lett.*, **1975b,** 665.

Lichtenthaler, F. W., T. Morino, W. Winterfeldt, and Y. Sanemitsu, *Tetrahedron Lett.*, **1975c,** 3527.

Lichtenthaler, F. W., T. Morino, and W. Winterfeldt, *Nucleic Acids Res.*, Spec. Publ. 1, S33 (1975d).

Lichtenthaler, F. W., E. Cuny, T. Morino, and I. Rychlík, *Chem. Ber.*, **112,** in press (1979).

Lieff, B. D., S. K. Sharpless, and K. Schlesinger, *J. Comp. Physiolog. Psych.*, **90,** 773 (1976).

Menzel, H. M., and F. W. Lichtenthaler, *Nucleic Acids Res.*, Spec. Publ. 1, s155 (1975).

Miyamoto, M., Y. Kawamazu, M. Shinohara, K. Nakanishi, Y. Nakadaira, and N. S. Bhacca, *Tetrahedron Lett.*, **1964,** 2371.

Monro, R. E., and D. Vazquez, *J. Mol. Biol.*, **28,** 161 (1967).

Monro, R. E., R. Fernandez-Munoz, M. L. Clema, A. Jiminez, and D. Vazquez, *Prog. Antimicrob. Anticancer Ther.*, **2,** 413 (1970).

Morton, G. O., J. E. Lancaster, G. E. Van Lear, W. Fulmor, and W. E. Meyer, *J. Am. Chem. Soc.*, **91,** 1535 (1969).

Moss, D. R., D. E. Moss, and D. Fahrney, *Biochim. Biophys. Acta*, **350,** 95 (1974).

Moss, D. E., and D. Fahrney, *J. Neurochem.*, **26,** 1155 (1976).

Nagarajan, R., and S. M. Nash, 9th International Symposium on the Chemistry of Natural Products (IUPAC), Ottawa Canada, 1974.

Nair, V., and D. J. Emmanuel, *J. Am. Chem. Soc.*, **99,** 1571 (1977).

Nakamura, S., and H. Kondo, *Heterocycles*, **8,** 583 (1977).

Nakanishi, T., F. Tomita, and T. Suzuki, *Agr. Biol. Chem.*, **38,** 2465 (1974).

Nakanishi, T., T. Iida, T. Tomita, and A. Furuya, *Chem. Pharm. Bull.*, **24,** 2955 (1976).

Nakanishi, T., F. Tomita, and A. Furuya, *J. Antibiot.*, **30,** 743 (1977).

Nathans, D., in *Antibiot.* Vol. 1, D. Gottleib and P. D. Shaw, Eds., Springer-Verlag, Berlin, 1967, p. 259.

Nicholson, A. W., and B. S. Cooperman, *FEBS Lett.*, **90,** 203 (1978).

Paggi, P. and G. Toschi, *J. Neurobiol.*, **2,** 119 (1971).

Pestka, S., *Proc. Natl. Acad. Sci. U.S.*, **69,** 624 (1972a).

Pestka, S., *J. Biol. Chem.*, **247,** 4669 (1972b).

Pestka, S., *Methods Enzymol.*, **30,** Part F, 470 (1974a).

Pestka, S., *Methods Enzymol.*, **30,** Part F, 479 (1974b).

Pestka, S., H. Rosenfeld, R. Harris, and H. Hintikka, *J. Biol. Chem.*, **247,** 6895 (1972).

Sabol, S., M. A. G. Sillero, K. Iwasaki, and S. Ochoa, *Nature*, **228,** 1269 (1970).

Sakagami, Y., R. L. Chang, K. Watanabe, S. Ichikawa, and Y. S. Wang, Abstracts, 4th International Fermentation Symposium, Kyoto, Japan, 1972, p. 212.

Seal, S. N., and A. Marcus, *Biochem. Biophys. Res. Commun.*, **46,** 1895 (1972).

Seto, H., and H. Yonehara, *J. Antibiot.*, **30,** 1019 (1977).

Seto, H., N. Ōtake, and H. Yonehara, *Agr. Biol. Chem.*, **37,** 2421 (1973).

Seto, H., K. Furihata, and H. Yonehara, *J. Antibiot.*, **29,** 595 (1976).

Sikorski, M. M., J. Černá, I. Rychlík, and A. B. Legocki, *Biochem. Biophys. Res. Commun.*, **475,** 123 (1977).

Stringer, E. A., P. Sarkar, and U. Maitra, *J. Biol. Chem.*, **252,** 1739 (1977).

Subramanian, A. R., and B. D. Davis, *Nature*, **228,** 1273 (1970).

Subramanian, A. R., and B. D. Davis, *Proc. Natl. Acad. Sci. U.S.*, **68,** 2453 (1971).

Subramanian, A. R., and B. D. Davis, *J. Mol. Biol.*, **74,** 45 (1973).

Subramanian, A. R., B. D. Davis, and R. J. Beller, *Cold Spring Harbor Symp. Quant. Biol.*, **34,** 223 (1969).

Suhadolnik, R. J., *Nucleoside Antibiotics*, Wiley, New York, 1970.

Sundaralingam, M., and S. K. Arova, *J. Mol. Biol.*, **71,** 49 (1972).

Symons, R. H., R. J. Harris, L. P. Clarke, J. F. Wheldrake, and W. H. Elliot, *Biochim. Biophys. Acta*, **179,** 248 (1969).

Thompson, H. A., and K. Moldave, *Biochemistry*, **13,** 1348 (1974).

Tomita, K., Y. Uenoyama, K. I. Fujisawa, and H. Kawaguchi, *J. Antibiot.*, **26,** 765 (1973).

Townsend, L. B., in *Handbook of Biochemistry and Molecular Biology, Nucleic Acids*, Vol. I, 3rd ed. G. D. Fasman, Ed., CRC Press, Cleveland, 1975, p. 271.

Uchida, K., *Agr. Biol. Chem.*, **40,** 395 (1976).

Uchida, K., and H. Wolf, *J. Antibiot.*, **27,** 783 (1974).

Uchida, K., and B. D. Das, 8th International Symposium on the Chemistry of Natural Products (IUPAC), New Delhi, India, 1972.

Uchida, K., T. Ichikawa, Y. Shimauchi, T. Ishikura, and A. Ozaki, *J. Antibiot.*, **24,** 259 (1971).

Umezawa, H., S. Takasawa, M. Okanishi, and P. Utahara, *J. Antibiot.*, **21,** 81 (1968).

Vazquez, D., *FEBS Lett.*, **40,** S63 (1974).

Vince, R., and S. Daluge, *J. Med. Chem.*, **20,** 930 (1977).

Vince, R., and R. G. Isakson, *J. Med. Chem.*, **16,** 37 (1973).

Vince, R. R. G. Almquist, C. L. Ritter, F. N. Shirota, and H. T. Nagasawa, *Life Sci.*, **18,** 345 (1976).

Vince, R., S. Daluge, and M. Palm, *Biochem. Biophys. Res. Commun.*, **46,** 866 (1972).

Vuilhorgne, M., S. Ennifar, B. C. Das, J. W. Paschal, R., Nagarajan, E. W. Hagaman, and E. Wenkert, *J. Org. Chem.*, **42,** 3289 (1977).

Watanabe, K. A., E. A. Falco, and J. J. Fox, *J. Am. Chem. Soc.*, **94,** 3272 (1972).

Weisblum, B., and J. Davies, *Bacteriol. Rev.*, **32,** 493 (1968).

Weissbach, H., and S. Pestka, Eds., *Molecular Mechanisms of Protein Biosynthesis*, Academic Press, New York, 1977.

Wittmann, H. G., *Eur. J. Biochem.*, **61,** 1 (1976).

Wu, J., C. P. Cheung, and R. J. Suhadolnik, *Biochem. Biophys. Res. Commun.*, **78,** 1079 (1977).

Wulff, V. J., *Pharmacol. Biochem. Behav.*, **1,** 177 (1973).

Yamaguchi, I., H. Shibata, H. Seto, and T. Misato, *J. Antibiot.*, **28,** 7 (1975).

Yonehara, H., and N. Ōtake, *Antimicrob. Agents Chemother.*, **1965,** 855 .

Zamir, A., *Methods Enzymol.*, **46,** 621 (1977).

Chapter 3 Inhibition of RNA Synthesis

3.1 CORDYCEPIN 118

Biosynthesis 118

Biochemical Properties 118

Contribution of Cordycepin to the Elucidation of the Steps Leading to Functional Mammalian Cytoplasmic mRNA 119

Inhibition of Methylation of Nuclear RNA 121

In vivo and *in vitro* Studies on RNA and Poly (A) Synthesis with Mammalian Cells in Culture, Normal Tissue, and Tumor Tissue 122

Selective Inhibition of Free and Chromatin-Associated Poly(A) Synthesis by Cordycepin 5′-Triphosphate 123

Effects of 3′dATP, ATP, β-AraATP, and α-Ara-ATP on Terminal Deoxynucleotidyltransferase (TdT) 123

Inhibition of Newly Synthesized Globin mRNA 124

Inhibition of Poly(A) Synthesis by Cordycepin 5′-Triphosphate in Isolated Rat Liver Mitochondria 124

Nascent Nuclear RNA and Poly(A) Synthesis in Regenerating Rat Liver and Optic Nerves 124

Increased Therapeutic Efficiency of Cordycepin by Adenosine Deaminase Inhibitors 125

Effect of Cordycepin on rRNA Synthesis and Enzyme Induction in Rat Liver 125

Elucidation of Iron-Stimulated Ferritin Synthesis by Cordycepin 127

Effects of Cordycepin on Embryonic Development 127

Effect of Cordycepin on Plant Tissue 127

Effect of Cordycepin on Viruses 128

Effect of Cordycepin on Cellular Kinase Reactions 129

Effect of the Absence of the 3′-Hydroxyl Group on AMP Aminohydrolase and AMP Kinase 129

Effect of Cordycepin in the Dehydrogenases and in DNA Synthesis in Isolated Nuclei by 3′dNAD$^+$ and ADP Ribosylation 129

Use of Cordycepin in the Determination of "Chemical Proofreading" for Protein Synthesis 131

Determination of Subclasses of HnRNA 133

Inhibition of RNA and DNA Synthesis by 3′dATP in Toluene-Treated *E. coli* Cells and with *E. coli* RNA Polymerase 133

3.2 5-AZACYTIDINE 135

Biochemical Properties 136
Action and Metabolism 136
Effect of 5-Azacytidine on Bacteria 137
Effects on rRNA, mRNA, tRNA, and Protein Synthesis 137
Effect of 5-Azacytidine on DNA Synthesis 139
Chromosomal Breakage 139
Effect on Induced Liver Enzymes 141
Effect on Polyamine Biosynthesis 141
Effect on the Cell Cycle 141
Inhibitory Action of 5-Aza-2′-deoxycytidine 141
Effect of 5,6-Dihydro-5-azacytidine 142
Toxicity of 5-Azacytidine 143
Clinical Use of 5-Azacytidine: Effect on Human Acute Myelogenous Leukemia 143
Effect on Solid Tumors 144
Resistance of Leukemic Cells to 5-Azacytidine 145
Treatment of Osteogenic Sarcoma or Melanoma 145

3.3 2′-AMINOGUANOSINE 145

Biochemical Properties 145

3.4 3′-AMINOADENOSINE 146

Biosynthesis of 3′-Aminoadenosine 146
Biochemical Properties 146

3.5 ARISTEROMYCIN 147

Biochemical Properties 147
Effect on Mammalian Cells in Culture and RNA Polymerase 147
Inhibition of Transmethylases 147
Inhibition of Plants 148
Effect of Aristeromycin Derivatives and Aristeromycin 5′-Diphosphate on Polynucleotide Phosphorylase 149
Activation of Rat Brain, Adrenal, or Liver cAMP-Dependent Protein Kinase 149

3.6 BREDININ 149

Discovery, Production, and Isolation 150
Physical and Chemical Properties 150
Structural Elucidation 150
Inhibition of Growth 150
Chemical Synthesis 151
Enzymatic Synthesis of Bredinin 151
Biochemical Properties 151
    Antiarthritic Activity of Bredinin, An Immunosuppressive Agent 151
    Immunosuppressive and Antirheumatoid Arthritic Properties 152
    Action of Bredinin on Mammalian Cells and Bredinin-Induced Chromosome Aberrations 152
    Effect of Bredinin and Its Aglycon on L51784 Cells in Culture 152
    Effect of Bredinin with GMP on L5178Y Mouse Leukemia Cells 153
    Cytotoxicity of Bredinin 5′-Monophosphate 153

3.7 PENTOPYRANINES 154

Discovery, Production, and Isolation 154
Physical and Chemical Properties 154
Structural Elucidation 156
Chemical Synthesis of Pentopyranines 157
    Pentopyranine A 157
    Pentopyranine C 157
    Pentopyranamine D 157
    Pentopyranic Acid 157
Biochemical Properties 157

3.8 PYRROLOPYRIMIDINE NUCLEOSIDE ANALOGS 158

Biosynthesis 159
Biochemical Properties of Tubercidin 161
    Effect of Tubercidin on Prokaryotes and Eukaryotes 161
    Inhibition of the Elongation Step in RNA Synthesis 162
    Methylase Inhibitors 162
    Effect of NAD^+ Containing Tubercidin in Place of Adenosine in Dehydrogenase Reactions and ADP Ribosylation 165
    Clinical and Anthelmintic Properties of Tubercidin 165
    Interferon Production 165
    Inhibition of Viruses 166

Biochemical Properties of Toyocamycin 166
Effect on Eukaryotes 166
Effect on Appearance of RNA in the Cytoplasm and rRNA Transcription 166
Effect on Viruses 167
Biochemical Properties of Sangivamycin 169

3.9 PYRAZOLOPYRIMIDINE NUCLEOSIDE ANALOGS 169

Chemical Synthesis 170
Prototype Tautomerism of Formycin and Fluorescent Derivatives 170
Formycin Analogs 172
Biosynthesis of Formycin 172
Biochemical Properties of Formycin 172
N-Methylformycins 173
Formycin 5′-Triphosphate and Anhydroformycins with RNA Polymerase and Adenosine Deaminase 173
Inhibition of AMP Nucleosidase by Formycin 5′-Phosphate and Formycin 174
Purine Nucleoside Metabolism in Humans with Severe Combined Immunodeficiency and Adenosine Deaminase Deficiency Induced with Coformycin 174
Formycin and Purine Nucleoside Phosphorylase in Human Erythrocytes 177
Effect of Formycin on the Processing of tRNA Precursors 177
Influence of Formycin in NAD^+ with Dehydrogenases 179
Cytokinin Activity of Adenosine Analogs 180
Effect of Formycin on Blood Platelet Aggregation 180
Effect on Bovine Adrenal Estrogen Sulfotransferase 181
Biochemical Properties of Formycin B 181
Effect on ADP Ribosylation of Chromatin Proteins, Proinsulin Synthesis, and Development 181
Interferon Induction 182
Synthesis of a C-Analog of AICAR from Formycin B 182
Oxoformycin B 182

3.10 THURINGIENSIN 183

Discovery, Production, and Isolation 184
Physical and Chemical Properties 185
Structural Elucidation 185
Chemical Synthesis of Thuringiensin 186
Inhibition of Growth 186

Toxicity 188
Biosynthesis of Thuringiensin 188
Biochemical Properties 188
Thuringiensin and Its Analogs: Effect on RNA, DNA, and Protein Synthesis 188
Effect of DNA-Dependent RNA Polymerase in Nuclei Isolated from Rat Liver Pretreated with Thuringiensin 189
Inhibition of DNA-Dependent RNA Polymerases in Nuclei Isolated from Rat Liver of Untreated Animals 189
Effect on Maturation of Mouse Nuclear RNA Synthesis 190
Inhibition of RNA Synthesis by Thuringiensin with *E. coli* DNA-Dependent RNA Polymerase 191
Effect of Thuringiensin on the DNA-Dependent RNA Polymerase Isolated from *B. thuringiensis* 191
Effect of Thuringiensin on Normal and Hormone-Stimulated RNA Synthesis in Isolated Nuclei from the Larvae and Adult *Sarcophaga bullata* 192
Effect of Thuringiensin on Polyphenylalanine Formation 192
Inhibition of Mitotic Spindle Formation by Thuringiensin 192
Inhibition of Adenyl Cyclase by Thuringiensin 193
Bacterial Membrane Transport of Thuringiensin 193
Effect of Thuringiensin on Nucleolar Morphology 194
Differentiation of *E. coli* RNA Polymerase from Phage T3 RNA Polymerase by Thuringiensin and α-Amanitin 194
Application of Thuringiensin in Insect Control 195

3.11 PUROMYCIN AMINONUCLEOSIDE 195

SUMMARY 196

REFERENCES 198

The primary inhibitory action of the naturally occurring nucleoside analogs reviewed in this chapter is their effect on RNA synthesis. However, as with most nucleoside analogs (including some of the analogs reviewed in this chapter), there are many other cellular processes that are affected. These additional inhibitory processes are discussed along with the inhibition of the RNA synthetic processes. The nucleosides reviewed in this chapter are cordycepin, 5-azacytidine, 2′-aminoguanosine, 3′-aminoadenosine, aristeromycin, bredinin, the pentopyranines, tubercidin, toyocamycin, sangivamycin, formycin, formycin B, oxoformycin B, thuringiensin, and puromycin aminonucleoside. Because bredinin, the pentopyranines, and thuringiensin are recently discovered analogs, their physical, chemical, and biochemical properties are reviewed in detail.

3.1 CORDYCEPIN

Biosynthesis

The biosynthesis of cordycepin by *Cordyceps militaris* has been shown to proceed by the direct conversion of [U-^{14}C;3′-^{3}H]adenosine to cordycepin without loss of the tritium at C-3′ (Suhadolnik et al., 1964; Lennon and Suhadolnik, 1976). The reduction of C-3′ of adenosine proceeds without loss of the 3′-hydrogen. This is similar to the biosynthesis of the 2′-deoxynucleoside 5′-di- or 5′-triphosphates from the corresponding ribonucleotides by ribonucleotide reductase.

Biochemical Properties

Cordycepin (3′-deoxyadenosine) (Fig. 3.1) was the first naturally occurring nucleoside antibiotic to be isolated. It is isolated from the culture filtrates of *Cordyceps militaris* and *Aspergillus nidulans* (Cunningham et al., 1951). Although cordycepin is a cytostatic agent and an isomer of 2′-deoxyadenosine, it is a functional analog of adenosine. The toxicity of cordycepin in eucaryotes and procaryotes is reversed by adenosine, but not by 2′-deoxyadenosine.

The first reports on the mode of action of cordycepin indicated that it is rapidly converted to its 5′-mono-, 5′-di-, and 5′-triphosphates by Ehrlich ascites tumor cells (Klenow, 1961). Rottman and Guarino (1964) demonstrated that cordycepin 5′-monophosphate competitively inhibits phosphoribosyl pyrophosphate amidotransferase from either *B. subtilis* or pigeon liver, which makes it a potent feedback inhibitor of purine biosynthesis *de novo*. It was shown to inhibit RNA synthesis (Klenow and Overgaard-Hansen, 1964; Rich et al., 1965). DNA and protein synthesis are not affected. Small amounts of [^{14}C]- or [^{3}H]cordycepin were found in the 3′-terminus of RNA (Cory et al., 1965). The lack of a 3′-hydroxyl group on cordycepin suggested that this adenosine analog was acting as a chain terminator for the 3′-end of growing RNA chains (Shigeura and Boxer, 1964). It was these initial studies with cordycepin plus the many reports since 1969 that finally led to elucidation of the posttranscriptional modifications on the

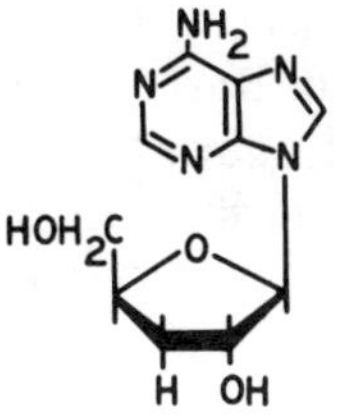

Figure 3.1 Structure of cordycepin.

3′- and 5′-terminus [i.e., poly(A) and methylation to form the "5′-cap"] needed for the formation of functional cytoplasmic mRNA.

Contribution of Cordycepin to the Elucidation of the Steps Leading to Functional Mammalian Cytoplasmic mRNA. Although it would have been difficult to address the questions of the biosynthesis, structure, and function of eukaryotic mRNA a few years ago, it is now possible to discuss these processes in considerable detail. For those mRNA molecules that require poly(A) at the 3′-terminus and a 5′-cap at the 5′-terminus, cordycepin has contributed markedly to our knowledge of these important posttranscriptional modifications. This section covers the time from the first reports of rich AMP fractions in mammalian RNA (Hoyer et al., 1963; Salzman et al., 1964; Henshaw et al., 1965) to the most recent reports by Darnell and coworkers in which the "splicing" process is presented to show how mRNA is synthesized (Nevins & Darnell, 1978; Darnell, 1978, 1979).

The discovery of poly(A) in rat liver began with the report by Georgiev and Mantieva (1962) on the DNA-like RNA in rat liver. Sibatani et al. (1962) subsequently reported on a unique physical feature of mRNA. Hadjivassiliou and Brawerman (1966, 1967) described the first procedure showing that RNA, very rich in adenine, could be isolated from rat liver cytoplasm. The finding of poly(A) as a natural cellular component was reported by Edmonds and Abrams (1960, 1963) in thymus nuclei and by Gottesman et al. (1962) and August et al. (1962) in *E. coli*. Subsequently, Siev et al. (1969) demonstrated that low concentrations of cordycepin in HeLa cells for short periods of time caused a significant depression of completed rRNA, ribosomal precursor (45 S) RNA, and tRNA. It was Lim and Canellakis (1970) who provided the first evidence that poly(A) exists in mRNA. Kates (1970) observed that vaccinia virus cores contained poly(A) sequences about 100 nucleotides long. Simultaneously, Penman et al. (1970) reported that cordycepin inhibited the transport of mRNA from the nucleus to the cytoplasm of mammalian cells while having little effect on the formation of HnRNA. DNA and protein synthesis was not inhibited. The location of poly(A) on the 3′-end of the majority of mRNA molecules in mammalian cells was reported by several laboratories (Mendecki et al., 1972; Weinberg,

1973; Darnell et al., 1971a, b; Edmonds et al., 1971; Lee et al., 1971; Kates, 1970; Lim and Canellakis, 1970). Finally, the elegant studies of Darnell and coworkers (Darnell et al., 1971b; Philipson et al., 1971) showed that cordycepin preferentially interfered with the posttranscriptional addition of poly(A) to preformed chains of HnRNA and mRNA in eukaryotic cells. This inhibition of poly(A) formation interfered with the maturation of mRNA in the nucleus.

Evidence that cordycepin may act at the level of the initiation of poly(A) synthesis and not the elongation process was provided by Diez and Brawerman (1974). The elongation process is relatively insensitive to cordycepin, wheras poly(A) synthesis *de novo* is very sensitive to this drug (Table 3.1).

Mendecki et al. (1972) and Niessing (1975) suggested that cordycepin, although inhibitory to poly(A) formation, is not incorporated into the growing poly(A) chain and therefore does not act as an inhibitor for poly(A) synthesis by means of chain termination. Although cordycepin inhibits poly(A) synthesis in mammalian cells in culture, HnRNA synthesis is not affected (Penman et al., 1970; Mendecki et al., 1972; Darnell et al., 1973; Nakazato et al., 1974; Rizzo et al., 1972). In *P. polycephalum* Fouquet et al. (1975) showed that, while cordycepin does not impair nuclear poly(A) synthesis, cordycepin does inhibit all species of RNA. With the exception of histone mRNA the evidence obtained indicated that nuclear RNA destined to become cytoplasmic mRNA must be polyadenylated. This poly(A) segment is a signal for the release of mRNA into the nucleus (Penman et al., 1970; Darnell et al., 1971b; Mendecki et al., 1972). The effect of cordycepin on nuclear synthesis and terminal turnover of poly(A) has been studied (Jelinek et al., 1973; Sawicki et al., 1977). There is a marked decrease in the labeling of nuclear poly(A) in 2 min; the ratio of internal AMP to terminal adenosine of the poly(A) from the 2 min experiment of the cordycepin-treated HeLa cells was only 10.8:1 (Fig. 3.2). This compares to a ratio of

Table 3.1 Effect of Cordycepin on *de novo* Synthesis and on Elongation of the Poly(A) Sequence

	De novo Synthesis (control cells)		Elongation (actinomycin-treated cells)	
Addition	Nuclei	Polysomes	Nuclei	Polysomes
None	33,800	8000	4200	1300
Cordycepin	3,100	2300	4050	840
Percent inhibition	91	71	3	35

Cpm [^{3}H]ads in poly(A). Reprinted with permission from Brawerman, 1976.

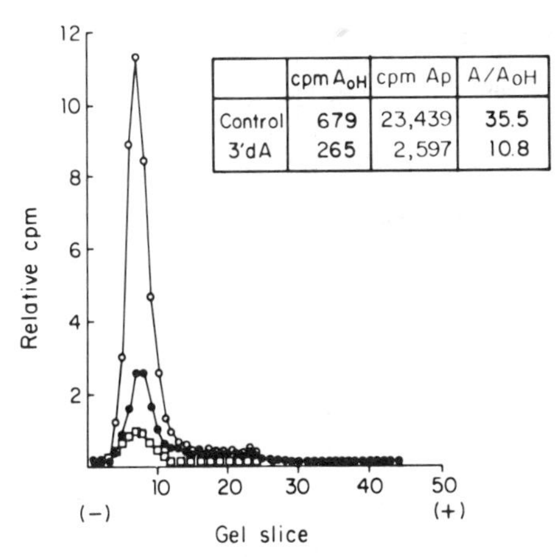

	cpm A_{OH}	cpm Ap	A/A_{OH}
Control	679	23,439	35.5
3′dA	265	2,597	10.8

Figure 3.2 Labeling of nuclear poly(A) in the presence of 3′-deoxyadenosine. A total of 3×10^8 HeLa cells were concentrated to 3×10^6 cells/ml and divided. Half of the culture was pretreated for 3 min with 100 μg 3′-deoxyadenosine/ml. At 30 s and 2 min after adding [^{3}H]adenosine, samples of both cultures were fractionated and the labeled nuclear poly(A) was assayed after purification. The ratio of radioactivity incorporated into Ap (i.e., internal AMP) compared to 3′-terminal adenylate was determined on the poly(A). Samples obtained from the 2 min pulse levels, which were recovered from slices 4–11 of gels run in parallel to those illustrated above. The cpm of the 2 min control and cordycepin samples were normalized with respect to the 30 s samples. Labels: 2 min [^{3}H]adenosine (□); 30 s [^{3}H]adenosine (●); control (○). Reprinted with permission from Sawicki et al., 1977.

35.5:1 for the control cells. Sawicki et al. (1977) summarized their findings on the effect of cordycepin as follows: (i) the nucleus is the site of *de novo* synthesis of poly(A), (ii) there is a nuclear and cytoplasmic 3′-addition of poly(A), (iii) only those molecules bearing poly(A) of 230 AMP residues or longer exit from the nucleus to the cytoplasm, (iv) nuclear terminal addition is much more rapid than is cytoplasmic terminal addition.

Nevins and Darnell (1978) and Darnell (1978, 1979) have demonstrated the sequence of events in the leaders-spacer-coding sequence-poly(A) that are required for the synthesis of mature mRNA. Three leader sequences are at the 5′-end of the primary transcript beginning at 16 on the map. These leaders in the mature RNA represent about 180–200 ribonucleotides. The excess nucleotides in the spacer region are removed and the leaders are spliced. mRNA processing involves the following steps: capping, poly(A) addition, methylation of internal AMP residues, and RNA:RNA splicing (Darnell, 1978, 1979). Cordycepin inhibits the formation of mature mRNA by the inhibition of poly(A) polymerase (Fig. 3.3).

Inhibition of Methylation of Nuclear RNA. The previous section described the role of cordycepin in poly(A) formation on the 3′-end of HnRNA and mRNA. This section is concerned with the newly discovered role of

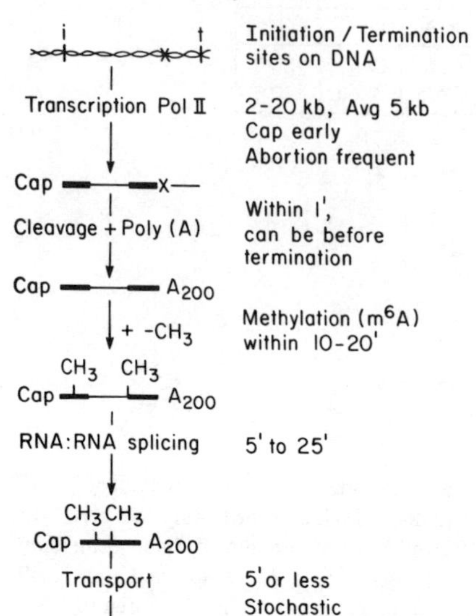

Figure 3.3 Processing steps in mRNA biosynthesis. Data from adenovirus type 2; kb = kilobase. From Darnell, personal communication.

cordycepin on the methylation of rRNA and of the 5′-cap. The methylation of nuclear rRNA and mRNA is one prerequisite for the formation of functional cytoplasmic rRNA and mRNA. For example, precursor 45 S rRNA requires 2′-*O*-methylation for the endonucleolytic cleavage to form 28 and 18 S rRNA (Weinberg and Penman, 1970; Perry and Kelley, 1972; Dabeva et al., 1976; Caboche and Bachellerie, 1977; Wolf and Schlessinger, 1977). mRNA requires methylation at the 5′-end to form the 5′-cap to allow mRNA to form a stable initiation complex for translation (Rao et al., 1975; Muthukrishnan et al., 1975; Both et al., 1975; Furuichi et al., 1977; Shimotohno et al., 1977). Although cordycepin inhibits the posttranscriptional processing of poly(A) on the 3′-end of mRNA, Glazer and Peale (1978) have also demonstrated that cordycepin (2.5×10^{-4} *M*) is a very effective inhibitor of 2′-*O*-methylation of adenosine, guanosine, and cytidine. The potent inhibition by cordycepin of rRNA and the 5′-cap of HnRNA may explain the inhibitory role that this analog plays in the synthesis of rRNA and mRNA.

In vivo *and* in vitro *Studies on RNA and Poly(A) Synthesis with Mammalian Cells in Culture, Normal Tissue, and Tumor Tissue.* Müller and coworkers have reported a series of logical, concise experiments on the effects of cordycepin and its 5′-triphosphate either on intact mouse L5178Y cells in culture or with partially purified RNA and poly(A) polymerases. Müller et al. (1977) observed that the primary inhibitory effects of [G-^{3}H]cordycepin were directed toward RNA and protein synthesis. DNA synthesis was not inhibited. Cordycepin was incorporated into the 3′-terminus of RNA and was found in different RNA species, but the incorporation was not uniform. Tritium-labeled cordycepin was found in the 28, 10, 5, and 4 S RNA. Müller et al. also showed that whereas DNA polymerase-α and polymerase-β from mouse lymphoma cells were not inhibited by 3′dATP, RNA polymerases I, II, and III from normal mouse liver are moderately inhibited by 3′dATP. For normal mouse liver RNA polymerase II, K_m/K_i is 0.60. Müller et al. (1977) also showed that only one 3′dAMP residue is incorporated into the 3′-end of oligo(ApA). Therefore, they demonstrated that 3′dATP is an initiator and substrate for poly(A) polymerase.

The above findings by Müller et al. (1977) on the incorporation of cordycepin into 28, 10, 5 and 4 S RNA may explain the 65% reduction in 18 S rRNA and the 35–40% reduction in 28 S rRNA (Fredericksen and Klenow, 1964). With normal rat liver and hepatoma nuclei, Rose and Jacob (1976) showed that the level of poly(A) polymerase is elevated in hepatoma relative to liver nuclei. Both enzymes are inhibited by 3′dATP.

Selective Inhibition of Free and Chromatin-Associated Poly(A) Synthesis by Cordycepin 5′-Triphosphate. Rose, Bell & Jacob (1977a) reported that 3′dATP is a competitive inhibitor of rat liver nuclear chromatin-free poly(A) polymerase; with chromatin-associated poly(A) polymerase, 3′dATP is a noncompetitive inhibitor. About 75 times more 3′dATP is needed for 50% inhibition of free nuclear poly(A) polymerase as for chromatin-associated enzyme. The inhibition of DNA-dependent RNA synthesis exhibited a dose response similar to that of free poly(A) polymerase (Rose et al., 1977b). Their findings offer a mechanism for the selective inhibition of initial polyadenylation of HnRNA *in vivo* by cordycepin and provides a satisfactory explanation for the indiscriminate effect of 3′dATP on "free" poly(A) and RNA polymerase. In a more recent study, Rose et al. (1977c) showed that the bound and free poly(A) polymerase exists as the same enzyme species in two different functional states.

In contrast to the above studies, Niessing (1975) reported that there are three distinct poly(A) polymerases in rat liver. Although all three polymerases were inhibited by 3′dATP, the nucleotide analog was not incorporated. On the basis of their studies Maale et al. (1975) suggested that the inhibition of poly(A) polymerase from HeLa cells and maize seedlings may be due to an incorporation of 3′dAMP. The poly(A) polymerase was no more sensitive to 3′dATP than to 2′dATP.

Effects of 3′dATP, ATP, β-AraATP, and α-AraATP on Terminal Deoxynucleotidyltransferase (TdT). TdT catalyzes the primer-dependent, but template-independent, DNA polymerization of deoxynucleoside 5′-phosphates (Bollum, 1962; Chang and Bollum, 1971). Bollum and coworkers subsequently showed that TdT is only present in thymus, bone marrow, and blood lymphoblasts of patients with acute lymphoblastic leukemia (Coleman et al., 1974, 1976). TdT has been reported in nonthymic cells and germinating wheat embryo (Srivastava, 1974; Brodniewicz-Proba and Buchowicz, 1976).

The activity of TdT in incubations of oligo[d(pA)$_3$] with 3′dATP or ATP is inhibited, and only tetranucleotide products are formed (Müller et al., 1977; Bhalla et al., 1977; Dicioccio and Srivastava, 1977). With 700-fold enriched TdT from thymus, Müller et al. (1978) have demonstrated that 3′dATP is a competitive inhibitor. ATP and 3′dATP act as chain terminators in that incubations of oligo[d(pA)$_3$] with 3′dATP or ATP form only tetranucleotide products.

Inhibition of Newly Synthesized Globin mRNA. Depending on the type of treatment of the cells with cordycepin, various effects on RNA synthesis are observed. For example, Beach and Ross (1978) reported that in cultured mouse fetal liver erythroid cells (MFL) cordycepin inhibits mRNA transcription and causes the accumulation of globin mRNA sequences rather than inhibition of poly(A). The data were obtained by hybridization of radioactive globin-specific RNA with cordycepin-treated and control erythroid cells with excess unlabeled globin cDNA. When MFL cells were exposed to cordycepin for 40 min at 20 μg/ml, the inhibition of tRNA was as effective as was the inhibition of newly synthesized mRNA. Cordycepin inhibited the synthesis of globin mRNA by 73%, whereas the inhibition of total RNA synthesis was only 43%.

Inhibition of Poly(A) Synthesis by Cordycepin 5′-Triphosphate in Isolated Rat Liver Mitochondria. Rose and Jacob (1976) showed that purified rat liver mitochondria synthesize poly(A) *in vitro*. Poly(A) synthesis required ATP. The average chain length of the poly(A) synthesized *in vitro* was 23 nucleotides. Cordycepin 5′-triphosphate (60 μM) is an effective inhibitor of poly(A) synthesis.

Nascent Nuclear RNA and Poly(A) Synthesis in Regenerating Rat Liver and Optic Nerves. Kann and Kohn (1972) and Siev et al. (1969) demonstrated a preferential inhibition by cordycepin of precursor nuclear rRNA in L1210 cells. Darnell et al. (1971a, b) and Mendecki et al. (1972) reported on a preferential inhibition of mRNA synthesis in nuclei of HeLa and S180 cells by the posttranscriptional polyadenylation of mRNA precursors. With regenerating rat liver, cordycepin shows a different effect. Glazer (1975, 1978) demonstrated that cordycepin was equally effective in its inhibition of the major species of nuclear RNA, that is, rRNA, non-poly(A) HnRNA, and poly(A) HnRNA, and poly(A) synthesis in regenerating rat liver.

Ingoglia (1978) reported that cordycepin blocks retinal ribosomal but not 4S RNA synthesis during nerve regeneration. Cordycepin also blocks the export of 4S RNA into regenerating optic axons.

Increased Therapeutic Efficiency of Cordycepin by Adenosine Deaminase Inhibitors. A serious difficulty encountered with 6-amino analogs of adenosine that are substrates for adenosine deaminase is the removal of the amino group and the formation of inactive analogs. This is the case with cordycepin. For example, *E. coli* is not inhibited by cordycepin because of its deamination to 3′-deoxyinosine, which is not toxic. Similarly, fetal calf serum, which is used for human epithelial (H. Ep.) #1 cells, has a very active adenosine deaminase that deaminates cordycepin (Cory et al., 1965). Therefore, it was not possible to calculate the exact concentration of cordycepin to which the H. Ep. #1 cells were exposed. To overcome this difficulty, Plunkett and Cohen (1975) increased the therapeutic efficiency of cordycepin by the addition of the adenosine deaminase inhibitor, *erythro*-9-(2-hydroxy-3-nonyl)adenine (EHNA), to mouse L cells in culture. Compared to cordycepin alone, cordycepin plus EHNA markedly reduced the incorporation of uridine and thymidine into RNA and DNA, respectively. The inhibition of DNA synthesis was attributed to the inhibition of RNA primer synthesis. Johns and Adamson (1976) described the enhanced antitumor activity of cordycepin in cell culture systems and in mice bearing P388 ascites leukemia by the adenosine deaminase inhibitor, 2′-deoxycoformycin (dCF) (see page 266). Additional detailed studies on the effect of adenosine deaminase inhibitors on cordycepin, ara-A, and formycin are reviewed in Chapter 5 and are not duplicated here.

Effect of Cordycepin on rRNA Synthesis and Enzyme Induction in Rat Liver. Siev et al. (1969) and Truman and Fredericksen (1969) reported that cordycepin inhibits ribosome biosynthesis by causing premature termination of 45 S ribosomal RNA precursor. Webb and coworkers (Rizzo and Webb, 1968, 1969; Rizzo et al., 1971, 1972) have demonstrated that cordycepin inhibits the accumulation of newly synthesized ribosomes in the cytoplasm of the resting cells of rat livers by 85% at 10 mg cordycepin/kg within 2.5 h of administration. They also showed that ribosome biosynthesis is controlled at the level of transcription of the 45 S precursor in the nucleus as well as at the level of processing in the nucleus and transport of the completed ribosomal subunit to the cytoplasm. The decrease in the 45 S rRNA is not due to an inhibition of RNA synthesis, but rather to a shift in the mean size distribution of the RNA to a lighter species. Rizzo et al. (1972)

concluded that cordycepin affects the premature termination of the transcription of the 45 S ribosomal precursor.

To determine the effect of cordycepin on the synthesis of specific proteins, Rizzo et al. (1972) studied the effect of cordycepin on the corticosteroid-mediated induction of hepatic tyrosine transaminase. This enzyme is induced about 4.5 h after a single dose of hydrocortisone. Although 5-azacytidine and 8-azaguanine do not inhibit the corticosteroid-mediated induction of this enzyme, cordycepin inhibits the induction by 60% at 200 mg/kg. These latter results are consistent with the observation by Darnell et al. (1971b) that cordycepin inhibits the formation of poly(A), which is involved in the processing of mRNA.

The specificity of action of cordycepin on L-1210 cells and the type of RNA affected has also been studied by electrophoresis of agarose–urea gels (Glazer et al., 1978). Cordycepin-treated cells predominantly inhibited nuclear rRNA synthesis by the equal suppression of labeling of 28 and 18 S RNA, which are the main rRNA species detected following a 30 min labeling period (Fig. 3.4*A*). Non-poly(A) HnRNA was heterodisperse and of low molecular weight (Fig. 3.4*B*); poly(A) HnRNA appeared to be restricted to the 11 S poly(A) species (Fig. 3.4*C*). These data suggest that the main pharmacological action of cordycepin, responsible for its antineoplastic

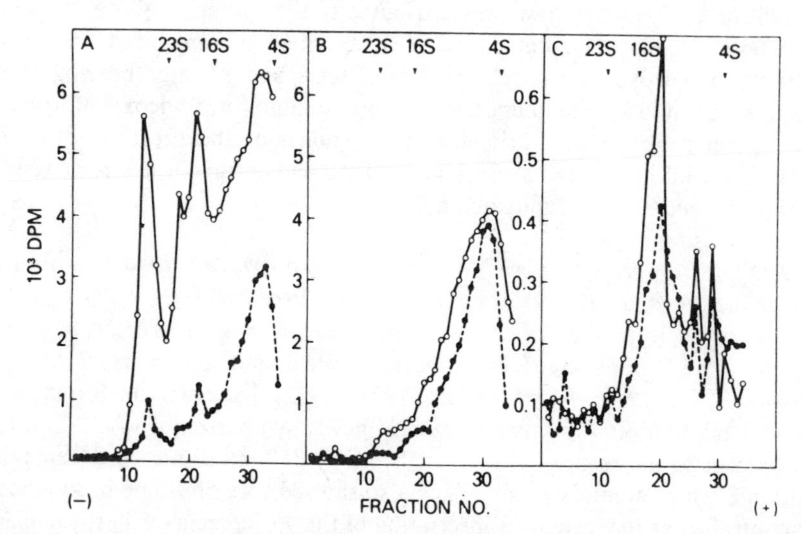

Figure 3.4 Agarose gel electrophoresis of rRNA after treatment of L1210 cells with cordycepin. L1210 cells (5×10^7 cells per flask) were incubated for 30 min with (●) and without (○) 2.5×10^{-4} *M* cordycepin. RNA was labeled by an additional 30 min incubation with [^{3}H]uridine and was fractionated and subjected to electrophoresis. (*A*) rRNA; (*B*) non-poly(A) HnRNA; (*C*) poly(A) HnRNA. Reprinted with permission from Glazer et al., 1978.

activity, is the inhibition of rRNA (Glazer et al., 1978). The sensitivity of rRNA to cordycepin may be due to the reported inhibition of the posttranscriptional processing of nuclear rRNA (Glazer and Peale, 1978).

Elucidation of Iron-Stimulated Ferritin Synthesis by Cordycepin. Cordycepin has been used to elucidate the mechanism by which iron increases ferritin synthesis. Ferritin is the major intracellular iron storage protein in many tissues (Linder-Horowitz et al., 1970). Iron increases the level of ferritin in the liver by increasing the amount of ferritin mRNA in the polyribosome–ferritin–mRNA complex (Drysdale and Munro, 1966; Zahringer et al., 1975). Concentrations of cordycepin reported by Tilghman et al. (1974) to block mRNA in the cytoplasm failed to prevent the iron-induced twofold increase in the polysomal ferritin-mRNA content. These findings have been subsequently confirmed by Zahringer et al. (1976). Apparently, iron exerts its stimulatory action on the synthesis of the ferritin in the cytoplasm by removing the inhibition of the translation of ferritin mRNA.

Effects of Cordycepin on Embryonic Development. The inhibition of rRNA synthesis, polyadenylation of mRNA, and the transfer of mature mRNA to cytoplasmic polysomes in eukaryotes by cordycepin (Siev et al., 1969; Penman et al., 1970; Darnell et al., 1971b; Philipson et al., 1971; Lavers et al., 1974; Harris and Dure, 1974; Fouquet et al., 1975) has now found wide application in developmental systems. Rowinski et al. (1975) demonstrated that cordycepin can block the development of inner cell mass derivatives in postimplantation mouse embyros in culture. Levey and Brinster (1977) studied the effect of cordycepin on development and macromolecular synthesis in mouse embryos at different stages of preimplantation development. This approach was based on the knowledge that actinomycin D blocks incorporation of RNA and protein precursors and inhibits embryonic development in culture (Mintz, 1974; Thomson and Biggers, 1966; Skalko and Morse, 1969; Tasca and Hillman, 1970; Monesi et al., 1970). They showed that the addition of cordycepin (10 μg/ml) to mouse embryos explanted into culture at the two-cell morulae and blastocyst stage caused a dose-responsive inhibition of cleavage and blastulation of these embryos. There is a suppression of RNA synthesis of morulae and blastocysts.

Effect of Cordycepin on Plant Tissue. Hammett and Katterman (1975) reported that most of the polyadenylated mRNA is also located in the nuclei of cotton seeds. Up to 6 h of germination, cordycepin inhibits nuclear and polysomal poly(A) by 75%. It has little effect on protein synthesis.

After 6 h, cordycepin markedly inhibits protein synthesis, presumably because the maturation of mRNA is inhibited. Delseny et al. (1975) showed that cordycepin at 200 μg/ml totally inhibits RNA synthesis. At lower concentrations, rRNA and tRNA syntheses were selectively inhibited, but protein synthesis was not affected. Harris and Dure (1974) reported that cordycepin did not inhibit RNA polymerase II in germinating cotton cotyledons; however, RNA polymerase I and III and polyadenylation of mRNA are inhibited. Marcus and coworkers in their studies with the early germination of wheat embryos asked, "What is the importance of RNA and polyribosome formation in the presence of cordycepin?" (Spiegel and Marcus, 1975). Cordycepin inhibited RNA synthesis by 73%, but did not affect polysome formation.

Effect of Cordycepin on Viruses. It was the original observation by Darnell and coworkers (Philipson et al., 1971) that cordycepin preferentially inhibited the addition of poly(A) sequences to virus-specific RNA during adenovirus replication that led to the subsequent studies on the effect of cordycepin on viruses. Although the mechanism of inhibition of host protein synthesis following viral infection is unknown (Bablanian, 1975; Moss, 1974), it is clear that the inhibition of virus-specific RNA synthesis (i.e., addition of poly(A) to viral mRNA) of viral-infected cells is the target of cordycepin (Nair and Owens, 1973; Nair and Panicali, 1976; Nevins and Joklik, 1975). Cellular mRNA is not inactivated by cordycepin in vaccinia virus-infected cells (Person and Beaud, 1978). Wu et al. (1972) reported that low concentrations of cordycepin suppress leukovirus production from murine fibroblasts induced by iododeoxyuridine. In a subsequent study, Richardson et al. (1975) demonstrated that the inhibitory effect of cordycepin results in a reduction of the number of cells producing virus. Cordycepin preferentially inhibits viral replication more than transformation of normal rat kidney cells by murine sarcoma virus. It is likely that cordycepin preferentially inhibits viral replication by inhibiting poly(A) and viral RNA, which in turn interferes with the maturation of viral mRNA.

In chick embryos infected with influenza virus, cellular RNA synthesis is inhibited 75% by cordycepin, whereas viral replication is not inhibited (Mahy et al., 1973). Cordycepin inhibits poly(A) parainfluenza mRNA synthesis. There is an 80% inhibition of the synthesis of parainfluenza RNA. Similar findings were observed when cordycepin was added to KB cells infected with adenovirus. Van Oortmerssen et al. (1975) reported that cordycepin strongly inhibited virus-associated small molecular weight RNA, but had little effect on viral HnRNA synthesis. Apparently, the synthesis of viral RNAs involves either two different RNA polymerase activities or two different RNA species in the *in vivo* transcription of the adenovirus genome. O'Brien and Boone (1977) reached the same conclusion in studies with cordycepin with cultured feline leukemia virus (FeLU) and FeLU-associated cell surface antigens (FeLU-CSA). Cordycepin does not affect either the size or the relative proportions of the 6–27 S and 18–22 S Newcastle disease virus mRNA species of embryonated hen's eggs infected with Newcastle disease virus (Weiss and Bratt, 1975). Cordycepin does not affect poly(A) associated with RNA. This is in contrast to the inhibition of poly(A)-associated HnRNA in the eukaryotic cell. With human rhinovirus and poliovirus, viral replication is completely inhibited by cordycepin in which 3′dATP is a competitive inhibitor and acts as a chain terminator of RNA (Nair and Panicali, 1976; Panicali and Nair, 1978).

Effect of Cordycepin on Cellular Kinase Reactions. A question concerning the mode of action of an inhibitor related to one specific reaction in the cells is always difficult to answer because of pleotropic activity. Cordycepin is a potent inhibitor of a nucleoside-stimulated protein kinase from *Trypanosoma cruzi* (Walter and Ebert, 1977). Glazer and Kuo (1977) reported that cordycepin competitively inhibits cAMP-dependent and cAMP-independent protein kinase from bovine heart and rat liver. These observations strongly suggest that cordycepin may affect transcription by interfering with phosphorylation of nonhistone chromosomal proteins (Legraverend et al., 1978). More recently cordycepin was reported to cause a threefold stimulation of translation of wheat germ embryo extracts and lysed rabbit reticulocytes for *in vitro* protein synthesis directed by myeloma mRNA and TMV mRNA (Leinwand and Ruddle, 1977) (Fig. 3.5). The speculation of Leinwand (private communication) is that cordycepin stimulates protein synthesis by increasing the initiation sites on mRNA. This would increase the rate of translation. Studies by Wu and Suhadolnik (unpublished results) show that 2′, 3′-dideoxyadenosine is more stimulatory for protein synthesis than is cordycepin.

Effect of the Absence of the 3′-Hydroxyl Group on AMP Aminohydrolase and AMP Kinase. The role of the 3′-hydroxyl group of adenosine in cellular reactions has been studied by Hampton and Sasaki (1973). They used AMP and several AMP analogs to study aminohydrolase, snake venom 5′-nucleotidase, and AMP kinase to elucidate the adenine–ribose torsion angle of enzyme-bound AMP. Structural changes in the nucleotides impair catalysis. The anti-type adenine–ribose torsion angle is such that the H-8 is oriented in the vicinity of C-4′.

Effect of Cordycepin in the Dehydrogenases and in DNA Synthesis in Isolated Nuclei by 3′dNAD$^+$ and ADP Ribosylation. Another utilization of cordycepin as a biochemical probe has been the synthesis of 3′dNAD$^+$, in which the binding of NAD$^+$ and the binding of 3′dNAD$^+$ to the coenzyme

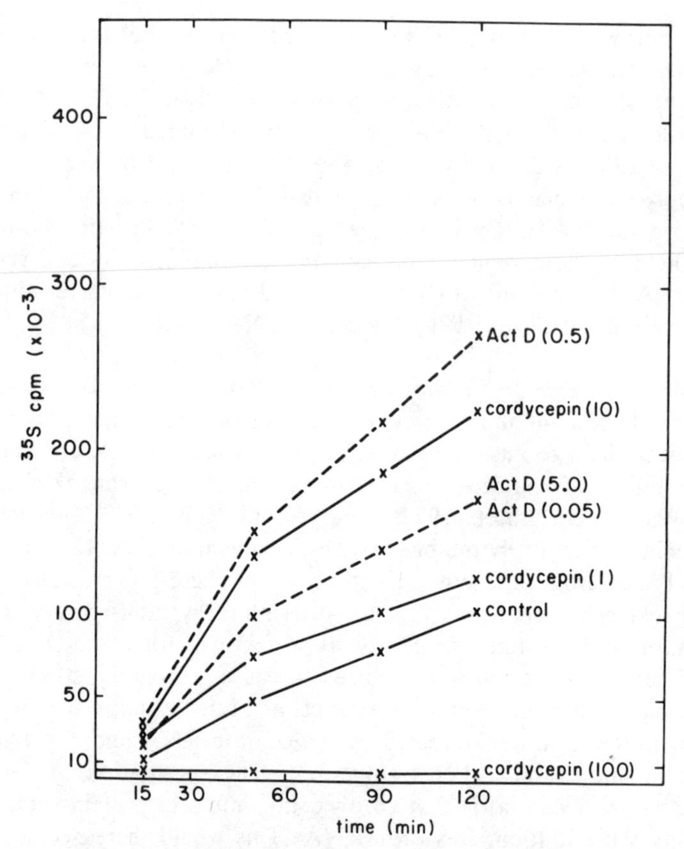

Figure 3.5 Incorporation of ^{35}S-labeled methionine into protein by myeloma mRNA translated in wheat germ embryo extracts containing 1 μg of myeloma poly(A)-containing RNA in 25 μl assays. Reprinted with permission from Leinwand and Ruddle, *Science*, Vol. 197, pp. 381–383, 1977. Copyright 1977 by the American Association for the Advancement of Science.

domain of the dehydrogenases have been compared (Suhadolnik et al., 1977a). The data show that the K_m does not change, but the V_{max} decreases by 80%. Therefore, these kinetic data support the X-ray diffraction data, which indicate that the 3′-hydroxyl group of the adenine-ribose of NAD^+ is essential for hydrogen bonding. However, the energy of the 3′-hydrogen bond is used for the proper conformation of either the dehydrogenase or the nicotinamide of NAD^+ to form a productive complex.

Another reaction related to the role of 2′- and 3′-hydroxyl groups of NAD^+ is in the addition of poly(ADP-ribose) to nuclear proteins. The ADP-ribose moiety of NAD^+ is covalently attached to the γ-carboxyl group of glutamate in histone-1 (Riquelme et al., 1977) and the homopolymer of ADP-ribose is linked by an $\alpha 1'' \rightarrow 2'$ glycosidic linkage (Chambon et al., 1966). Suhadolnik et al. (1977b) reported that 0.5 *M* NAD^+ inhibits DNA synthesis in nuclei isolated from rat liver by only 9%; however, $2'dNAD^+$ and $3'dNAD^+$ inhibit DNA synthesis by 90% (Fig. 3.6). With nuclei from Novikoff hepatoma and fetal rat liver, NAD^+ does not inhibit DNA synthesis; however, with $2'dNAD^+$ and $3'dNAD^+$, DNA synthesis in nuclei from Novikoff and fetal rat liver is inhibited (Suhadolnik et al., 1977b). Müller and Zahn (1975) reported the effect of the nucleoside antibiotics, cordycepin, ara-A, coformycin, tubercidin, showdomycin, and formycin B, on the activity of poly(ADP-ribose) synthase. Only formycin B and showdomycin were inhibitory. Cordycepin was without effect.

Use of Cordycepin in the Determination of "Chemical Proofreading" for Protein Synthesis. Initial studies indicated that the specificity of aminoacylation of tRNA was due to specific complex formation between a specific tRNA and the corresponding aminoacyl-tRNA synthetase (Loftfield, 1972; Söll and Schimmel, 1974; Kisselev and Favorova, 1974).

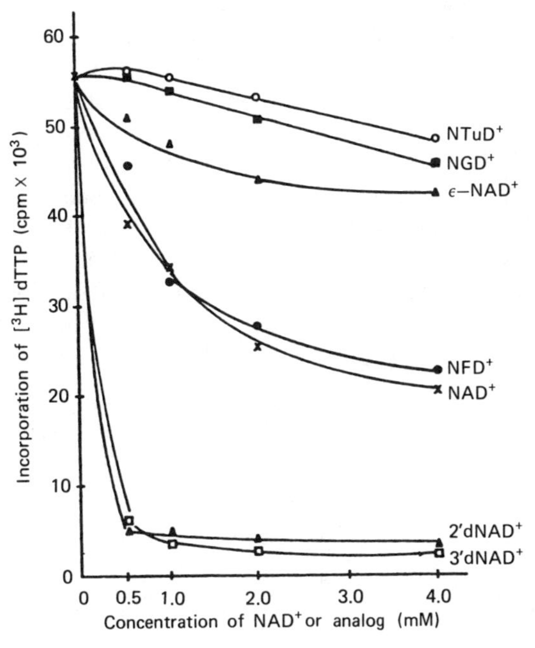

Figure 3.6 Effect of NAD^+ analogs on the inhibition of template activity for DNA synthesis in nuclei isolated from rat liver. Replacement of [^{3}H]dTTP with [^{3}H]dATP in the NAD^+ experiment gave essentially the same percent inhibition. Reprinted with permission from Suhadolnik et al., 1977b.

However, recent studies indicate that the recognition of tRNA is not a simple process. Instead, recognition of an individual tRNA is achieved by passing through a series of steps. The elucidation of this complicated series of events has been made possible by the studies of Cramer, Rich, Hecht and their coworkers, who have used adenosine analogs at the 3′-adenylate end of the —C—C—A of tRNA to study the mechanism of aminoacylation by incorporating 2′dATP, 3′dATP, 3′-amino-3′dATP, and formycin 5′-triphosphate into the 3′-end of tRNA with tRNA-nucleotidyl transferase (Fraser and Rich, 1973; Sprinzl and Cramer, 1973, 1977; von der Haar and Cramer, 1976; Hecht et al., 1974b; Ofengand and Chen, 1972). For a recent review, see Sprinzl and Cramer (1978); for structural requirements of the terminal adenosine of tRNA see formycin, page 177.

The investigators listed above have shown how the activated aminoacyl-AMP adds to tRNA to form an aminoacyl tRNA that is the central repetitive reaction in protein synthesis. The $tRNA^{Phe}$-3′-deoxyadenosine, charged in the 2′-hydroxyl position, is not a substrate for poly(U)-directed polyphenylalanine synthesis (Sprinzl and Cramer, 1973). With 3′-amino-3′-deoxyadenosine at the 3′-terminal end of tRNA, Fraser and Rich (1973) showed that this modified tRNA is capable of accepting an *N*-Ac-phenylalanine from the donor site of the ribosome. However, the ribosome is not able to cleave the amide bond at the tRNA-3′-amino position. Therefore, the $tRNA^{Phe}$-CC-3′-amino-3′-deoxyadenosylate has acceptor, but not donor, activity in protein synthesis. These data lead to the assumption that native $tRNA^{Phe}$-CCA is first acylated on the 2′-hydroxyl of adenosine. The carboxyl group of the amino acid then undergoes a rapid acyl group migration from the 2′- to the 3′-hydroxyl position (Fig. 3.7). Only the 3′-aminoacyl tRNA is active as a peptide acceptor in the peptidyl transferase reaction.

Von der Haar and Cramer (1976) have shown that the hydrolysis of an amino acid on tRNA occurs only if the nonaccepting hydroxyl is present.

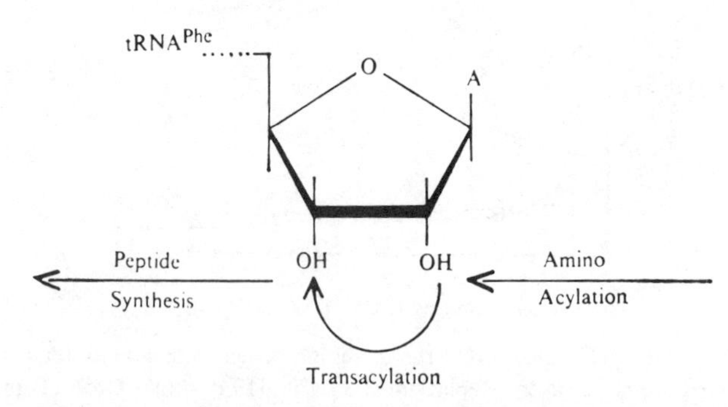

Figure 3.7 Proposed migration of the aminoacyl group on $tRNA^{Phe}$—C—C—A from the 2′- to the 3′-position during protein biosynthesis. Reprinted with permission from Sprinzl and Cramer, 1973.

Therefore, a mismatched amino acid (i.e., valine with isoleucyl-tRNA synthetase) is transferred to the improper tRNA. This mismatched amino acid is removed by hydrolysis. For example, valine (which has five carbons) allows a water molecule to be placed isosterically in the space where the methyl group of the normal substrate, isoleucine, would be found. The valyl group of $tRNA^{Ile}$-CCA is then split to give free valine and free $tRNA^{Ile}$. Von der Haar and Cramer (1978b) have extended their approach to the analysis of the selection of noncognate versus cognate tRNA. They studied the mischarging of tRNA and compared the data with competitive binding of tRNA under true equilibrium conditions. They were able to show that "recognition of individual tRNA is achieved by passing through a cataract of several steps." An error fraction as high as 0.016 in mischarging is greatly reduced in the presence of the cognate tRNA by competing with the noncognate tRNA. Fersht and Dingwall (1979) have confirmed the misacylation/deacylation for the valyl tRNA synthetases. However, cysteinyl tRNA synthetase does not need an editing mechanism. From studies with methionyl tRNA synthetase, they concluded that the $2' \rightarrow 3'$acyl transfer is not a general phenomenon for editing.

Determination of Subclasses of HnRNA. To determine the subclasses of HnRNA, cordycepin and 5,6-dichloro-β-D-ribofuranosylbenzimidazole (DRB) have been used. Darnell and Tamm have reported that DRB inhibits the initiation of a subclass of HnRNA by inhibiting the initiation of RNA chains, but it does not inhibit nuclear poly(A) synthesis (Sehgel et al., 1976a, b). More recently, Dreyer and Hausen (1978) demonstrated that DRB triphosphate inhibits RNA polymerase B. Winicov (1979) showed that 30% of the >12 S HnRNA sequences transcribed *in vitro* are sensitive to DRB.

Inhibition of RNA and DNA Synthesis by 3′dATP in Toluene-Treated* E. coli *Cells and with* E. coli *RNA Polymerase. *E. coli* rapidly deaminates cordycepin, thereby making it an inactive analog; Gumport et al. (1976) overcame this difficulty by using 3′dATP and toluene-treated *E. coli* cells. They showed an inhibition of the ATP-dependent DNA replicative apparatus when 3′dATP was added. The inhibition of 3′dATP was competitive with ATP, but not with 2′dATP (Fig. 3.8). Gumport et al. state that "one possible mechanism by which this analog may interfere with DNA synthesis is at the initiation step involving the synthesis of the primer RNA."

Bruzel et al. (1977, 1978) have supplied additional evidence that [α-^{32}P]3′dAMP is incorporated into the 3′-end of RNA as catalyzed by RNA polymerase from *E. coli* and RNA polymerase II from mouse plas-

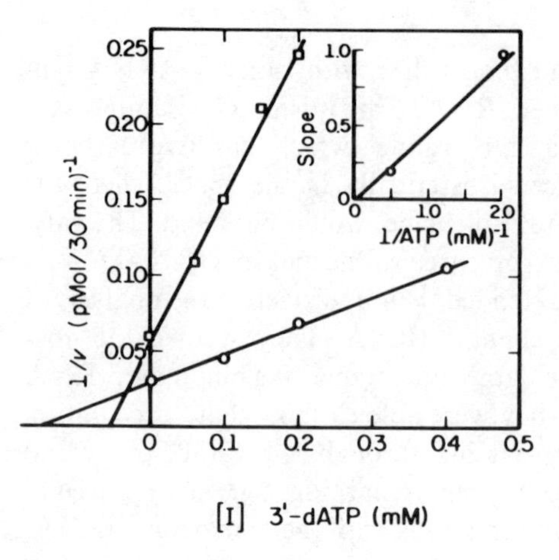

Figure 3.8 A Dixon plot of the reciprocal of the velocity of DNA synthesis at two ATP concentrations as a function of 3′dATP concentration. Reaction mixture: 100 μl; ATP concentration: 0.5 mM (□), or (○) 2.0 mM. Reaction mixtures contained indicated amounts of 3′-dATP. The specific activity of the [^{3}H]dTTP was 500 cpm/pmol. Lines were fit to the data points by a least-squares analysis. The insert is a plot of the slopes of the two curves versus the reciprocal of the ATP concentration. Reprinted with permission from Gumport et al., *Biochemistry*, **15**, 2804–2809 (1976). Copyrighted by the American Chemical Society.

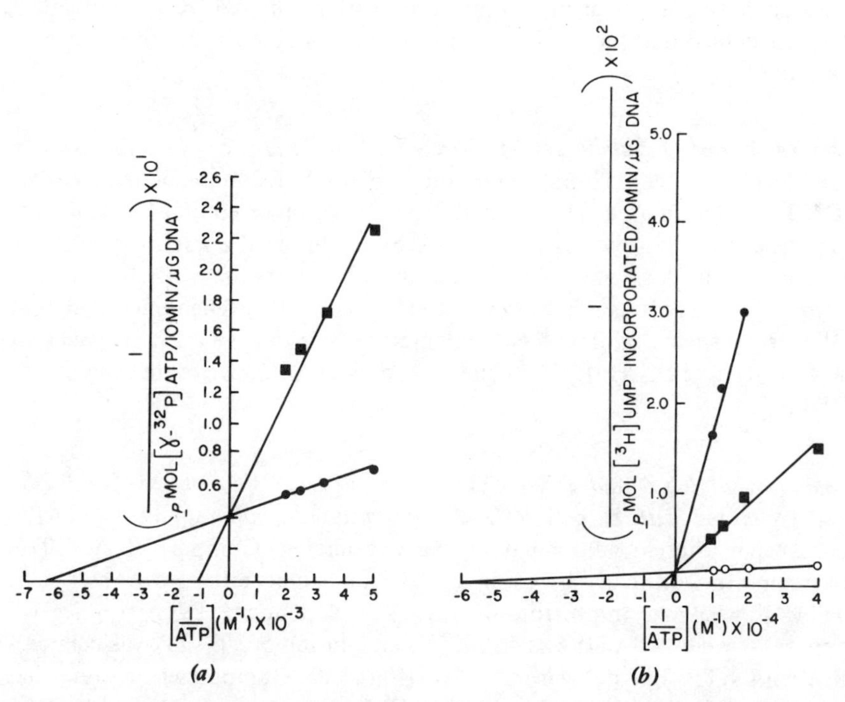

Figure 3.9 Effect of 3′dATP on the reactions catalyzed by *E. coli* RNA polymerase with respect to the (*a*) initiation and (*b*) elongation steps. In the intiation reaction, the K_m for ATP is 164 μM, K_i for 3′dATP is 104 μM. In the elongation reaction, the K_m for ATP is 18 μM, K_i for 3′dATP is 1.2 μM. (*a*) Minus 3′dATP (●); plus 500 μM 3′dATP (■). (*b*) minus 3′dATP (●); plus 10 μM 3′dATP (■); plus 50 μM 3′dATP (○) (Bruzel et al. 1978).

macytoma (MOPC 315) and acts as a chain terminator. This inhibition is competitive with ATP. The effect of cordycepin on the initiation and elongation steps of RNA synthesis is shown in Fig. 3.9. 3′dATP acts primarily at the elongation step (Fig. 3.9*B*). The RNA chains labeled with either [G-^{3}H]3′dAMP or [α-^{32}P]3′dAMP were hydrolyzed chemically and enzymatically to the position of the labeled 3′dAMP at the 3′-terminus. Using *E. coli* RNA polymerase, the powerful inhibitory effect of 3′dATP is demonstrated by the competitive inhibition of ATP binding ($K_{m\,ATP}$ = 21 μM); with MOPC 315 RNA polymerase II, $K_{m\,ATP}$ = 770 μM; $K_{i\,3'dATP}$ = 50 μM.

Because 3′dATP can partially substitute for ATP in the energy-requiring steps in DNA replication, Suhadolnik and Uematsu (1978) synthesized 3′-deoxyuridine by the transribosylation of benzoylated cordycepin with bissilyluracil. Reichenbach et al. (1980) have reported that 3′-deoxyuridine 5′-triphosphate (3′dUTP) inhibits RNA and DNA synthesis in nuclei isolated from HeLa cells. They demonstrated that, unlike 3′dATP, 3′dUTP does not contribute to the energy-requiring steps in DNA replication. Therefore, interpretation of data is not complicated by this variable activity with 3′dUTP. The inhibition of 3′dUTP with highly purified *E. coli* and MOPC 315 RNA polymerase II and DNA polymerase-α from HeLa cells shows that 3′dUTP only inhibits RNA synthesis.

3.2 5-AZACYTIDINE

Pískala and Šorm (1964) reported on the synthesis of the *s*-triazin ribonucleoside analog, 5-azacytidine, 4-amino-1-β-D-ribofuranosyl-1,3,5-triazin 2-one (Fig. 3.10). The naturally occurring nucleoside was subsequently isolated from the culture filtrates of *S. ladakanus* by Haňka et al. (1966) and Bergy and Herr (1966). The 2′-deoxy derivative of 5-azacytidine has also been synthesized (Fig. 3.10). The physical and chemical properties of 5-azacytidine have been reviewed in depth elsewhere (Suhadolnik, 1970; Veselý and Čihák, 1978). This section describes the mechanism of action of 5-azacytidine in intact normal cells, intact abnormal cells, and *in vitro* studies.

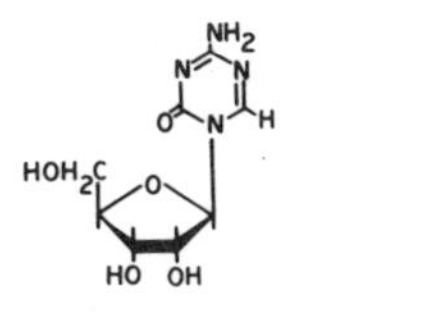

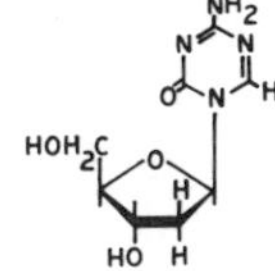

Figure 3.10 Structures of 5-azacytidine and 5-aza-2′-deoxycytidine.

Biochemical Properties

Action and Metabolism. 5-Azacytidine affects a number of reactions in the cell. It is a cytotoxic analog of cytidine that is rapidly deaminated (Čihák and Šorm, 1965). Because 5-azacytidine interferes with many cellular metabolic processes, the action is assumed to be polyvalent (Suhadolnik, 1970; Veselý and Čihák, 1975; Von Hoff et al., 1976). Although the mechanism of action of 5-azacytidine is not completely understood, this analog blocks the incorporation of orotic acid in the liver of leukemic AKR mice. The net effect is an inhibition of RNA, DNA, and protein synthesis (Li et al., 1970a; Jurovčík et al., 1965; Kalousek et al., 1966; Weiss and Pitot, 1975).

Cellfree extracts of leukemic livers phosphorylate 5-azacytidine to its 5′-monophosphate, which decreases the activity of orotidine 5′-phosphate decarboxylase (Veselý and Čihák, 1978). Čihák (1974) demonstrated that the administration of 5-azacytidine immediately after partial hepatectomy decreases orotate phosphoryl transferase and orotidine 5′-phosphate decarboxylase activities. The inhibition of the decarboxylase activity is due to the newly formed 5-azacytidine 5′-phosphate. The orotate phosphoribosyl transferase is inhibited by the OMP accumulated because of the inhibition of the OMP decarboxylase (Čihák, 1974). The analog is phosphorylated to the 5′-mono-, 5′-di-, and 5′-triphosphates by Ehrlich ascites tumor cells (Jurovčík et al., 1965). With L1210 cells, Li et al. (1970a) reported that 5-azacytidine occurs primarily as the 5′-triphosphate, is reduced to the deoxyribonucleotide, and is incorporated into DNA. Uridine kinase, but not cytidine kinase, phosphorylates 5-azacytidine (Lee et al., 1974; Coleman et al., 1975). Cytidine inhibits the phosphorylation of 5-azacytidine (Veselý et al., 1968; Li et al., 1970a), whereas the analog does not inhibit the phosphorylation of cytidine and uridine (Lee et al., 1974). Partially purified calf thymus RNA polymerase incorporated 5-azacytidine 5′-triphosphate into RNA (Lee and Momparler, 1977). The K_m of 5-azacytidine 5′-triphosphate is eighteenfold greater than the K_m for CTP; therefore, the analog is a poor competitive inhibitor for CTP. With the wild-type strain of *E. coli* and a cytidine deaminase mutant, the metabolism of 5-azacytidine is markedly different (Doskočil and Šorm, 1971). The analog is deaminated by the wild strain and incorporated into RNA as 5-azauracil. The 5-azauracil is degraded to a ring-opened form (Čihák et al., 1964). Human leukemia cells also deaminate 5-azacytidine (Chabner et al., 1973). With the cytidine deaminase-deficient cells, the incorporated 5-azacytidine can be recovered intact. Tetrahydrouridine, an inhibitor of cytidine deaminase (Camiener, 1968), inhibits deamination of 5-azacytidine.

5-Azacytidine is incorporated into RNA of normal and AKR mice with lymphatic leukemia that are sensitive or resistant to the analog

(Suhadolnik, 1970). Furthermore, 5-azacytidine and cytidine are competitive precursors for incorporation into RNA (Čihák et al., 1966). The hydrogen bonding of the aglycon of the analog does not change (Pitha et al., 1966).

Effect of 5-Azacytidine on Bacteria. Doskočil and Šorm (1970a, b, 1971) studied the effect of 5-azacytidine in *E. coli*. The primary target site of the drug is the inhibition of protein synthesis. With wild strains of *E. coli*, the drug is deaminated to 5-azauridine. The ring of the aglycon of the 5′-monophosphate of 5-azauridine, following incorporation into mRNA, is unstable and opens, giving rise to a nonfunctional protein by miscoding. In *E. coli* mutants deficient in cytidine deaminase, 5-azacytidine, in which the aglycon is stable, is a weak inhibitor of protein synthesis even though the drug is incorporated into RNA (Doskočil and Šorm, 1970a). For additional detailed review of earlier studies of 5-azacytidine in *E. coli* see Suhadolnik (1970) or Veselý and Čihák (1978).

Effects on rRNA, mRNA, tRNA, and Protein Synthesis. The incorporation of 5-azacytidine into RNA and the subsequent effect on protein synthesis has been studied in many laboratories (Pačes et al., 1968; Raška et al., 1966; Čihăk et al., 1966, 1967a, b, 1968; Zain et al., 1973). Mammalian cell lines have been used to study the effect of 5-azacytidine on the transcription of a specific mRNA and its translation into protein. The incorporation of 5-azacytidine into mRNA and tRNA modifies the RNA such that it becomes nonfunctional in normal protein synthesis. These RNAs show different elution patterns on DEAE cellulose (Shutt and Krueger, 1972; Lee and Momparler, 1976; Reichman and Penman, 1973; Lee, 1973; Momparler et al., 1976; Lee and Karon, 1976).

The maturation of rRNA is also inhibited by 5-azacytidine. The formation of 28 and 18 S, but not 38 S, RNA is completely inhibited (Fig. 3.11) (Weiss and Pitot, 1974a; Reichman et al., 1973; Čihák et al., 1974). Novikoff hepatoma cells show a 50% decrease in 28 and 18 S RNA following a 1 h exposure to 5-azacytidine (Fig. 3.11*c*); after 2 h, labeling of 28 and 18 S RNA continues to be inhibited in the 5-azacytidine-treated cells. The degradation of polysomes is increased by 5-azacytidine. The result is an accumulation of monosomes and a marked inhibition of protein synthesis (Čihák et al., 1968, 1973a, 1974; Levitan and Webb, 1969; Reichman and Penman, 1973). Čihák et al. (1973a) reported that there is a decreased binding of leucine-charged tRNA during polysome breakdown. In addition, HeLa cell tRNA isolated from 5-azacytidine-treated cells is 2.5 times less active in supporting protein synthesis than in normal cells (Lee, 1973; Momparler et al., 1976). This decrease is probably due to abnormal tRNA

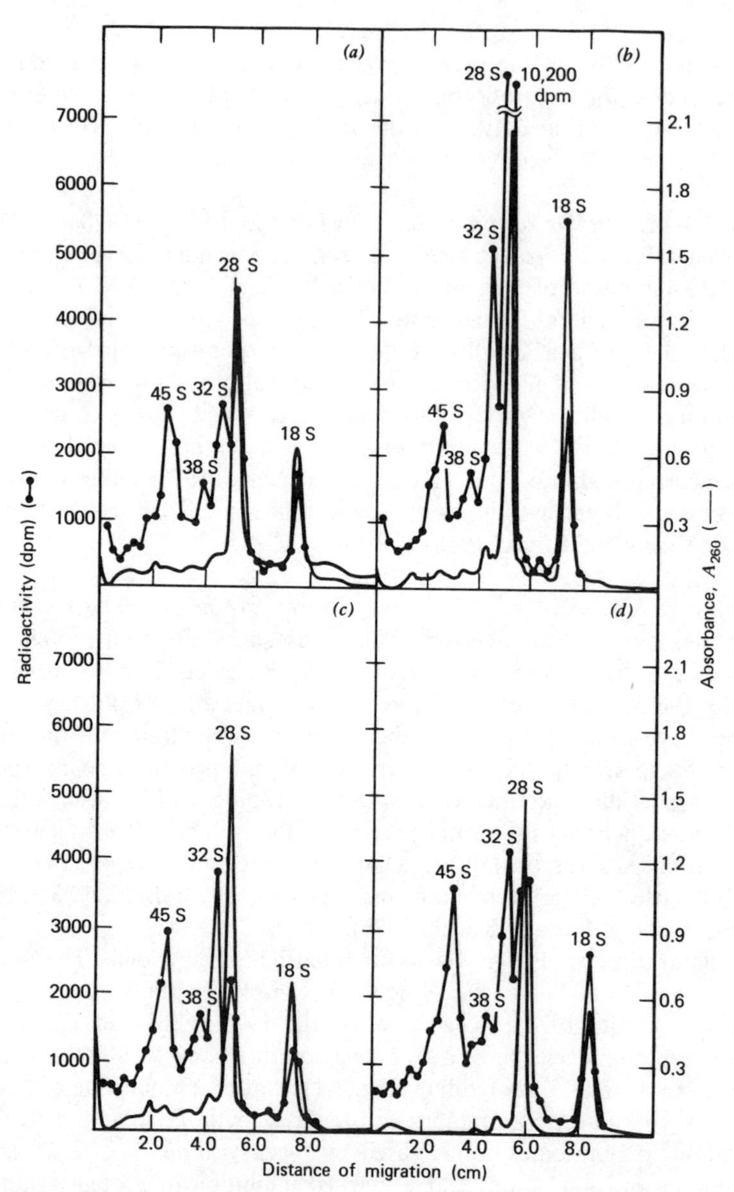

Figure 3.11 Effect of 5-azacytidine on [^{3}H]guanosine labeling of ribosomal RNA. Cells were treated with 5×10^{-4} *M* 5-azacytidine or 5×10^{-4} *M* cytidine (control) in the presence of [^{3}H]guanosine (0.2 μCi/ml, 6.1 Ci/mmol). Total-cell RNA samples were prepared after 1 and 2 h of labeling and were analyzed by acrylamide–agarose gel electrophoresis. Absorbance at 260 nm, A_{260}, and radioactivity, (dpm) are plotted versus distance of migration (cm). (*a*) Control, 1 h; (*b*) control, 2 h; (*c*) 5-azacytidine-treated, 1 h; (*d*) 5-azacytidine treated, 2 h Reprinted with permission from Weiss and Pitot, 1974a.

and not to analog incorporation at the CCA end of tRNA. 5-Azacytidine does not compete for the incorporation of UTP into RNA. McGuire et al. (1978) recently studied (i) the degree and duration of alteration of DNA and RNA synthesis in normal and tumor tissue after exposure to 5-azacytidine, (ii) differences in the antimetabolic effects of 5-azacytidine in normal and tumor tissue in scheduling subsequent drug doses, (iii) the ability of 5-azacytidine to inhibit the growth of a tumor cell line that has known resistance to cytosine arabinoside. Their data show that in the L1210 tumor system in mice, TdR incorporation into DNA is more affected in leukemic cells than in bone marrow or in the G.I. tract when 5-azacytidine is administered intraperitoneally. The inhibition of RNA synthesis is maximal at 12 h with full recovery after 24 h (Fig. 3.12).

Effect of 5-Azacytidine on DNA Synthesis. 5-Azacytidine inhibits the incorporation of thymidine into DNA more than it inhibits the incorporation of uridine into RNA in L1210 cells and ascites cells (Li et al., 1970a). Uridine and cytidine reverse the inhibition of DNA synthesis; deoxyuridine and deoxycytidine are ineffective.

The effects of 5-azacytidine on DNA synthesis in intact and regenerating rat liver have been studied in detail. Čihák et al. (1976) reported that repeated doses of 5-azacytidine resulted in a tenfold increase of thymidine incorporation into DNA in intact rat liver. Although there was a decrease in dTMP, thymidine and thymidylate kinase activities did not change. With regenerating rat liver, the analog interferes with DNA synthesis by blocking thymidine incorporation when administered immediately after partial hepatectomy (Čihák and Veselý, 1969). Protein synthesis was inhibited 2–8 h after administration of 5-azacytidine; however, thymidine and thymidylate kinase activities were inhibited for 30 h (Čihák and Veselý, 1972). This sequence of inhibitors suggests that 5-azacytidine blocks RNA synthesis that occurs before the synthesis of induced enzymes and DNA replication. Čihák et al. (1972) and Čihák and Rabes (1974) showed that if the analog is given 24–40 h before partial hepatectomy, there is an enhanced uptake of thymidine into the DNA of 24 h regenerating rat liver. This increase in thymidine incorporation into DNA can be explained by a synchronization of growing hepatocytes in the regenerating liver.

Chromosomal Breakage. Karon and Benedict (1972) reported that chromosomal breakage occurs in the S and G_2 phases of the cell cycle following the incorporation of 5-aza-2′-deoxycytidine monophosphate into DNA. The explanation offered for the chromosomal breakage is that the incorporation of this analog into DNA results in a less stable secondary structure (Čihák et al., 1967a, 1976; Fučík et al., 1970; Karon and Benedict, 1972; Zadražil

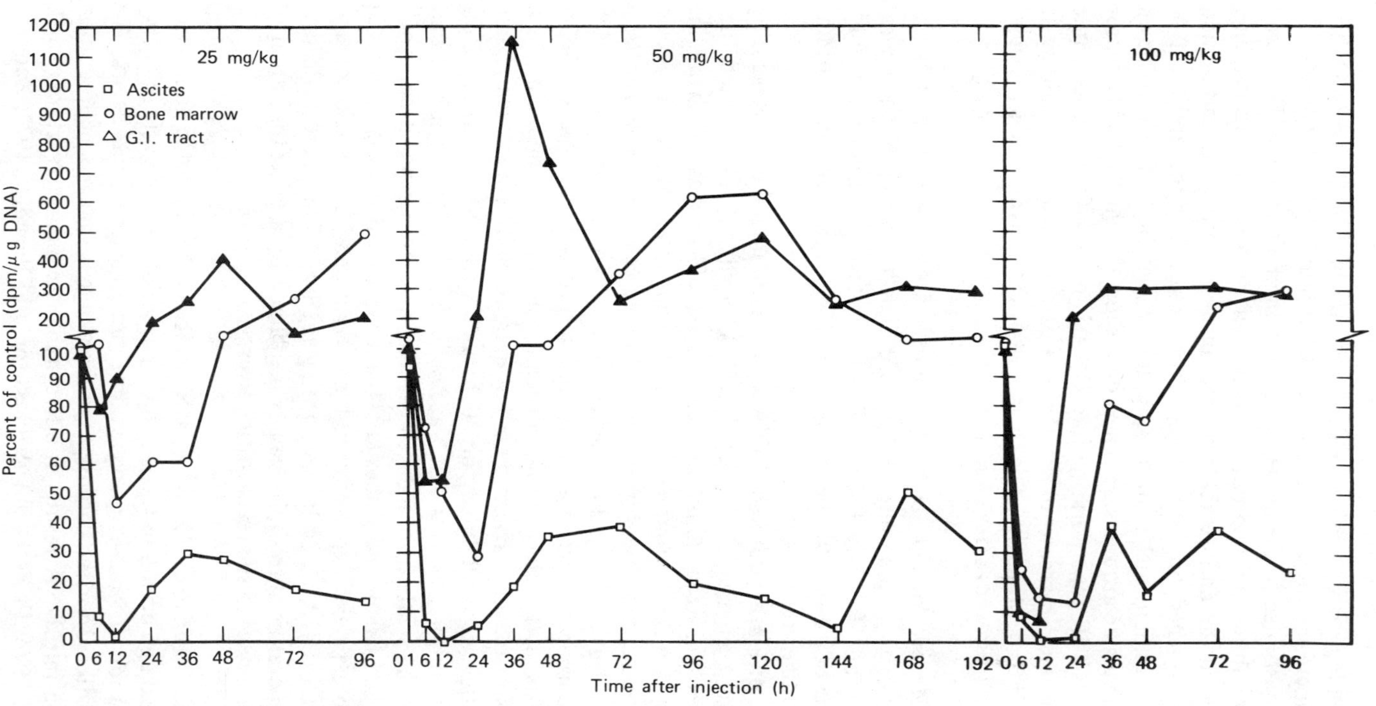

Figure 3.12 Alterations in [^{3}H]TdR incorporation into DNA of L1210 ascites tumor, bone marrow, and gastrointestinal mucosa due to 25, 50, and 100 mg/kg of 5-azacytidine i.p. in BDF mice. Control incorporation rates were 70,000 ± 12,500 dpm/μg of DNA for ascites tumor, 950 ± 100 dpm/μg of DNA for bone marrow, and 500 ± 40 dpm/μg of DNA for gastrointestinal mucosa. Reprinted with permission from McGuire et al., *Biochemical Pharmacology*, **27**, 745–750 (1978), "Alteration in [^{3}H]Thymidine Incorporation into DNA and [^{3}H]Uridine Incorporation into RNA Induced by 5-Azacytidine *in vivo*." Copyright 1978, Pergamon Press, Ltd.

et al., 1965). When the 5-aza-2′-deoxycytidine is incorporated into DNA, the DNA has a lower molecular weight and a lower melting point (Zadražil et al., 1965).

Effect on Induced Liver Enzymes. 5-Azacytidine changes the synthesis and activity of induced enzymes of liver. The amino acid-metabolizing enzymes are especially affected. When 5-azacytidine is given simultaneously with L-tryptophan, there is a 60% inhibition of substrate-induced enzyme (Čihák et al., 1976). However, tryptophan oxygenase of liver was higher in animals pretreated with 5-azacytidine 18–24 h prior to L-tryptophan when compared to control animals (Čihák et al., 1969).

The sensitivity to 5-azacytidine with liver and hepatoma 5123D tyrosine aminotransferase differs from that with tryptophan oxidase (Levitan et al., 1971; Čihák et al., 1973b). The initial induction of liver tyrosine aminotransferase by cAMP or glucagon is not inhibited by 5-azacytidine. However, McNamara and Webb (1973, 1974) demonstrated that once induction occurs by glucagon or cAMP, the corticosteroid-mediated induction of tyrosine aminotransferase is inhibited by 5-azacytidine.

Effect on Polyamine Biosynthesis. All the enzymes in the polyamine biosynthetic pathway in L1210 leukemic mice are decreased by 5-azacytidine, and polyamine accumulation in leukemic mice is inhibited (Heby and Russell, 1973a). When the administration of 5-azacytidine is stopped, polyamine synthesis is restored to normal (Heby and Russell, 1973a, b). Heby and Russell (1973a) reported that the putrescine-dependent *S*-adenosyl-L-methionine decarboxylase activity in the spleens of mice increased within the first day of tumor inoculation. In the 5-azacytidine-treated mice, the enzyme activity decreased. 5-Azacytidine also blocked the stimulation of ornithine and *S*-adenosyl-L-methionine decarboxylase activities even when the inducer, 12-*O*-tetradecanoylphorbol-13-acetate, was administered to mice (O'Brien, 1976).

Effect on the Cell Cycle. Li et al. (1970b) and Bhuyan et al. (1972) studied the phase specificity of 5-azacytidine. The analog drastically inhibited DNA synthesis and prevented hamster cells from entering the S-phase. Tobey (1972) showed that S-phase cells have difficulty in going into mitosis. In some cases the effect of 5-azacytidine is phase independent. Nitschke (1974) showed that the analog affected the cell by interfering with the cell membrane when phytohemagglutinin-induced blastogenesis of lymphocytes was used as the model.

Inhibitory Action of 5-Aza-2′-deoxycytidine. When 5-aza-2′-deoxycytidine (Fig. 3.10) is added to a mutant of *E. coli* deficient in cytidine deaminase, the analog is still deaminated and cleaved to yield the aglycon, 5-azauracil (Doskočil and Šorm, 1970c; Čihák and Veselý, 1977). With bacteria, which lack deoxycytidine kinase, the only mode of uptake of the deoxynucleoside is deamination and transamination. Because bacteria do not have a kinase to phosphorylate 5-aza-2′-deoxycytidine, it cannot be incorporated into bacterial DNA (Doskočil, 1972; 1974). The loss of sensitivity to the analog by bacteria has been shown by Doskočil (1974) and Doskočil and Holý (1974) to be due to the loss of a specific nucleoside-transporting system. For additional studies by Doskočil and Holý on the nucleoside-transporting systems and inducible nucleoside permease in *E. coli* with showdomycin see page 56. 5-Azauracil then enters the cell and reacts with 5-phosphoribosyl-1-pyrophosphate to give 5-azauridine 5′-monophosphate (Čihăk et al., 1964). In AKR mouse leukemic cells, 5-aza-2′-deoxycytidine lowers the level of the acid-soluble pool of 2′-dCMP and inhibits the incorporation of dCMP into the DNA Veselý and Čihák, 1977, 1978; Čihák, 1978). 5-Aza-2′-deoxycytidine is also deaminated by mouse tissue and Ehrlich ascites cells and results in the formation of 5-azauracil; deamination is prevented by O^4,4,5,6-tetrahydrouridine (Čihák, 1978; Čihák and Veselý, 1979). [^{3}H]-5-Aza-2′-deoxycytidine is localized in the nuclei and incorporated into DNA of mouse leukemic cells (Veselý and Čihák, 1977). In normal mice the analog is preferentially incorporated into the spleen and thymus (Čihák and Veselý, 1979), whereas in AKR mice with lymphatic leukemia, the drug is localized in the lymphatic tissue (Čihák, 1978).

The incorporation of dCMP into DNA in mice is almost completely inhibited at a dose of 50 mg/kg of 5-aza-2′-deoxycytidine (Čihák and Veselý, 1979). In long term treatment, there is about an 80% inhibition of incorporation of thymidine into DNA. The deoxy analog does not affect thymidine phosphorylation.

5-Aza-2′-deoxycytidine is rapidly deaminated in mice and Ehrlich ascites cells (Čihák and Veselý, 1979). The level of the 2′-deoxy analog increases in tetrahydrouridine-treated cells (Čihák, 1978).

Effect of 5,6-Dihydro-5-azacytidine. Because 5-azacytidine can be deaminated to the unstable 5-azauridine (Chabner et al., 1973), the design of a suitably stable analog that would have equal or better therapeutic effects with less toxic properties has been undertaken. Beisler et al. (1976, 1977a, b) have described the synthesis of 5,6-dihydro-5-azacytidine, which has biological antagonistic properties similar to the action of 5-azacytidine against L1210 leukemic cells, but its cytostatic action requires a concentration tenfold higher than that of 5-azacytidine. The dihydro analog has a tenfold greater affinity for adenosine deaminase from HeLa cells than does 5-azacytidine (Voytek et al., 1977). Tetrahydrouridine in combination with the dihydro analog (both at 10^{-4} *M*) completely inhibits HeLa cells, whereas the dihydro analog alone is without effect.

Toxicity of 5-Azacytidine. The toxicity of 5-azacytidine is most easily detected in beagles. There is a decrease in leukocytes and necrosis of lymphatic organs when 5-azacytidine is administered (Palm et al., 1973; Palm and Kensler, 1970). Cytidine reverses the toxicity of 5-azacytidine. The i.v. administration of 5-azacytidine to beagles results in excretion of 5-azacytidine, 5-azacytosine, 5-azauracil, urea, and guanidine-like compounds (Coles et al., 1974). In mice, O^4-4,5,6-tetrahydrouridine, ara-A, vincristine, or prednisone causes a sixfold increase in the amount of 5-azacytidine in the urine. Man can tolerate much higher doses of 5-azacytidine than can beagles, rodents, or monkeys (Von Hoff et al., 1975). Although 5-azacytidine is incorporated into the DNA of bacteria and mammalian tissues (Zadražil et al., 1965; Jurovčík et al., 1965; Townsend, 1975), Troetel et al. (1972) reported that 5-azacytidine is incorporated into RNA, but not into mammalian DNA. There is no metabolism to carbon dioxide (Israili et al., 1976).

There are two unusual properties of 5-azacytidine with L1210 leukemic cells (Presant et al., 1975). First, there is a biphasic dose response and second, there is an antileukemic effect that lasts many days following administration.

Clinical Use of 5-Azacytidine: Effect on Human Acute Myelogenous Leukemia. 5-Azacytidine is active against human adult acute myelogenous leukemia and the blastic phase of chronic myelogenous leukemia (Burchenal et al., 1972; Li et al., 1970a; Vadlamudi et al., 1970; Karon et al., 1973; McCredie et al., 1973; McGuire et al., unpublished results). Šorm, Čihák, and Veselý have reported that 5-azacytidine shows preferential affinity for the lymphatic system and inhibits various forms of experimental neoplasias (Šorm et al., 1964; Čihák and Veselý, 1979; Šorm and Veselý, 1968).

Hrodek and Veselý (1971) showed that 5-azacytidine was active in inducing partial remissions in clinical trials with 19 children with acute lymphatic leukemia and one child with acute myelogenous leukemia when administered intramuscularly (25–75 mg/m^2, daily for 1–4 weeks). In all cases prednisone therapy was started 2 weeks after the initiation of 5-azacytidine administration. With this combination, there was 88% complete remissions (Hrodek and Veselý, 1971).

Of patients with chronic myelogenous leukemia, 36% responded to 5-azacytidine (Saiki et al., 1977; Vogler et al., 1976; McCredie, 1975; Tan et al., 1973; Levi and Wiernik, 1975; Bateman, 1975). Combination therapy improved the response to 68% (Levi and Wiernik, 1975). Patients with acute lymphocytic leukemia, when treated with 5-azacytidine showed a partial remission (Tan et al., 1973; Quagliana et al., 1974; Cunningham et al., 1974; Bellet et al., 1974; Moertel et al., 1972). More recently, Saiki et al. (1977) reported that 5-azacytidine has significant activity in acute nonlymphoblastic leukemia. Scheduling of therapy with 5-azacytidine has ranged from empiric to a regimen of i.v. infusion at 8 h intervals for 5 days (McGuire et al., 1978). The difficulty of the present empiric daily infusion of 5-azacytidine for 5 days might be remedied by the method recently described by McGuire et al. (1978) in which the administration of 5-azacytidine was dictated by following the incorporation of [^{3}H]thymidine into human myeloblasts isolated from a patient in the blastic phase of myelogenous leukemia following incubation with 5-azacytidine. This same patient had sustained partial remission on 5-azacytidine therapy (McGuire et al., 1978). Studies have been done in patients with ovarian ascites tumor after i.v. methotrexate therapy (Chabner and Young, 1973). Toxicity to 5-azacytidine included moderate to severe nausea, diarrhea, stomatitis, myelosuppression, and neurological side effects.

Effect on Solid Tumors. 5-Azacytidine shows less encouraging results in solid breast tumors, lung, colon, rectum, malignant myeloma, and miscellaneous tumors (Von Hoff et al., 1976; Quagliana et al., 1974; Cunningham et al., 1974; Bellet et al., 1974; Moertel et al., 1972; Vogler et al., 1976; Omura, 1977a, b; Weiss et al., 1972; Lomen et al., 1975). Two separate studies by Weiss et al. (1972, 1977) evaluated solid tumors in patients who received 5-azacytidine by rapid and slow i.v. infusion. Nausea was severe by rapid i.v. infusion. An antitumor effect of 17 and 21% of patients with breast and malignant lymphoma, respectively, was seen. It was concluded that 5-azacytidine was of minimal value as a single agent in the treatment of solid tumors. Bellet et al. (1972a, b, 1973, 1974) evaluated a variety of advanced solid malignant tumors by the subcutaneous injection of 5-azacytidine for 10 successive days to a total dose of 275–280 mg/m^2. Objective remission was achieved in one out of four ovarian and one out of four mammary tumors. Five patients suffered hepatic toxicity, three of whom died of hepatic coma. Histologic evaluation of needle biopsies of liver in eight patients did not show any differences between pretreatment and posttreatment samples.

To increase the therapeutic efficiency of 5-azacytidine, tetrahydrouridine is administered simultaneously (Neil et al., 1975). Because 5-azacytidine and ara-C are inhibitors of acute myelogenous leukemia, their combination has been used with hamster fibrosarcoma cells in S-phase. Maximal killing effect of 5-azacytidine occurred when the analog was given 6 h after ara-C.

When the analog was added to S-phase cells, the maximal killing effect was produced by ara-A added 6 h later (Momparler et al., 1975). A significant antagonism was observed when the analog and ara-C were added to S-phase cells simultaneously (Neil et al., 1976). Presant et al. (1977) studied the toxicity of 5-azacytidine and adriamycin. Adriamycin produced additive toxicity when administered 16 h before or after 5-azacytidine.

Resistance of Leukemic Cells to 5-Azacytidine. Veselý et al. (1966) reported that AKR leukemic mice became resistant to 5-azacytidine. This resistance was due to an impaired uptake of cytidine and 5-azacytidine by the leukemic cell. There is a progressive decrease of uridine kinase activity with increased resistance. With Novikoff ascites tumor, the two isozymic forms of uridine kinase increase twice and then decline to about 30% of the control by the tenth generation (Keefer et al., 1975).

Treatment of Osteogenic Sarcoma or Melanoma. Hanka and Clark (1975) have proposed that 5-azacytidine could be more effective in the treatment of isolated tumors with well-defined blood supplies such as osteogenic sarcoma or melanoma. By the administration of the proper ratio of 5-azacytidine and cytidine, the authors propose that it should be possible to increase the level of the drug in the area of the tumor while reversing the inhibitory activity of 5-azacytidine in the general circulation.

3.3 2′-AMINOGUANOSINE

Biochemical Properties

2′-Aminoguanosine (Fig. 2.16) is discussed on pages 93–96 as an inhibitor of protein synthesis. Figure 2.18 (page 96) shows that 2′-aminoguanosine also inhibits RNA synthesis. Eighty-six percent of the tritium-labeled 2′-aminoguanosine taken up by *E. coli* is found in the acid-soluble fraction as the nucleotide. A small amount of tritiated 2′-aminoguanosine was isolated from the RNA. Nakanishi et al. (1977) favor the idea that 2′-aminoguanosine is converted to its 5′-triphosphate, which then serves as a GTP analog and is incorporated into mRNA, tRNA, and rRNA. This incorporation produces nonfunctional RNAs that inhibit protein synthesis. Furthermore, Hobbs and Eckstein (1977) reported that poly(2′-aminouridylate) and poly-(2′-aminocytidylate) are not hydrolyzed by ribonuclease.

3.4 3′-AMINOADENOSINE

Biosynthesis of 3′-Aminoadenosine

The biosynthesis of 3′-aminoadenosine from adenosine has been described by Suhadolnik (1970). Adenosine is the direct precursor for the biosynthesis of 3′-aminoadenosine.

Biochemical Properties

3′-Aminoadenosine is a naturally occurring purine nucleoside antibiotic that is isolated from the culture filtrates of *Cordyceps militaris*, *Aspergillus nidulans*, and *Helminthosporium*. This nucleoside antibiotic has been reviewed elsewhere (Suhadolnik, 1970). The biochemical studies since 1970 are described here.

3′-Aminoadenosine (Fig. 3.13) inhibits RNA polymerase but not DNA polymerase (Truman and Klenow, 1968). It has also been used to study the aminoacylation step in protein synthesis. Fraser and Rich (1973) reported that 3′-aminoadenosine replaced the adenylyl residue at the 3′-terminus of tRNA. Phenylalanine was covalently bound to the 3′-amino group of this modified tRNA. Although the phenylalanyl-(3′-aminoadenosyl)-tRNA was bound to the ribosomes, the abnormal amide bond involving the 3′-amino group of 3′-aminoadenosine and the carboxyl group of phenylalanine was not cleaved. Therefore, the phenylalanyl-(3′-aminoadenosyl)-tRNA has acceptor activity, but it does not have the donor activity essential for protein synthesis. The inability to be a donor for protein synthesis is attributed to the amide linkage, which is considerably more stable than the ester bond formed in aminoacyl-tRNA. For review, see Sprinzl and Cramer (1978). Although puromycin and cordycepin are both analogs of 3′-aminoadenosine, the *p*-methoxytyrosine on the 3′-amino group of puromycin provides an analog with different biological properties. Puromycin, as an inhibitor of cellular reactions, acts as an amino acid antimetabolite, but is not an analog of adenosine, whereas 3′-deoxyadenosine (cordycepin) and 3′-aminoadenosine are analogs of adenosine.

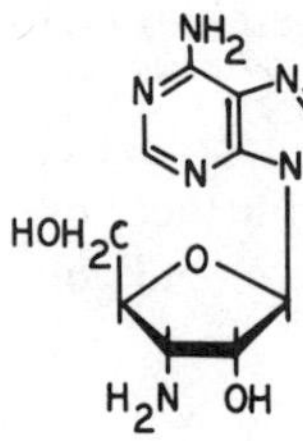

NH_2 N N N N HOH_2C O H_2N OH

Figure 3.13 Structure of 3′-aminoadenosine.

3.5 ARISTEROMYCIN

Aristeromycin (Fig. 3.14) is a carbocyclic derivative of adenosine. The racemic mixture was first synthesized by Shealy and Clayton (1969). The naturally occurring nucleoside antibiotic was isolated by Kusaka et al. (1967) from *Streptomyces citricolor*. The structure has been established as 9-[(1*R*,2*S*,3*R*,4*R*)-2,3-dihydroxy-4-(hydroxymethyl)cyclopentyl] adenine. The physical, chemical, and biochemical properties of aristeromycin have been reviewed elsewhere (Suhadolnik, 1970).

Biochemical Properties

Effect on Mammalian Cells in Culture and RNA Polymerase. Bennett et al. (1968) reported that H. Ep. #2 cells phosphorylate aristeromycin. Aristeromycin is not incorporated into either RNA or DNA by H. Ep. #2 cells. The nonphosphorylated aristeromycin is inhibitory because aristeromycin is toxic to H. Ep. #2 cells deficient in adenosine kinase. Adenosine reverses the toxicity of aristeromycin (Kusaka, 1971). Although aristeromycin inhibits AMP synthesis, it does not interfere with the phosphorylation or deamination of adenosine. The 5′-triphosphate of aristeromycin cannot replace ATP in the synthesis of NAD^+ (Kusaka, 1971, 1972). Adenosine reverses the inhibition of aristeromycin (Fig. 3.15). Although aristeromycin is toxic to fibroblast cells such as mouse 3T3 at concentrations of more than 1 μg/ml, it induces DNA synthesis and cell growth at a low concentration of 0.4 μg/ml when added to the confluent monolayer culture (M. Suno, K. Tsukamoto, and Y. Sugino, personal communication).

Inhibition of Transmethylases. The transmethylation reactions are affected by aristeromycin. Borchardt and coworkers (Borchardt and Wu, 1976; Pugh et al., 1977) reported that *S*-aristeromycinyl-L-homocysteine (SAmH) inhibits *S*-adenosylmethionine-dependent catechol-*O*-methyltransferase, phenethanolamine *N*-methyltransferase, histamine *N*-methyltransferase, and hydroxyindole-*O*-methyltransferase. Because of the sensitivity of the transmethylases to the *S*-aristeromycinyl analog of SAH, the "capping"

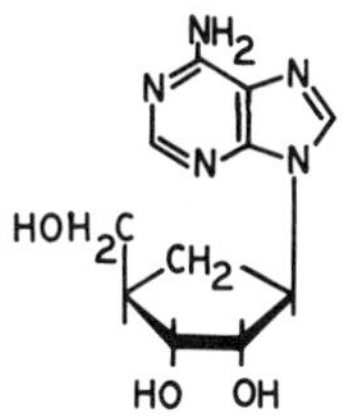

Figure 3.14 Structure of aristeromycin.

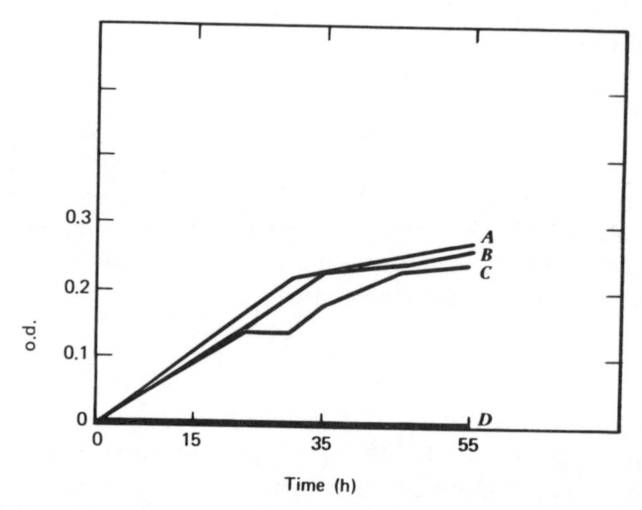

Figure 3.15 Reversal of aristeromycin inhibition by adenosine. Mixtures containing 0.25 ml of adenosine solution [final concentration (*D*) 0, (*C*) 0.05, (*A*) 0.1, or (*B*) 0.5 m*M*], 0.25 ml of aristeromycin solution [final concentration 0.005 m*M* (*A–D*)], 4.5 ml of XO medium, and 0.1 ml of seed suspension were incubated and o.d. was determined. Reprinted with permission from Kusaka, 1971.

reaction of viral and eukaryotic mRNAs by the eukaryotic or viral mRNA methyltransferase would be inhibited (for review, see Shatkin, 1976). The methylated structure, m^7GpppN(m), is essential for efficient translation and the binding of mRNA to the ribosomes (Both et al., 1975; Muthukrishnan et al., 1975a, 1975b). Although SAmH shows limited inhibitory properties against guanine-7-methyltransferase, from Newcastle disease, this transferase is strongly inhibited by *S*-tubercidinyl-L-homocysteine (Pugh et al., 1977).

By using 2′-deoxy- and 3′-deoxy-derivatives of adenosine and aristeromycin (in which the 0-4 of the ribose of adenosine has been replaced by a methylene group), it has been found that these SAmH analogs have low inhibitory properties (Shapiro and Schlenk, 1965; Coward and Slisz, 1973). Therefore, the methylases have a very high degree of specificity.

Inhibition of Plants. In an attempt to find a regulator of the growth of grasses and plants, Hagimoto et al. (1969) reported that 50 μM aristeromycin inhibits the growth of rice leaf, rice root, and many grasses. Aristeromycin inhibits cell division and cell elongation. The significance of these findings is that aristeromycin at 2 g/are controls the growth of grass.

Effect of Aristeromycin Derivatives and Aristeromycin 5′-Diphosphate on Polynucleotide Phosphorylase. Since aristeromycin is very cytotoxic to mammalian cells, several derivatives have been synthesized to overcome this toxicity, namely, 2′-deoxy-, 3′-deoxy-, 3′-amino-, 3′-aminoarabinofuranosyl-, 6-hydroxy-, 6-mercapto-, 8-bromo-, and 8-hydroxyaristeromycin and aristeromycin 3′,6′-cyclic phosphate (Marumoto et al., 1975, 1976; Daluge and Vince, 1976). 3′-Aminoaristeromycin inhibits herpes simplex virus and vaccinia virus (Daluge and Vince, 1978). Aristeromycin 5′-diphosphate and 8,2′-anhydro-8-mercapto-9-β-D-arabinofuranosyl 5′-diphosphate have been used by Ikehara and coworkers to show that there are two or more binding sites for polynucleotide phosphorylase (Ikehara and Fukui, 1973; Ikehara et al., 1969). They demonstrated that aristeromycin 5′-diphosphate, although a poor substrate by itself, polymerizes when ADP is added to the assays to form copolymers of poly(A, aristeromycin). This suggests that aristeromycin binds to the polymerization site and is a good substrate, but has a lower affinity to bind tightly to the initiation site.

Activation of Rat Brain, Adrenal, or Liver cAMP-Dependent Protein Kinase. Aristeromycin 3′,6′-cyclic phosphate is slightly less effective than cAMP with respect to the protein kinase. Introduction of the bromine atom to the 8-position of purine base resulted in an increase of the activity, which suggests that the oxygen in the ribofuranose ring would have little effect on the activity (Marumoto et al., 1979).

3.6 BREDININ

Bredinin, 4-carbamoyl-1-β-D-ribofuranosylimidazolium-5-olate (Fig. 3.16), is a novel imidazole nucleoside analog isolated by Mizuno et al. (1974) from *Eupenicillium brefeldianum* M2166. Bredinin is a derivative of 5-amino-4-imidazole carboxamide nucleoside (AICA nucleoside), an intermediate of purine nucleotides, and is structurally similar to pyrazofurin (page 282), and the synthetic nucleoside analog virazole (page 286). Bredinin does not

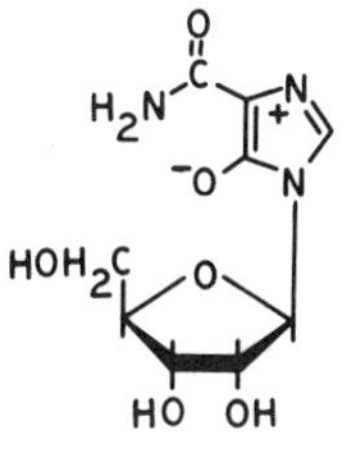

Figure 3.16 Structure of bredinin.

inhibit bacteria, but it does inhibit the growth of *Candida albicans*, vaccinia virus, and the multiplication of several mammalian cell lines in culture. It is cytotoxic to L5178Y cells and is a potent immunosuppressive agent (Mizuno et al., 1974). It also affects the secondary lesions that occur from adjuvant injection in rats causing suppression of primary and secondary immune responses. These findings strongly suggest the possibility of a clinical application for bredinin in rheumatoid arthritis. Bredinin blocks the conversion of IMP to XMP to GMP. It also inhibits nucleic acid synthesis, but is not incorporated into either RNA or DNA. The growth inhibitory effect of bredinin can be reversed by GMP, guanosine, or guanine in L5178Y cells (Sakaguchi et al., 1976a).

Discovery, Production, and Isolation

Mizuno et al. (1974) isolated bredinin from the fermentation broth of *Eupenicillium brefeldianum* from a soil sample in Hachijo, Tokyo as a part of a screening program for antifungal agents. The production medium and isolation are described by Mizuno et al. (1974). Bredinin is crystallized from acetone–water at 5°C.

Physical and Chemical Properties

The molecular formula of bredinin is $C_9H_{13}N_3O_6$; mol wt 259; mp 200°C (decomp.); $[\alpha]_D^{27}$ $-35°$ (c 0.8, H_2O); ultraviolet spectral properties: λ_{max}^{H2O} 245 nm (ϵ = 6300), 279 nm (ϵ = 14,200) (Hayashi et al., 1975); λ_{max}^{HCl} 245 nm ($E_{1\,cm}^{1\,percent}$ 260), 281 nm ($E_{1\,cm}^{1\,percent}$ 495); λ_{max}^{NaOH} 277 nm ($E_{1\,cm}^{1\,percent}$ 660) (Mizuno et al., 1974). Nmr spectra (DMSO): δ 3.58 (2H), 3.90 (H), 4.08(H), 4.40 (H), and 5.53 (H) for the protons on C-5′, C-4′, C-3′, C-2′, and C-1′ of ribose, respectively. Bredinin is a water-soluble and weakly acidic nucleoside.

Structural Elucidation

Hydrolysis of bredinin in 6 N HCl, 105°C, 20 h resulted in the isolation of glycine. The chemical structure of bredinin was elucidated by X-ray analysis (Yoshioka et al., 1975).

Inhibition of Growth

Bredinin is inhibitory against *C. albicans* and vaccinia virus and is cytotoxic to lymphoma cell line L5178Y. It does not inhibit bacteria, fungi, Ehrlich ascites tumors, or P388 tumors, nor is it effective against mice inoculated with leukemia L1210 (Mizuno et al., 1974).

Chemical Synthesis

The chemical synthesis of bredinin was accomplished by the condensation of the trimethylsilyl derivative of **II** with **III** to give **IV** (Fig. 3.17). Deacylation of **IV** (methanolic NH_3) gave bredinin (Hayashi et al., 1975).

Enzymatic Synthesis of Bredinin

Mizuno and coworkers have reported on the enzymatic synthesis of bredinin from cellfree extracts of *E. coli* on incubation of the aglycon with ribose 1-phosphate (Mizuno et al., 1975). Adenine, but not guanine, is a competitive inhibitor. Because purine nucleoside phosphorylase is an important enzyme in purine salvage synthesis, this enzyme is probably involved in the above synthesis of bredinin. A large number of organisms were able to use the salvage pathway for the production of bredinin when the aglycon was added to the medium. Bredinin was synthesized from the aglycon by resting cells of *E. brefeldianum*.

Biochemical Properties

Antiarthritic Activity of Bredinin, An Immunosuppressive Agent. Bredinin is safer than 6-mercaptopurine, azathiopurine, and 8-azaguanine as an immunosuppressive agent because of its restricted side effects, especially with respect to bone marrow damage and possible irreversible pancytopenia (Kusaka et al., 1968). Iwata et al. (1977) reported that bredinin has beneficial effects on experimental rheumatoid arthritis. They suggested that suppression of developing adjuvant polyarthritis in rats by bredinin is due to the inhibition of antibody formation and/or multiplication of sensitized

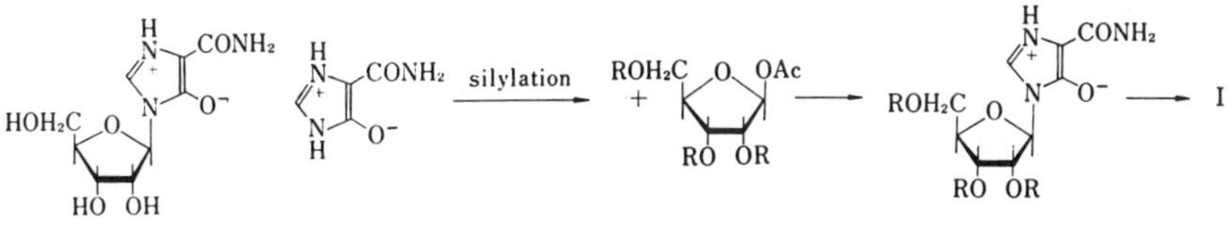

I   II   silylation   IIIa: R=acetyl   IIIb: R=benzoyl   IVa: R=acetyl   IVb: R=benzoyl

Figure 3.17 Chemical synthesis of bredinin (**I**). Reprinted with permission from Hayashi et al., 1975.

lymphocytes by a possible antigen. Bredinin does not possess antiinflammatory activity against carrageenan edema.

Immunosuppressive and Antirheumatoid Arthritic Properties. Bredinin is an excellent immunosuppressive nucleoside. It slightly decreases the peripheral leukocytes. When mice received 5000 mg/kg i.p. or 1500 mg/kg i.v., bredinin was not lethal for 9 days (Mizuno et al., 1974). The immunosuppressive activity of bredinin was studied by comparing its activity with that of azathiopurine (Imuran), the major drug used in combination with corticosteroids in immunosuppressive therapy (Kuroyanagi, 1971). Bredinin was more active than azathiopurine and showed potent activity following oral administration. Bredinin suppressed the primary and secondary hemolysin production in mice after the antigen stimulation of sheep red blood cells.

Action of Bredinin on Mammalian Cells and Bredinin-Induced Chromosome Aberrations. Bredinin has a cytotoxic effect on L5178Y cells and inhibits multiplication of several cells lines (Sakaguchi et al., 1975a). GMP antagonized the cytotoxic growth-inhibitory effect of bredinin (Sakaguchi et al., 1975a). GMP ($5 \times 10^{-5}M$) completely reversed the inhibition of bredinin ($1 \times 10^{-6}\ M$). Bredinin ($2 \times 10^{-5}\ M$) causes chromosomal abberrations with exposure for 1 h to L5178Y cells in culture (Sakaguchi et al., 1975a, 1976b). GMP prevented chromosome breakage. Sakaguchi et al., (1975a) suggest that bredinin blocks the conversion of either IMP to XMP or of XMP to GMP. The inhibition of GMP synthesis by bredinin (as its 5′-monophosphate) is similar to the inhibition of XMP aminase by psicofuranine and decoyinine (page 280) and that by the synthetic nucleoside virazole (page 287), which inhibits IMP dehydrogenase as reported by Robins and coworkers (Streeter et al., 1973).

Effect of Bredinin and Its Aglycon on L5178Y Cells in Culture. Sakaguchi et al. (1975b) studied the functional relationship of bredinin and its aglycon in L5178Y cells. The aglycon was as cytotoxic as bredinin. GMP, guanosine, or guanine reversed the cytotoxicity of bredinin and its aglycon. The aglycon and bredinin inhibited the incorporation of thymidine and uridine into DNA and RNA; the incorporation of leucine into protein was unaffected. Adenine, but not adenosine or AMP, reverses the growth inhibition by the aglycon. With bredinin, adenine does not reverse the inhibitory effect. The aglycon is converted to bredinin by adenine phosphoribosyltransferase (Sakaguchi et al., 1975b). Following oral administration to rats, most of the aglycon was recovered as bredinin from the serum and urine.

Effect of Bredinin with GMP on L5178Y Mouse Leukemia Cells. Because GMP does not completely reverse bredinin inhibition, the possibility existed that bredinin was acting on another site in the cell. Sakaguchi et al. (1976a) observed that bredinin (1.2×10^{-5} *M*) strongly inhibits the growth of L5178Y; GMP reverses this inhibition. Bredinin (5×10^{-5} *M*) in the presence of excess GMP was cytostatic. The cytostatic effect of bredinin could be reversed by GMP when cAMP was added. Other cyclic nucleotides were ineffective. Cell survival was not observed in the absence of GMP. It appears that cAMP reverses the efffect of bredinin as a function of GMP concentration because cAMP prevents the action of bredinin at a site other than that protected by GMP (Fig. 3.18). Evidence that bredinin and GMP operate by reducing intracellular cAMP, which is required for cell division, is provided by Sakaguchi et al. (1976a). [^{14}C]Bredinin is not incorporated into either RNA or DNA (Hayashi et al., 1975).

Cytotoxicity of Bredinin 5′-Monophosphate. Because the 5′ phosphates of the naturally occurring nucleoside analogs (i.e., tubercidin, toyocamycin, sangivamycin, formycin, ara-A) are the active forms of these inhibitors (Mizuno and Miyazaki, 1976), the 5′-monophosphate of bredinin was synthesized (Mizuno and Miyazaki, 1976). The nucleotide was not active

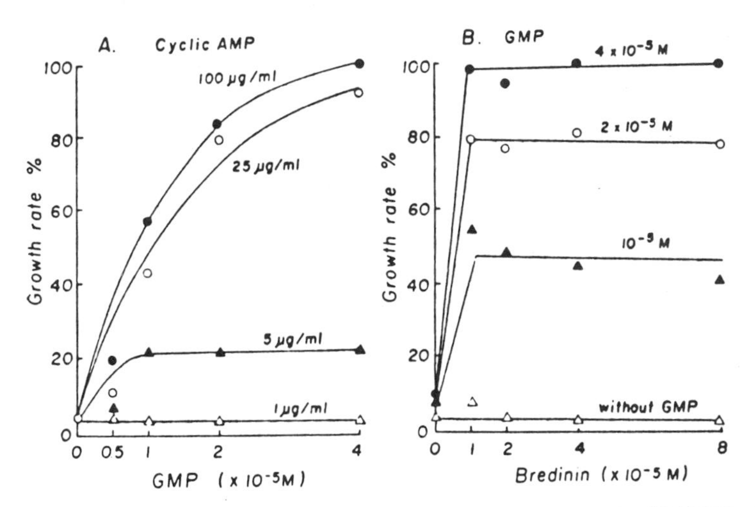

Figure 3.18 Effect of GMP and cyclic AMP on the inhibition of growth of L5178Y cells by bredinin. L5178Y cells were incubated for 40 h with bredinin (8×10^{-5} *M*) and various concentrations of GMP and cyclic AMP (*A*). The effects of various concentrations of bredinin and GMP with 100 µg/ml cyclic AMP were studied similarly (*B*). Reprinted with permission from Sakaguchi et al., 1976b.

against *C. albicans*, but it was cytotoxic against L5178Y cells. Like the uptake of ara-AMP by mammalian cells in culture (page 226) bredinin 5′-monophosphate can also enter the mammalian cell. The dose responses of bredinin and bredinin 5′-monophosphate for anti-L1210 activity in mice were the same. Apparently, bredinin 5′-monophosphate is converted to bredinin in the animal.

3.7 PENTOPYRANINES

Nine cytosine nucleoside analogs have been isolated from the culture medium of *Streptomyces griseochromogenes* by Seto and coworkers (Seto et al., 1976a; Seto and Yonehara, 1975; Seto, 1973). Of the six pentopyranines (A–F), pentopyranines A, C, and E are α-L-pentopyranosylcytosines, while pentopyranines B, D, and F are β-D-pentopyranosylcytosines (Fig. 3.19). The related group of β-D-hexopyranosylcytosine derivatives that have been isolated are pentopyranamine D, pentopyranine G, and pentopyranic acid. The structures of pentopyranines A, C, E, F, and G, pentopyranamine D, and pentopyranic acid have been confirmed by total chemical synthesis or comparison with known nucleosides (Chiu et al., 1973; Watanabe et al., 1974, 1976a, b; Fox and Goodman, 1951; Hilbert and Jansen, 1936). Hilbert and Jansen (1936) reported the synthesis of 1-(β-D-glucopyranosyl)cytosine, which turned out to be pentopyranine G. Pentopyranamine D is the nucleoside moiety of blasticidin H (Watanabe et al., 1976a). Although the ribo- and deoxyribonucleosides occur naturally as β-D-isomers, the only naturally occurring nucleoside found with the α-D-configuration is 5,6-dimethyl-1-α-D-ribofuranosylbenzimidazole (found in vitamin B_{12}).

Discovery, Production, and Isolation

The culture medium for the production of the pentopyranines and their isolation is as described by Seto and Yonehara (1975) and Seto et al. (1968, 1976a).

Pentopyranines A and C are the major nucleosides; A and C are crystallized from methanol; B and D are from ethanol. The following yields of crystalline pentopyranines were obtained from 500 liters of medium: A, 420 mg; B, 50 mg; C, 1250 mg; D, 80 mg.

Physical and Chemical Properties

The physical and chemical properties of pentopyranines A–D are shown in Table 3.2 (Seto, 1973).

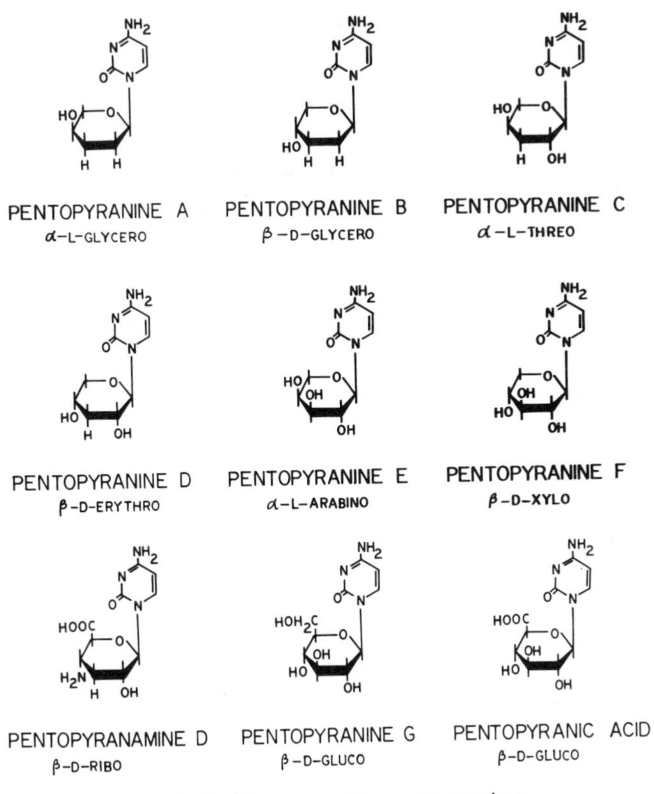

Figure 3.19 Structures of the pentopyranines.

Table 3.2 Physical and Chemical Properties of Pentopyranines A—D

	Molecular Formula	Mass Spectrum M^+: m/e	Mp (°C)	$[\alpha]^{21}$ (c, H_2O)	λ_{max} (ϵ) (nm) In 0.01 N HCl	λ_{max} (ϵ) (nm) In 0.01 N NaOH
A	$C_9H_{13}O_3N_3$	211	258 (decomp.)	+31.5 (2.0)	278 (13,100)	270 (8850)
B	$C_9H_{13}O_3N_3$	211	242 (decomp.)	+12.0 (4.2)	278 (11,300)	270 (8290)
C	$C_9H_{13}O_4N_3$	227	143~145	+20.0 (1.2)	278 (12,200)	270 (8450)
D	$C_9H_{13}O_4N_3$	227	261 (decomp.)	+13.7 (1.3)	278 (12,000)	270 (8370)

Reprinted with permission from Seto, 1973.

Structural Elucidation

The structure and nmr parameters of pentopyranine C and acylated derivatives of pentopyranines C and A are shown in Fig. 3.20 (Seto et al., 1972). The uv spectral data indicate that 1-substituted cytosine is the chromophore of all pentopyranines. The nmr spectrum of pentopyranine C shows signals of an anomeric proton, four methylene protons, and two ill-resolved protons attached to carbon bearing an oxygen atom. The steric relation between H-1′ and H-2′ was found to be trans diaxial as determined by the nmr spectra of pentopyranine C (I) and its tribenzoate (III). The absolute configuration of the 3-deoxypentose as L-threo was ascertained using the 2,4-dinitrophenylosazone. The chemical and physical properties of the sugar isolated from pentopyranine C and the authentic synthesized deoxy sugar were the same. Therefore, the structures of pentopyranines A and C are established as 1-α-(2′,3′-dideoxy-L-*glycero*-pentopyranosyl)cytosine and

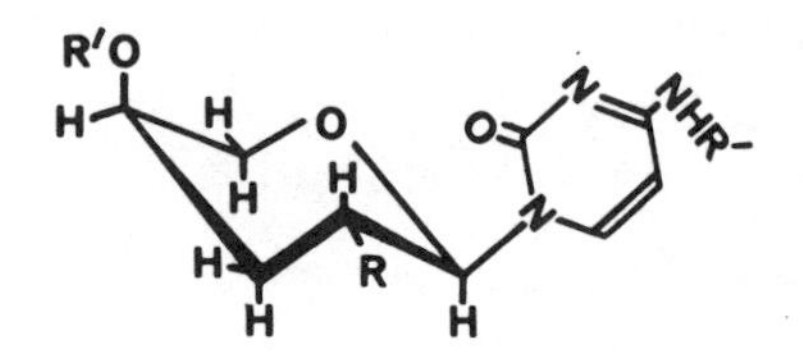

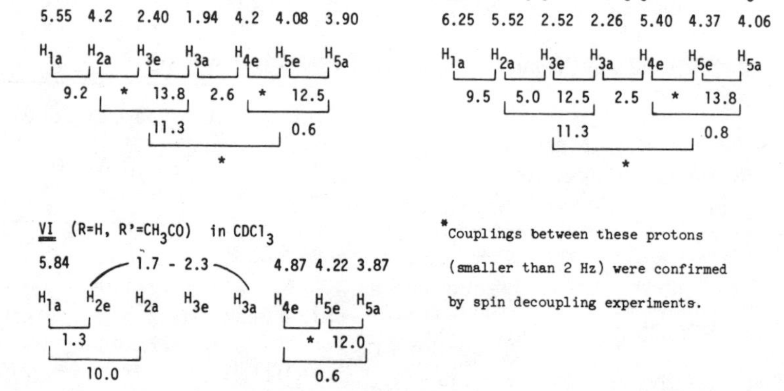

Figure 3.20 Structure and nmr parameters of pentopyranine C (**I**), its tribenzoate (**III**), and pentopyranine A (diacetate) (**VI**). Asterisks indicate cases where couplings between protons (smaller than 2 Hz) were confirmed by spin decoupling experiments. Reprinted with permission from Seto et al., *Tetrahedron Letters*, p. 3991 (1972), "Isolation of Pentopyranines A, B, C, and D." Copyright 1972, Pergamon Press, Ltd.

1-α-(3′-deoxy-L-*threo*-pentopyranosyl)cytosine, respectively (Seto et al., 1973). The proof of structures of pentopyranines E, F, and G as 1-(α-L-arabinopyranosyl)-, 1-(β-D-xylopyranosyl) and 1-(β-D-glucopyranosyl)-cytosine, respectively, has been reported (Watanabe et al., 1976a; Seto, personal communication).

Seto et al. (1976b) reported on the isolation of pentopyranic acid from the culture medium of *S. griseochromogenes*. Based on the physical and chemical data, the structure of pentopyranic acid is 1-(β-D-glucopyranosyluronic acid) cytosine (Seto et al., 1976a; Seto and Yonehara, 1975; Hilbert and Jansen, 1936). Pentopyranic acid has a structure closely related to C substance, the nucleoside obtained from the degradation of gougerotin (Watanabe et al., 1972).

Chemical Synthesis of Pentopyranines

Pentopyranine A. Pentopyranine A and its diacetate were synthesized (Watanabe et al., 1974) from 1-(α-L-arabinopyranosyl)cytosine (pentopyranine E) by a multistep procedure involving the introduction of the 2′,3′-unsaturated linkages. The synthetic nucleoside and its diacetate were identical with natural pentopyranine A and its diacetate.

Pentopyranine C. Chiu et al. (1973) described the synthesis of pentopyranine C from 3-deoxy-di-*O*-isopropylidene-α-D-*xylo*-hexofuranose. The final step in the synthesis of pentopyranine C involves the removal of the acetyl groups from the triacetate. The physical and chemical properties of these synthetic nucleosides were the same as the natural product and its triacetate.

Pentopyranamine D. The total synthesis of pentopyranamine D, the nucleoside moiety of blasticidin H, has been reported by Watanabe et al. (1976a). The synthesis is accomplished in eight steps starting from 1-(3-deoxy-β-D-*xylo*-hexapyranosyl)cytosine.

Pentopyranic acid. Pentopyranic acid was synthesized by condensation of bis(trimethylsilyl)-N-acetylcytosine and methyl tetra-O-acetyl-β-D-glucopyranouronate in the presence of stannic chloride followed by hydrolytic removal of protecting groups (Watanabe et al., 1976b).

Biochemical Properties

Although the pentopyranines are not potent inhibitors of eukaryotes or prokaryotes, Seto (1973) reported that the pentopyranines A–D inhibited

Table 3.3 Inhibitory Effect of Pentopyranines A, B, C and D on the Incorporation of [^{3}H]Uridine Into RNA of Ehrlich Ascites Cells

	Inhibition (%) at		
Compound	1000 μg/ml	166 μg/ml	28 μg/ml
A	76	88	91
B	92	87	101
C	18	66	84
D	96	94	94

Reprinted with permission from Seto, 1973.

the incorporation of [^{3}H]uridine into the RNA of *Ehrlich ascites* tumor cells (Table 3.3).

3.8 PYRROLOPYRIMIDINE NUCLEOSIDE ANALOGS

The pyrrolopyrimidine nucleoside antibiotics, tubercidin, toyocamycin, and sangivamycin (Fig. 3.21), have stimulated considerable interest because they are powerful antibacterial, antifungal, and cytotoxic agents. They have also found use against some forms of human cancer, such as in the treatment of cutaneous neoplasms (Acs et al., 1964; Bisel et al., 1970; Bloch et al., 1967; Grage et al., 1970; Smith et al., 1970). These three nucleosides, all analogs of adenosine, are highly cytotoxic to mammalian cells in culture. Modification of the cyano group at C-5 and/or the amino group at C-4 (see numbering, Fig. 3.21) markedly changes the antineoplastic and antiviral activities of these pyrrolopyrimidine nucleosides. Tubercidin is rapidly taken up by

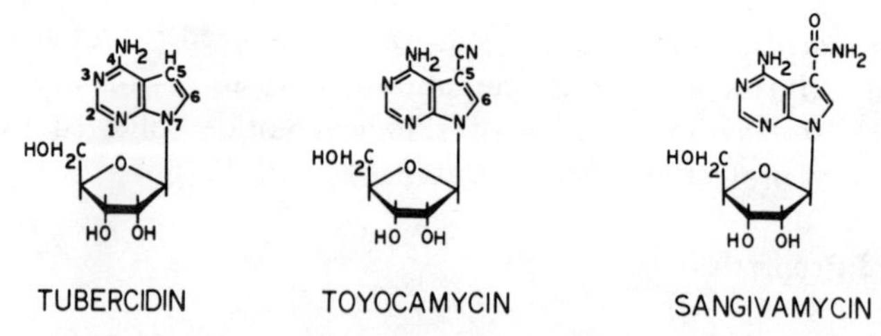

Figure 3.21 Structures of the pyrrolopyrimidine nucleoside analogs, tubercidin, toyocamycin, and sangivamycin.

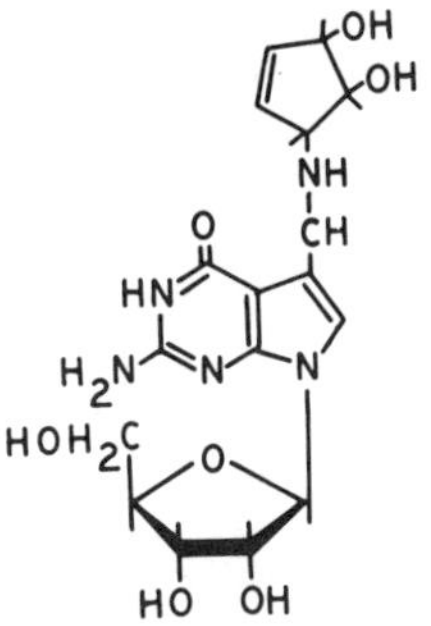

Figure 3.22 Structure of nucleoside Q.

red blood cells and remains inside the cells as 5′-phosphates (Smith et al., 1970).

Although the isolation of the pyrrolopyrimidine nucleoside antibiotics was previously limited to the *Streptomyces*, Nishimura and coworkers have now shown that the pyrrolopyrimidine ring is found in tRNA in eukaryotes and prokaryotes. The nucleoside they isolated, nucleoside Q (Fig. 3.22), occupies the first position of the anticodon of *E. coli* tRNA^{Tyr}, tRNA^{His}, tRNA^{Asn}, and tRNA^{Asp} (Kasai et al., 1975a, 1975b). The structure of nucleoside Q is 2-amino-5-(4,5-*cis*-dihydroxy-1-cyclopenten-3-yl-*trans*-aminomethyl(7-β-D-ribofuranosyl)pyrrolo[2,3-*d*]pyrimidin-4-one. Based on the biosynthetic studies of the pyrrolopyrimidine nucleoside antibiotics by Suhadolnik and Uematsu (1970), Nishimura and coworkers (Kuchino et al., 1976) reported that guanine also serves as the precursor for part of the carbon–nitrogen skeleton for the pyrrolopyrimidine aglycon of nucleoside Q. Of interest is the report of Townsend and coworkers (Cheng et al., 1976) in which they described the synthesis of 2-amino-5-cyano-7-(β-D-ribofuranosyl)pyrrolo-[2,3-*d*]pyrimidin-4-one from toyocamycin as a potential precursor of the biosynthesis of nucleoside Q. Okada and Nishimura (1979a) have recently synthesized base Q.

Biosynthesis

Although tubercidin, toyocamycin, and sangivamycin have the pyrrolopyrimidine aglycon, Smulson and Suhadolnik (1967) reported that ^{14}C from [2-^{14}C]adenine was incorporated into tubercidin. Uematsu and Suhadolnik (1970) provided direct evidence by chemical degradation that ^{14}C from [2-^{14}C]adenine of the purine ring was directly incorporated into the equivalent carbon atom in toyocamycin. In their determination of the biosynthetic origin of the two pyrrole carbon atoms and the cyano groups of toyocamycin, Suhadolnik and Uematsu (1970) found evidence that the intermediate in the biosynthesis of the pyrrolopyrimidine nucleoside analogs required the addition of a second ribose to N-9 of the purine ring once the ureido carbon (C-8) was removed. This was accomplished by using ^{14}C labeled adenosine in which 4% of the ^{14}C was in the adenine and 96% resided in the ribose. The percent distribution of ^{14}C in the aglycon, in the cyano group and in the ribose of toyocamycin firmly established that the pyrrole ring formation does not occur until a second ribose is attached to N-9. Proof that GTP and not ATP was the immediate precursor for the biosynthesis of the pyrrolopyrimidine nucleosides is given by Elstner and Suhadolnik (1971), who reported on the isolation and partial purification of GTP-8-formylhydrolase. This enzyme catalyzes the production of formic acid from the ureido carbon of GTP. The role of GTP as the common precursor for the biosynthesis of important biochemical cofactors and vitamins is shown in Fig. 3.23. Hirasawa and Isono (1978) have also shown that guanine is the carbon–nitrogen precursor for the biosynthesis of 8-azaguanine. C-8 of guanine is eliminated and is replaced by a nitrogen.

The importance of the pyrrolopyrimidine ring has now expanded from the

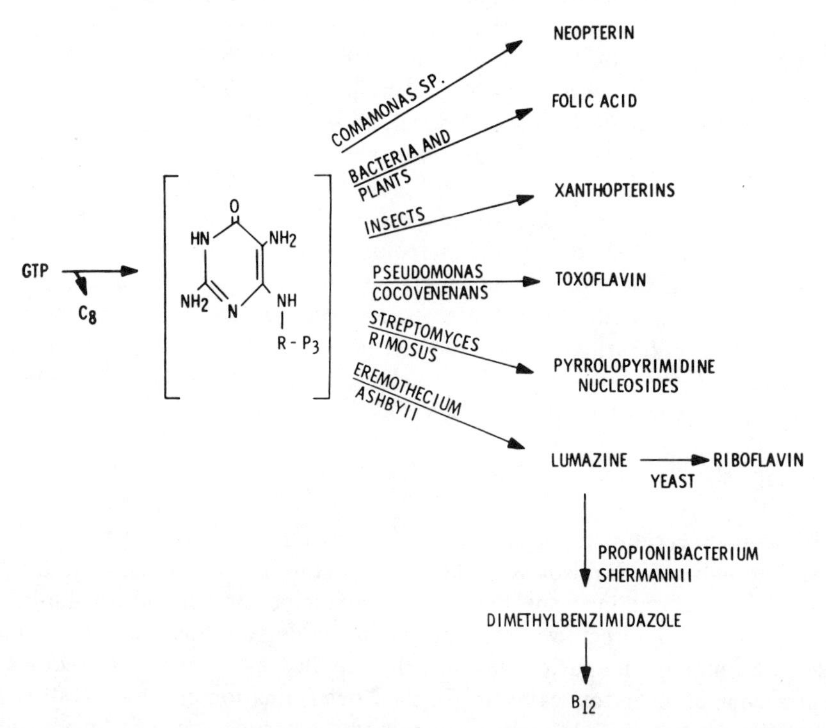

Figure 3.23 The role of GTP in the biosynthesis of the pteridine ring, the azapteridine ring, the benzimidazole ring, and the pyrrolopyrimidine ring. Reprinted with permission from Elstner and Suhadolnik, 1971.

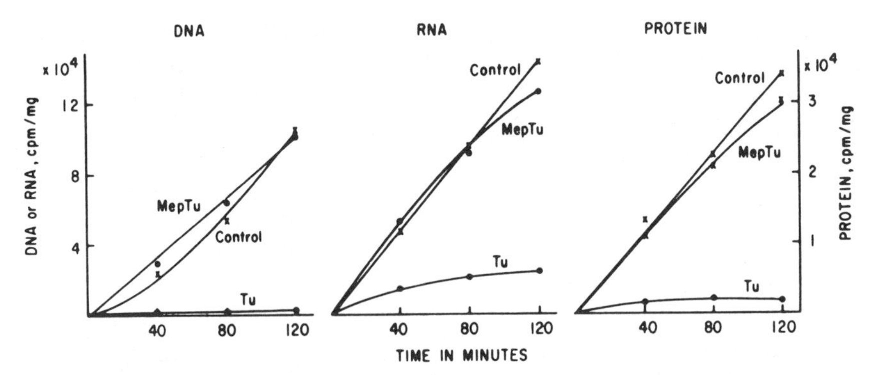

Figure 3.24 Effect of tubercidin (Tu) and the methyl ester of tubercidin 5′-phosphate (MepTu) (both at 20 μg/ml) on macromolecule synthesis in KB cells. Reprinted with permission from Smith et al., *Advances in Enzyme Regulation*, **5**, 121 (1973), "Biochemical and Biological Studies with Tubercidin (7-Deazaadenosine), 7-Deazainosine and Certain Nucleotide Derivatives of Tubercidin." Copyright 1973, Pergamon Press, Ltd.

Streptomyces to mammalian tissue with the discovery of the pyrrolopyrimidine aglycon in the 5′-terminal position of the anticodon in asparagine, aspartate, histidine, and tyrosine tRNA by Nishimura and coworkers (Kasai et al., 1975b). Kuchino et al. (1976) have reported that carbon-14 from [2-^{14}C]guanine, but not from [8-^{14}C]guanine, is incorporated into base Q. The posttranscriptional modification of tRNA occurs by the removal of guanine in the first position of the anticodon followed by the insertion of either 7-(aminomethyl)-7-deazaguanine or 7-(cyano)-7-deazaguanine into tRNA (Okada et al., 1979b). The enzyme that catalyzes the exchange of guanine has been isolated and purified to homogeneity (Okada and Nishimura, 1979).

Biochemical Properties of Tubercidin

Effect of Tubercidin on Prokaryotes and Eukaryotes. Tubercidin blocks glucose utilization in bacteria and inhibits mitochondrial respiration in Ehrlich Lettre tumor cells (Bloch et al., 1967; Miko and Drobnica, 1975). Tubercidin inhibits purine synthesis *de novo*, the anabolism of adenosine (Parks and Brown, 1973; Stegman et al., 1973), rRNA processing, methylation of tRNA, and protein and nucleic acid synthesis and causes visible nuclear damage (Wainfan et al., 1973; Baxter and Byvoet, 1974; Wainfan and Landsberg, 1973; Bassleer et al., 1976). Tubercidin (20 μg/ml), but not the 5′-methyl ester of tubercidin 5′-phosphate, completely inhibits DNA, RNA, and protein synthesis following the addition to growing KB cells (Fig 3.24). The glycosyl bond of tubercidin is unusually flexible and shows no preference for either *anti* or *syn* (Evans and Sarma, 1975). Tubercidin, once incorporated into the nucleic acid structure, disrupts the traditional Watson-Crick type base pairing. This disruption probably places tubercidin in the *syn* conformation, which explains its cytotoxic properties in rat liver nuclei. Activity of cAMP-dependent protein kinase is inhibited by tubercidin; however, there is a threefold stimulation of protein kinase activity in *Trypanosoma cruzi* and *Trypanosoma gambiense* (Walter and Ebert, 1977; Hirsch and Martelo, 1976; Walter, 1976). Tubercidin is toxic in all phases of the cell cycle (Bhuyan et al., 1972).

Tubercidin is a competitive inhibitor of ATP for the ATP–pyrophosphate exchange catalyzed by Met-tRNA synthetase (K_i for tubercidin is 30 μM) (Lawrence et al., 1974). Tubercidin is not a substrate for either nucleoside phosphorylase (Bloch, 1975) or adenosine deaminase (Ikehara and Fukui, 1974), but is phosphorylated by red blood cells or microorganisms (Bloch et al., 1967; Agarwal et al., 1975; Smith et al., 1970).

The red blood cell converts tuberdicin to the 5′-triphosphate (Smith et al., 1970; Parks and Brown, 1973) and prevents adenine metabolism by inhibiting adenosine kinase, adenosine phosphoribosyltransferase, nucleoside phosphorylase, and other adenosine enzymes (Ross and Jaffee, 1977; Henderson et al., 1975). Bloch (1975) reported that the inhibition of bacterial cells by tubercidin can be attributed to faulty regulation of phosphofructokinase by tubercidin 5′-triphosphate. To explain the incorporation of tubercidin into DNA, Suhadolnik and coworkers reported that the 5′-diphosphate and 5′-triphosphate of tubercidin are substrates for ribonucleotide reductase (Suhadolnik et al., 1968a; Chassy and Suhadolnik, 1968).

Inhibition of the Elongation Step in RNA Synthesis. Ward and coworkers (Kumar et al., 1977) have used tubercidin to clarify the initiation, elongation, and termination of RNA synthesis. They demonstrated that tubercidin 5′-triphosphate can effectively replace ATP as an RNA-initiating nucleoside triphosphate; however, tubercidin 5′-triphosphate can moderately replace ATP as an RNA chain-elongating nucleotide.

Methylase Inhibitors. During the past 10 years, *S*-adenosylmethionine (SAM) has been established as the most important donor in transmethylation reactions and the alkyl donor in the biosynthesis of spermidine (for reviews, see Tabor and Tabor, 1976; Salvatore et al., 1977; Paik, 1979; Cantoni, 1975; Shapiro and Schlenk, 1965). Chemically, three bonds in the sulfonium of SAM are equivalent. Therefore, transfer reactions involving any one of these three substituents should be possible. Proof of this is the important discovery by Tabor et al. (1958) that SAM, following decarboxylation, can act as the alkyl donor in the biosynthesis of spermidine.

Methylase inhibitors have been extremely useful biochemical probes to define the methylnucleoside composition of the 5′-cap structure for cytoplasmic poly(A), mRNA (Kaehler et al., 1977), catecholamine methylation (Michelot et al., 1977), and tRNA methylation (Chang and Coward, 1975). The importance of SAH (**5a**) as an extremely potent product inhibitor of the methylation of catecholamines by SAM (**4a**) was firmly established by Coward et al. (1972) (Scheme 3.1). Coward and coworkers (Coward et al., 1974) were the first to describe the use of *S*-tubercidinylhomocysteine (**5b**) (STH) as a physiological, stable, potent inhibitor of SAM-dependent methylases. Their studies were aimed at the regulation of SAM-dependent methylation and their rationale for the synthesis of STH was the use of an analog of adenosine with a stable base–ribose bond that is resistant to phosphorylases. STH serves this purpose because the pyrrolopyrimidine–ribose bond is not hydrolyzed by either nucleoside phosphorylases or hydrolases. In addition, tubercidin is not a substrate for adenosine deaminase. Therefore, STH is an important SAH analog because cleavage of SAH in mammalian cells can proceed in several ways (Scheme 3.1). The 5′-thioether (**5a**) can be cleaved to give homocysteine (**6**) and adenosine (**1a**), which is the major hydrolytic pathway. In polyamine biosynthesis (i.e., synthesis of spermidine, **10**), a phosphorylase cleaves 5′-methylthioadenosine (**9a**) to adenine (**11a**) and 5-methylthioribose (**12**).

STH partially inhibits cytoplasmic poly(A)-RNA methylation of the 2′-*O*-methylnucleoside, m^7GpppN′, at the second 2′-*O*-methylnucleoside and at internal N^6-methyladenosine (Kaehler et al., 1977). Methylation of 7-methylguanosine is not affected. The results of these studies indicate that a completely methylated cap is not required for transport of mRNA into the cytoplasm. With STH, Kaehler et al. (1977) suggest that it may be possible to assess the relative time sequence of methylation *in vivo* at specific sites within mRNA. Similarly, 5′-methylthiotubercidin (**9b**), which is resistant to rat ventral prostate phosphorylase, competitively inhibits the hydrolytic cleavage of the purine–ribose bond in methylthioadenosine (**9a**) (Coward et al., 1977). The 7-deaza analog of decarboxylated *S*-adenosylmethionine catalyzes the transfer of the propylamino group from either **7a** or **7b** to putrescine (**8**) to form spermidine (**10**). 5′-Methylthiotubercidin (**9b**), unlike **9a**, is not further metabolized by hydrolytic cleavage of the pyrrolopyrimidine–ribose bond or by deamination of the amino group on the aglycon.

Methylation of tRNA of phytohemagglutinin-stimulated rat lymphocytes and rat liver tRNA is inhibited by STH (**5b**) (Chang and Coward, 1975; Coward et al., 1974). STH is a better inhibitor than the natural product inhibitor, SAH. Because the products of this degradation are adenosine (**1a**) and homocysteine (**6**), the adenosine from SAH (**5a**) stimulates DNA synthesis in rat lymphocytes, whereas STH inhibits DNA synthesis (Chang and

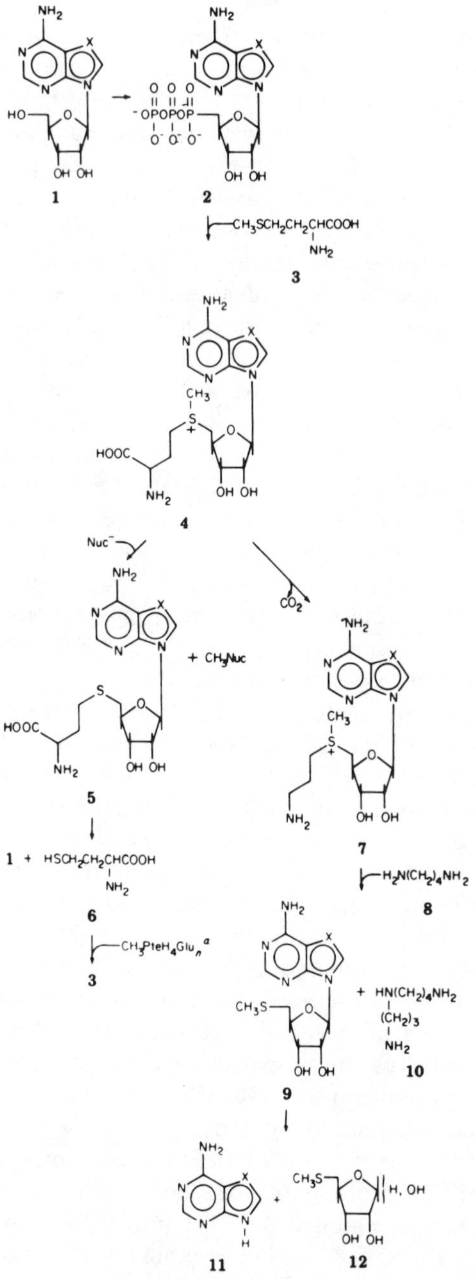

a, X = N; **b**, X = CH

Scheme 3.1 The role of S-adenosylmethionine (SAM, 4a) and its metabolites in cellular reactions. Reprinted with permission from Coward et al., 1977.

Coward, 1975). The fact that STH can enter the cell intact and is resistant to metabolic degradation allows studies on its transport and methylase reactions *in vivo* (for review, see Coward and Crooks, 1979).

Effect of NAD^+ Containing Tubercidin in Place of Adenosine in Dehydrogenase Reactions and ADP Ribosylation. On the basis of their studies on the role of the adenine and the adenine-ribose of NAD^+ in the coenzyme domain of the dehydrogenases, Suhadolnik et al. (1977a) reported that the K_m and K_D of the NAD^+ analog in which tubercidin replaced the adenosine moiety of NAD^+ are essentially the same as those of NAD^+; however, the V_{max} decreased markedly. Replacement of the adenosine of NAD^+ with tubercidin ($NTuD^+$) does not change the binding of this NAD^+ analog to the coenzyme domain of the dehydrogenases, but does decrease productive complex formation. $NTuD^+$ is also a substrate for the ADP ribosylation of elongation factor 2 by diphtheria toxin in lysed rabbit reticulocytes (Lennon and Suhadolnik, 1979; Lennon et al., 1976). In studies related to the role of NAD^+ in DNA synthesis in nuclei of eukaryotes, $NTuD^+$ is not effective as an inhibitor of DNA synthesis in nuclei isolated from rat liver (Suhadolnik et al., 1977b) (Fig. 3.6). Another reaction that markedly affects cellular processes is that of the 3′,5′-cyclic tubercidin monophosphate. It has much greater lipolytic activity than cAMP in adipose tissue (Blecher et al., 1971).

Clinical and Anthelmintic Properties of Tubercidin

Although studies with various types of advanced neoplastic diseases treated with tubercidin have not been encouraging, Klein and coworkers reported that tubercidin is very effective against basal cell carcinoma, actinic keratoses, mycosis fungoides, reticulum cell sarcoma, and squamous cell carcinoma (Burgess et al., 1974; Klein et al., 1975). The basal cell carcinoma did not reoccur.

One of the most promising clinical applications of tubercidin is as an anthelmintic agent. In the presence of tubercidin, adult bloodworms of *Schistosoma mansoni* cannot synthesize purine nucleotides and must rely on the purine salvage mechanisms for energy maintenance (Senft et al., 1972, 1973; Jaffee et al., 1971, 1975; Much et al., 1975).

Interferon Production. Witkop and coworkers have contributed significantly to our knowledge of interferon production. Poly(tubercidin) does not induce interferon production in the presence of poly(7-deazainosine), a potent inducer (DeClercq et al., 1974; Torrence et al., 1974). Poly(tubercidin) apparently inhibits interferon production by binding to a cellular receptor site (Torrence et al., 1974).

Inhibition of Viruses. 5-Bromotubercidin affects the acetylcholine receptors in muscle cells from chick embryos. It also inhibits SV-40 proliferation in African green monkey kidney cells and Rous sarcoma virus in chick embryo fibroblasts and it reversibly blocks the synthesis of cellular HnRNA, rRNA, and mRNA. The inhibition of viruses by 5-bromotubercidin is not related to the interference of nucleic acid precursor synthesis; its mode of action is at the level of nucleic acid polymerization (Brdar et al., 1973). Additional evidence that encourages the application of nucleoside analogs to the study of DNA viruses is the report by Darlix et al. (1971).

Biochemical Properties of Toyocamycin

Effect on Eukaryotes. Toyocamycin inhibits yeast (Venkov et al., 1977), virus (Brdar and Reich, 1976), and 45 S RNA synthesis in eukaryotes. The inhibition of 45 S RNA prevents the restarting of rRNA transcription in amino acid-starved cells once the cells are returned to a complete medium (Iapalucci-Espinoza et al., 1977) and in Rous sarcoma virus (RSV) (Brdar and Reich, 1976).

Suhadolnik et al. (1967) reported that toyocamycin is converted to its 5′-triphosphate and is incorporated into the RNA of Ehrlich ascites tumor cells, which seems to explain the mode of its antineoplastic activity. Tavitian et al. (1968, 1969) in a more complete study showed that its incorporation into the RNA of L cells selectively inhibits rRNA synthesis. The studies of Weiss and Pitot (1974b) on the effect of toyocamycin, tubercidin, and 6-thioguanosine on rRNA maturation in cultured Novikoff hepatoma cells showed that the processing of 45 S RNA to 38S RNA is not inhibited; however, the formation of mature 28 and 18S RNA is inhibited. The inhibition of rRNA maturation by tubercidin and toyocamycin may explain their antineoplastic activity. Weiss and Pitot (1974b) demonstrated that toyocamycin interferes with RNA metabolism by preventing polyadenylation and/or methylation of adenosine residues. Gotoh et al. (1974) showed that toyocamycin incorporated into HeLa cell 45 S precursor ribosomal RNA is not cleaved by *E. coli* ribonuclease III.

Toyocamycin can also replace adenosine in the RNA of *S. cerevisiae* (Venkov et al., 1977). While limited pre-rRNA synthesis occurs, the processing and maturation of pre-rRNA is inhibited and there is an accumulation of 27 and 20 S pre-rRNA. As the concentration of toyocamycin increases, the conversion of 27 S pre-rRNA, 25 S RNA, and 20 S pre-rRNA to 18 S rRNA is stopped.

Effect on Appearance of RNA in the Cytoplasm and rRNA Transcription. Phillips and Phillips (1971) studied the effect of toyocamycin and species differences with mouse and Chinese hamster cells in monolayer cultures. Toyocamycin did not increase the newly formed RNA in the nucleolus of Chinese hamster cells. It is not clear if this was due to a turnover of RNA or if nucleolar RNA synthesis simply stopped following incubation with toyocamycin. Toyocamycin inhibits the appearance of newly formed RNA in the cytoplasm. Toyocamycin-treated mouse and hamster cells show a gradual disappearance of the 150 Å granules from the particulate region in the nucleoli. Therefore, that region of the nucleolus that contains early rRNA precursor ultimately becomes fibrillar. Heine (1969) reported similar findings. Actinomycin D, an inhibitor of all RNA synthesis, caused a segregation and finally a disaggregation of nucleolar components. The effect of toyocamycin on rRNA transcription has been studied in much detail by Franze-de Fernandez and coworkers (Iapalucci-Espinoza et al., 1977). They demonstrated that toyocamycin affects rRNA transciption in cells that have been incubated in complete medium, but not in amino acid-deprived cells, even though toyocamycin inhibits rRNA maturation in cells incubated under both conditions. Because they found no inhibition of protein synthesis by toyocamycin, the explanation offered is that the inhibition of the processing of rRNA somehow affects transcription.

Effect on Viruses. Brdar and Reich (1976) reported that toyocamycin reversibly blocks Rous sarcoma virus-infected embryonic chick fibroblast multiplication. Toyocamycin (0.1 μg/ml) decreased the synthesis of cellular DNA and certain classes of RNA. Protein synthesis continued at disproportionately elevated rates for several days. Toyocamycin prevented the processing of ribosomal 45 S precursor to the mature 28 and 18 S RNA. Toyocamycin does not affect HnRNA synthesis. In the presence of toyocamycin, the quantities of virus-specific RNA of Rous sarcoma virus particle were normal (Fig. 3.25). To study the effect of toyocamycin on the antiviral action of interferon, Périès et al. (1974) reported that although toyocamycin inhibited the synthesis of ribosomal RNAs, it blocked the antiviral activity of interferon in encephalomyocarditis virus-infected L cells. The precursor–product relationship between rapidly sedimenting nuclear RNA (HnRNA) and the more slowly sedimenting polyribosome-associated RNA (mRNA) has been demonstrated in both uninfected and virus-infected mammalian cells (for a review see Perry, 1976). To further characterize the sequence of events in the posttranscriptional modifications at the 3′-terminus (polyadenylation) and the 5′-terminus (insertion of inverted dinucleotide and methylation) to make mature, functional mRNA, Swart and Hodge (1978) have isolated nuclear adenovirus from HeLa S_3 cells treated with toyocamycin. Their data imply that toyocamycin inhibits poly(A) additions (Fig. 3.26), but the introduction of the inverted nucleotides and methylations can occur at the 5′-end independently of polyadenylation.

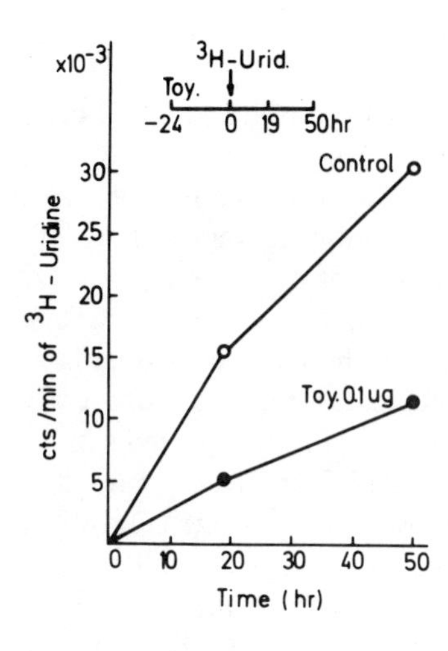

Figure 3.25 Effect of toyocamycin on RSV production in transformed cultures. Control (○); toyocamycin, 0.1 μg/ml (●). Reprinted with permission from Brdar and Reich, 1976.

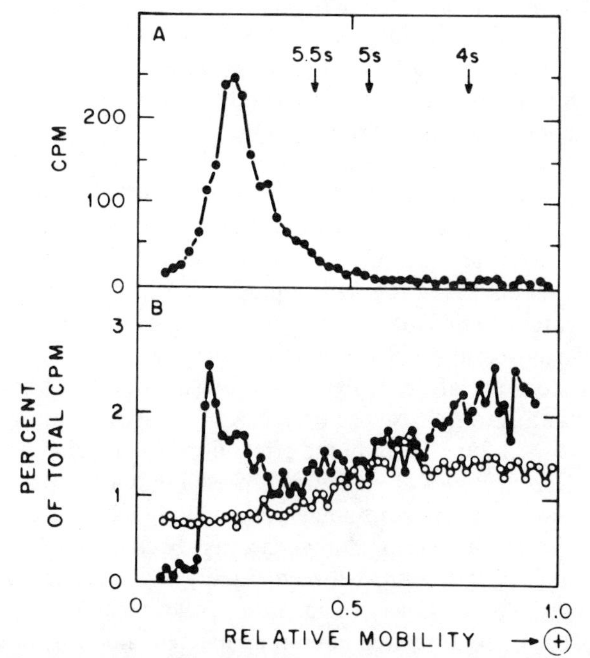

Figure 3.26 Poly(A) addition in the presence of toyocamycin. Adenovirus-specific nuclear RNA labeled with radioactive ^{32}P was obtained by selective extraction late in infection and poly(A) segments were prepared by enzymatic digestion. Prior to or after selection on oligo$(dT)_{10}$ cellulose, samples were fractionated in 10% cylindrical polyacrylamide gels by electrophoresis. (A) RNA fragments selected by oligo$(dT)_{10}$ cellulose synthesized in the absence of toyocamycin (●). (B) RNA fragments prior to selection by oligo$(dT)_{10}$ cellulose synthesized in the absence (●) and presence (○) of toyocamycin. Reprinted with permission from Swart and Hodge, 1978.

The isopentenyl, seleno, and alkylseleno derivatives of toyocamycin have been synthesized and studied for antitumor and antifungal activity (Lewis and Townsend, 1974; Schramm and Townsend, 1974; Schramm et al., 1975; Townsend and Milne, 1975).

Biochemical Properties of Sangivamycin

Sangivamycin is one of the few nucleosides that have been selected for clinical studies. It has strong antileukemic activity (Cairns et al., 1967). Suhadolnik et al. (1968b) reported that sangivamycin 5′-triphosphate competes with ATP, is a substrate for RNA polymerase from *Micrococcus lysodeikticus*, and is incorporated into RNA with various DNA templates. It is a substrate for tRNA adenyltransferase and is incorporated into the 3′-terminus of tRNA (Uretsky et al., 1968). This modified tRNA can function in the esterification of amino acids. When [^{3}H]sangivamycin was injected into mice, the 5′-mono-, 5′-di-, and 5′-triphosphates were isolated from the red blood cells (Hardesty et al., 1974). Sangivamycin is incorporated into the RNA and DNA of all tissues except the brain. In the brain, only the RNA contains sangivamycin. When incorporated into red blood cells, the half-life of sangivamycin is 50 h. This compares to a half-life of 43 h for tubercidin (Smith et al., 1970). The biochemical properties of cSMP are compared with those of cTuMP and cToMP on page 165.

3.9 PYRAZOLOPYRIMIDINE NUCLEOSIDE ANALOGS

The pyrazolopyrimidine nucleoside antibiotics, formycin (formycin A), formycin B (laurusin), and oxoformycin B (Fig. 3.27), have been isolated from the culture filtrates of *Nocardia interforma* (Hori et al., 1964), *Streptomyces lavendulae* (Aizawa et al., 1965), and *Streptomyces gunmaences*. Another important nucleoside isolated from *N. interforma* is coformycin (Fig. 5.1) (Sawa et al., 1967). Formycin, a cytotoxic analog of adenosine, is

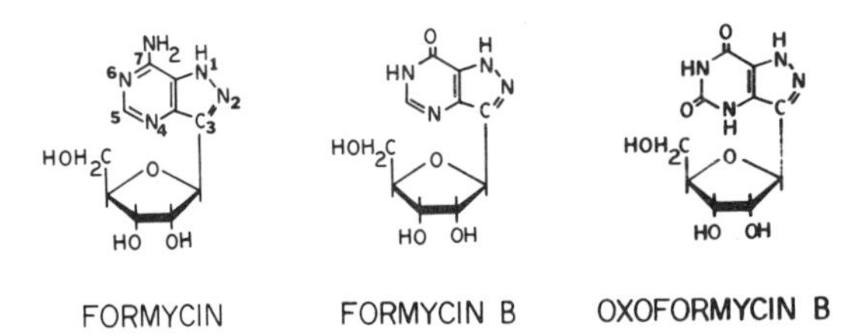

Figure 3.27 Structures of the pyrazolopyrimidine nucleoside analogs, formycin, formycin B, and oxoformycin B.

deaminated and easily phosphorylated by adenosine kinase; formycin B is an analog of inosine; oxoformycin B is an analog of xanthosine, but is not toxic either to eukaryotes or prokaryotes. Formycin has antineoplastic activity and inhibits bacteria, fungi, and viruses. Formycin B inhibits mouse sarcoma 180 cells and influenza A_1 virus and protects rice plants against *Xanthomonas oryzae* infections. Coformycin and its 2′-deoxy derivative are very potent inhibitors of adenosine deaminase and are immunosuppressive agents (see page 259).

The chemical, physical, and biochemical properties of the pyrazolopyrimidine nucleosides have been reviewed previously by Suhadolnik (1970), Townsend (1975), Bloch (1978), and Daves and Cheng (1976).

Chemical Synthesis

A very elegant method for the synthesis of the C-nucleoside analogs, showdomycin, formycin, and formycin B, has been described by Kalvoda (1976; 1978) (Scheme 3.2). Kalvoda found that the acylation of 3-cyano-2-alkenoic esters [ester-nitrile (**II**)] with HCN in the presence of *tert*-butyloxycarbonylmethylenetriphenylphosphorane (**I**) affords pyrazole **IV** on cycloaddition of diazoacetonitrile (**III**) with the simultaneous elimination of HCN. The *tert*-butyl group was split off compound **IV** by acid to form compound **V**, which was converted to the urethan (**VI**), by a modified Curtius degradation. The protecting trichloroethyl group was cleaved reductively to form the cyanoamine (**VII**), which on heating with formamidine acetate followed by alkali-catalyzed methanolysis yielded formycin (**VIII**).

Prototype Tautomerism of Formycin and Fluorescent Derivatives

Townsend and coworkers (Chenon et al., 1976; Townsend et al., 1974) have used ^{13}C nmr to study the prototype tautomerism of the aglycon of formycin. Formycin undergoes a temperature-dependent tautomeric equilibrium between the N(1)H and N(2)H forms. The N(1)H form accounts for 85% of the tautomer. In an attempt to determine if the rate of formycin tautomerization is faster than the turnover rate for many enzymes, Cole and Schimmel (1978) determined the tautomerization rate constants at 25°C. The rapid rate ($1/\lambda > 10^3\ s^{-1}$) means that even though the (1)H tautomer predominates, the quick tautomerization to the (2)H form occurs fast enough so as not to limit the rate of an enzymatic process that requires the (2)H species. Lactam–lactim tautomerism was not observed when there is a 7-oxo function (Chenon et al., 1976).

Chung and Zemlicka (1977), as a result of the current interest in fluorescent nucleosides, have obtained the fluorescence spectra of the for-

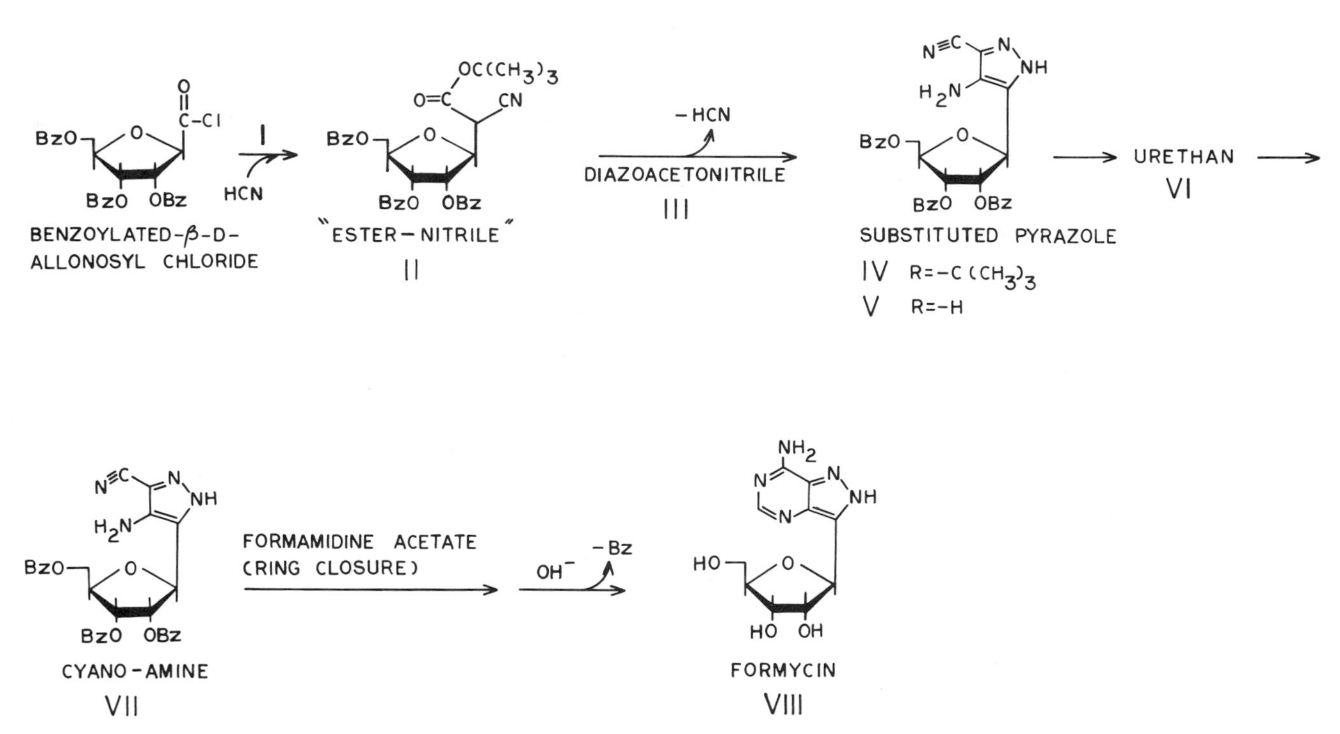

Scheme 3.2 Chemical synthesis of formycin (**VIII**). Reprinted with permission from Kalvoda, Collection of Czechol. Chemical Communication 43/1978, p. 4131, published by Academia, Publishing House of the Czechoslovak Academy of Sciences, Prague, Czechoslovakia.

mycin anhydronucleosides. The "frozen" 4,5′-anhydroformycin (*syn* form) and 2,5′-anhydroformycin (*anti* form) exhibit fluorescence emission two and four times as intense as that of formycin. Similarly, the deaminated "frozen" *anti* conformation also has a fluorescence intensity over the parent, deaminated formycin. There is also a bathochromic shift of the emission maximum. Although formycin has a lower *anti–syn* energy barrier, the rotation of the base is still not absolute.

Formycin Analogs

Townsend, Parks, and Giziewicz and coworkers have synthesized the N_1-, N_2-, N_4-, N_6-, N^7-, 2′-*O*-, 3′-*O*-methylformycins as possible substrates for adenosine deaminase and for antiviral activity studies with primary rabbit kidney (PRK) and L1210 cells in culture (Long et al., 1971; Milne and Townsend, 1972; Townsend et al., 1974; Lewis and Townsend, 1978; Crabtree et al., 1979; Giziewicz et al., 1975; Giziewicz and Shugar, 1977).

Biosynthesis of Formycin

Based on the ^{14}C-incorporation studies of Kunimoto et al. (1971), Ochi et al. (1974, 1975, 1976), and Ochi (personal communication), the biosynthesis of the carbon skeleton of formycin by *Streptomyces* (Fig. 1.34, page 56) appears to utilize the acetate–glutamate pathway. This suggestion is based on efficient incorporation of [^{14}C]glutamate into formycin. ^{14}C and ^{3}H labeled adenine and uridine were not incorporated into formycin, which demonstrates that purines or pyrimidines are not precursors for the pyrazolopyrimidine aglycon of formycin (Kunimoto et al., 1971). Pyrazofurin (see page 283 for structure), the naturally occurring pyrazole riboside, is not a precursor for formycin (Ochi et al., 1976). Furthermore, pyrazofurin inhibits the biosynthesis of formycin. Ochi et al. (1975) have also shown that formycin B is converted to formycin.

Biochemical Properties of Formycin

Formycin is deaminated to formycin B; formycin, but not formycin B, is enzymatically phosphorylated to the 5′-mono-, 5′-di-, and 5′-triphosphate; formycin inhibits purine *de novo* synthesis in tumor cells. At 10 μg/ml, formycin inhibits DNA synthesis; at 0.1 μg/ml, formycin inhibits protein synthesis by 10%; RNA synthesis is not affected. Cell division is inhibited 40%. FTP is incorporated into RNA *in vitro* and codes like ATP with bacterial and viral RNA polymerases; formycin polymers are very slowly hydrolyzed by spleen phosphodiesterase. This resistance to hydrolysis is due to the formycin residues existing between the *syn* and *anti* conformation. FTP is a substrate for aminoacyl-tRNA synthetase and tRNA-CCA pyrophosphorylase (Maelicke, 1974).

N-Methylformycins. Crabtree et al. (1979) have compared five *N*-methylformycins with formycin and adenosine with regard to their substrate activity with human erythrocytic adenosine deaminase, their ability to form intracellular nucleotides and their cytotoxicity to L1210 cells. 2-Methylformycin, which is predominantly the *syn* conformer, is deaminated; of the 1-methyl, 4-methyl, and N^7-methylformycins, which exist as *anti* conformers and which should be better substrates, only N^7-methyl is a substrate. Similarly, Giziewicz et al. (1975) reported that N_2-methylformycin is deaminated ten times more slowly than formycin; with calf intestinal adenosine deaminase, N_1-methylformycin is not a substrate. These data suggest that the conformation of an adenosine analog is not a major consideration in determining adenosine deaminase activity. With respect to 5′-phosphate formation, 1-methyl and 2-methylformycin, but not 4-methyl and 6-methylformycin, were converted to the nucleotide form by human erythrocytes. 2′-Deoxycoformycin increased the formation of the 5′-phosphates of 2-methyl and N^7-methylformycin. Formycin, 1-methyl, 2-methyl, and N^7-methylformycin, but not 4-methyl and 6-methylformycin, showed similar toxicity to L1210 cells. Giziewicz et al. (1975) reported that 1-methyl and 2-methylformycins are not toxic to PRK cells used for evaluation of antiviral activity; formycin and formycin B showed cytotoxicity. The differences between the findings of Crabtree et al. (1979) and Giziewicz et al. (1975) may be due to different cell lines or different assay conditions. From the observations presented above, Crabtree et al. (1979) raise three interesting questions that deserve further study: (i) can the antiviral or cytotoxic activity of 2-methylformycin be increased by concurrent treatment with dCF which increases the formation of analog nucleotides, (ii) is 1-methylformycin ineffective against vaccinia virus because of inactivity of its nucleotides with enzymes as ribonucleotide reductase or viral DNA polymerase, and (iii) how do formycin and *N*-methyl derivatives cause cytotoxicity?

Formycin 5′-Triphosphate and Anhydroformycins with RNA Polymerase and Adenosine Deaminase. Asano et al. (1971a, b) reported K_m values for formycin 5′-triphosphate (FTP) of 5.1×10^{-4} *M* and 1.4×10^{-4} *M* with calf thymus DNA and polyd(A-T), respectively; the K_m values for ATP are 7.0×10^{-5} *M* and 7.1×10^{-4} *M*. Ward and coworkers (Kumar et al., 1977) used adenosine analogs to elucidate the initiation, elongation, and termination sites of RNA polymerase from *Azotobacter vinelandii.* FTP reacts 17 times slower than ATP at the initiation site. However, in the elongation step, FTP is incorporated into RNA at the same rate as ATP. The rigid specificity at the initiation site explains the slow rate of incorporation by FTP. FTP shows an *anti–syn* conformation that compares to the *anti* conformation for ATP. The peculiar enzyme-related behavior of poly(F) is attributed to the existence of poly(F) as a right-handed helix (Ward and Reich, 1968). There is some evidence that poly(F)·poly(U) may exist in the *anti* form as a result of interconversion of the C—C glycosidic bond (Uesugi et al., 1975; Prusiner et al., 1973).

The absence of hydrogen at the 2-position of the aglycon of formycin (corresponding to C-8 of adenosine) and the longer glycosidic bond (1.50 Å for formycin versus 1.47 Å for adenosine) lower the energy barrier of rotation, which permits the *syn–anti* equilibrium (Prusiner et al., 1973). Whereas formycin has a lower *anti–syn* energy barrier, the free rotation of the base is not absolutely free (Chung and Žemlička, 1977). The purine nucleosides have a more hindered rotation because of the interaction of the C-2′ hydroxyl and the C-8 proton (Agarwal and Parks, 1975). Žemlička (1975) and Makabe et al. (1975) have reported that the sterically constrained cyclonucleoside derivative of formycin, 2,5′-anhydroformycin (*anti* conformation), is deaminated by adenosine deaminase, whereas 4,5′-anhydroformycin (*syn* conformation) is not deaminated. These anhydroformycins are inactive against mouse and L1210 ascites tumors (Makabe et al., 1975).

Inhibition of AMP Nucleosidase by Formycin 5′-Phosphate and Formycin. Formycin 5′-monophosphate is a very strong competitive inhibitor of AMP nucleosidase. The K_m/K_i ratio is greater than 2500, which led Schramm and coworkers to propose this as a transition state analog of the enzyme (Schramm et al., 1978). The tight binding is proposed to occur because of a preference for a *syn* conformation of the glycosidic torsion angle. Binding of AMP in the *syn* conformation would strain the glycosidic bond and is proposed to account for a portion of the catalytic rate enhanced for this enzyme. Formycin was found to inhibit with an inhibition constant of 4 μM compared to 0.04 μM for formycin 5′-PO_4.

Purine Nucleoside Metabolism in Humans with Severe Combined Immunodeficiency and Adenosine Deaminase Deficiency Induced with Coformycin. Agarwal et al. (1975) have reported that formycin is rapidly deaminated by erythrocytic adenosine deaminase. The K_m for formycin is 1000 μM; the K_m for adenosine is 25 μM; V_{max} for formycin is 800%; V_{max} for adenosine is 100%. The electronic factor (withdrawal of electrons) from the amino group of formycin allows a better nucleophilic displacement reaction. However, this is not true for calf intestine adenosine deaminase where formycin is a poorer substrate than adenosine.

To study the nucleotide pool in human erythrocytes and to compare the pool of humans with inherited deficiency of adenosine deaminase, Agarwal et al. (1976) used coformycin (see page 262). Coformycin did not alter the pattern of nucleotide incorporation in the adenosine deaminase-deficient cells. The level of ATP increased by 135% in the coformycin-treated cells. This compares with a 130–150% increase in ATP in the adenosine deaminase-deficient cells. This finding immediately led to a study of the phosphorylation of formycin in humans suffering from erythrocytic and lymphocytic adenosine deaminase deficiency (ADA-deficiency) (Agarwal et al., 1976). ADA deficiency is an inherited autosomal recessive trait that is associated with a severe combined immunodeficiency disease characterized by deficits in thymus-derived and bone marrow-derived cell-mediated immunity. Children with this condition usually succumb to infection early in life. However, at least two children have become immunocompetent following bone marrow transplantation (Parkman et al., 1975). ADA-deficient patients show an increased conversion of adenosine to ATP. Coformycin does not alter the nucleotide profiles in ADA-deficient cells because adenosine deaminase is already defective. An important finding of this study is that normal erythrocytes, incubated 4 h with formycin, do not show any FTP accumulation (Fig. 3.28*C*), but the FTP level in the ADA-deficient erythrocytes is three times as great as the ATP concentration in normal erythrocytes (Fig. 3.28*C*). Formycin, which in normal erythrocytes is deaminated eight times more rapidly than adenosine, but is not deaminated by the ADA-deficient patients, is converted to FTP in quantities three times as high as ATP. Therefore, incubations of erythrocytes with formycin should make it possible to detect those individuals suffering from this inherited disease.

The possibility that the concentration of adenosine accumulates in the ADA-deficient patients and either kills immunocompetent cells, prevents their response to antigenic stimulation, or is selectively toxic to lymphoid cells has been tested by abolishing adenosine deaminase activity with the inhibitors EHNA, CF, and dCF. Fox et al. (1975), Snyder et al. (1977), Wolberg et al. (1975), and Ballet et al. (1977) reported that EHNA did not affect mitogen-stimulated proliferation of normal human lymphocytes following destruction of cells sensitized *in vitro* by sensitized mouse lymphocytes. Hovi et al. (1976) observed erratic inhibition of mitogen-stimulated proliferation by CF. Ballet et al. (1977) showed that CF inhibited maturation of precursor lymphocytes, but did not inhibit the response of either T or B lymphocytes to mitogens. Carson and Seegmiller (1976)

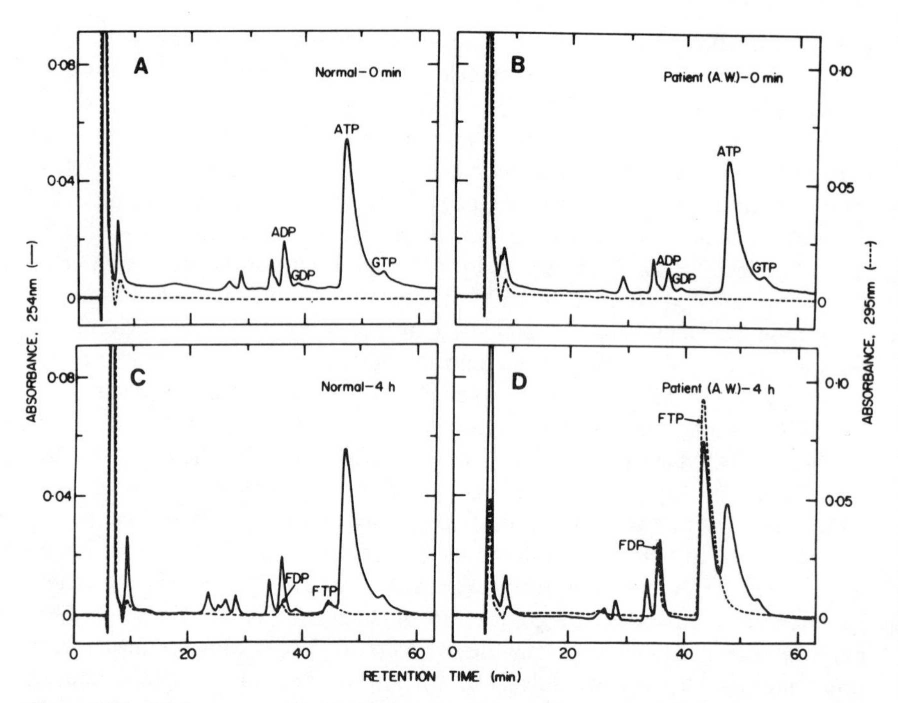

Figure 3.28 High pressure liquid chromatography profiles of erythrocytes incubated with formycin. (*A* and *C*) Profiles of erythrocytes from a normal patient that have been incubated with formycin (1 m*M*). The large adenosine deaminase levels in the normal subject prevented an increase of formycin nucleotides (FDP, FTP). (*B* and *D*) Profiles of erythrocytes from an adenosine deaminase-deficient patient. Reprinted from Agarwal et al., *The Journal of Clinical Investigation*, **57**, 1025–1035 (1976) by copyright permission of The American Society for Clinical Investigation.

reported that EHNA partially inhibited the incorporation of leucine into protein, but did not affect DNA synthesis. Most recently Burridge et al. (1977) studied the immunologic responsiveness in mice treated *in vivo* with dCF. Adenosine deaminase activity was completely abolished by dCF; however, there was no effect on the viability or responsiveness of either thymocytes or splenic cells. Evidence that the sensitivity of human splenic lymphoblasts is not due to toxicity by nucleotides was provided by using mutants from the W-I-L2 line of human lymphoblasts deficient in adenosine kinase or adenine phosphoribosyltransferase. These cells remained sensitive to growth inhibition caused by adenine and not due to toxicity by nucleotide accumulation (Hershfield et al., 1977).

Finally, it has been suggested that the functional inhibition of terminal deoxyribonucleotidyltransferase in the thymus due to high levels of ATP in patients suffering from adenosine deaminase deficiency may explain the observed immunodeficiency in this clinically known condition (Modak, 1978).

Formycin and Purine Nucleoside Phosphorylase in Human Erythrocytes. The activity of purine nucleoside phosphorylase in human erythrocytes is 1000 times as great as that of adenosine deaminase (Agarwal et al., 1975). Formycin B is not a substrate for purine nucleoside phosphorylase of normal human erythrocytes (Parks and Brown, 1973). The hydrolysis of inosine is competitively inhibited by formycin B. Because nucleoside phosphorylase can prevent purine nucleoside analogs from entering the nucleotide pool and thereby limit their effectiveness, laurusin and/or formycin B is an inhibitor of this enzyme (K_i 10^{-4} M) for studies on the effectiveness of adenosine or inosine analogs that would normally be hydrolyzed.

Effect of Formycin on the Processing of tRNA Precursors. The biosynthesis of tRNA in prokaryotes requires several nucleolytic cleavages of larger precursors for maturation to functional tRNA (Altman, 1975; Smith, 1976). In animal cells the initial transport of tRNA genes appears in the nucleus. Following specific cleavages, the tRNA is transported into the cytoplasm (Lon, 1977).

Because formycin is an analog of adenosine, it has been used to elucidate the ribosome–tRNA interaction by replacing the terminal adenosine nucleotide (Maelicke et al., 1974). The stacking of the terminal formycin to the penultimate cytidine in $tRNA^{Phe}$ —C—C—F is weaker than the stacking of the terminal adenosine in native $tRNA^{Phe}$ —C—C—A. With the protein synthesizing system from lysed rabbit reticulocytes, $tRNA^{Phe}$ —C—C—F and $tRNA^{N\text{-}Ac\text{-}Phe}$—C—C—F have donor activity for peptide bond formation; however, it is much slower than normal tRNA. The result is an obligatory lag that is not normally observed. In addition, the interaction of the N-AcPhe-$tRNA^{Phe}$—C—C—F with the ribosomal P-site is weaker than the normal —C—C—A end in tRNA. Thus the charged tRNA-C—C—F is a poorer substrate for the P-site, which makes the earlier stages of polypeptide synthesis slower. As the polypeptide chain increases, there is a more proper alignment of the peptidyl-tRNA in the P-site and the reaction proceeds at its normal speed (Baksht et al., 1975).

Because formycin can be incorporated into the —C—C—A end of tRNA, it has been used to study the regulation of synthesis of active tRNA molecules. The model system used for this study is the silk gland of the silkworm (Majima et al., 1977). During the development of the silk gland, tRNA pools for certain amino acids undergo massive production of fibroin and sericin. Formycin, an analog of adenosine, preferentially inhibits the synthesis of low molecular weight RNA, particularly tRNA precursors in the cytoplasm. These data imply that formycin-containing 4.5 S precursor RNA is responsible for the failure to process 4.5 S RNA into normal 4 S tRNA. Tsutsumi et al. (1978) have confirmed this suggestion by demonstrating that 4.5 S precursor RNA containing formycin in the silk gland is less susceptible to hydrolysis by a specific endonuclease that is required for the processing of tRNA precursors.

Adenosine analogs have also been used to define the initial sites of aminoacylation of the 3′-adenosine of tRNA (Sprinzl and Cramer, 1973). By using tRNA terminating in 2′- or 3′-deoxyadenosine, Alford and Hecht (1978) showed that *E. coli* alanyl- and lysyl-tRNA synthetases normally aminoacylate the 3′-hydroxyl group of adenosine, while yeast phenylalanyl-tRNA synthetase aminoacylates the 2′-hydroxyl group (for more details on transacylation and chemical proofreading involving changes from the adenosine on the 3′-end of tRNA to 3′-deoxyadenosine and 3′-aminoadenosine, see pages 131 and 146).

Von der Haar and Cramer (1976) have shown that tRNA, in which the adenosine at the 3′-end is replaced with formycin, accepts the amino acid from the aminoacyl-AA-tRNA on the 3′-hydroxyl and occupies a position that is not accessible, so the synthetase cannot make corrections for misactivated amino acids. Von der Haar and Cramer (1978a, b) have studied the triggering of five aminoacyl-tRNA synthetases by the invariant 3′-terminal adenosine. By substituting analogs of adenosine at the 3′-end of tRNA (i.e., formycin, formycin oxi-red, 2′-deoxyadenosine, 3′-deoxyadenosine, and 3′-aminoadenosine), they have demonstrated that (i) with phenylalanyl-tRNA synthetase the total ribose moiety is involved (Von der Haar and Gaertner, 1975), (ii) with seryl-, threonyl-, and valyl-tRNA synthetases the reacting-site triggering must be intimately related to the amino acid-accepting hydroxyl; (iii) the triggering in the isoleucyl system is related to a non-accepting 3′-hydroxyl group. In all systems studied, the 3′-terminal adenosine of tRNA induces a conformational change in the synthetase. The use of modified adenosine analogs in the —C—C—A end of tRNA produced evidence that isoleucyl-tRNA synthetase has two binding sites for tRNA per single chain of molecular weight 115,000.

More recently, Bhuta and Žemlička (unpublished results) showed that 2′(3′)-*O*-L-phenylalanyl derivatives of *anti* and *syn* anhydroformycins (**I** and **II**) (Fig. 3.29) can be used as tools for the investigation of conformational specificity of ribosomal peptidyltransferase from *E. coli* (Fig. 3.30). Thus *anti*-compound **I** is an excellent substrate for peptidyltransferase, whereas *syn*-compound **II** only functions as a substrate at higher concentrations. However, it is of interest to note that at higher concentrations the *anti*-product/*syn*-product ratio is about 4, which indicates a rather low con-

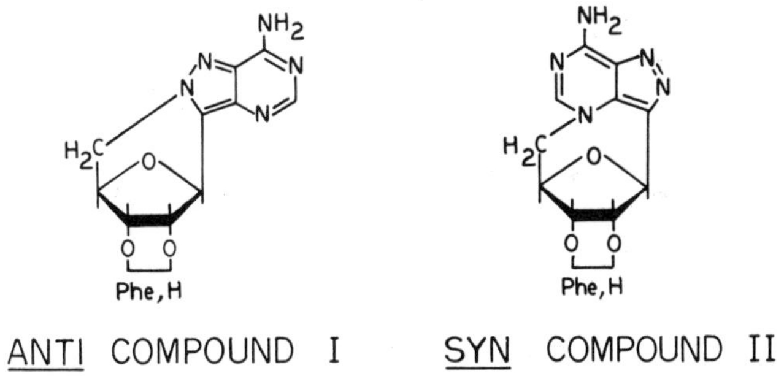

ANTI COMPOUND I SYN COMPOUND II

Figure 3.29 Structures of 2′(3′)-*O*-L-phenylalanyl-2,5′-anhydroformycin (*anti*-compound **I**) and 2′(3′)-*O*-L-phenylalanyl-4,5′-anhydroformycin (*syn*-compound **II**) (Bhuta and Zemlicka, unpublished results).

formational specificity. The result with *syn*-compound **II** is in accord with another *syn* model, 8-bromo-2′-(3′)-*O*-D-phenylalanyladenosine (Žemlička et al., 1975).

Influence of Formycin in NAD⁺ with Dehydrogenases. To elucidate the role of the adenosine moiety (i.e., the "nonworking" end) of NAD^+ in the coenzyme domain of the dehydrogenases, Suhadolnik et al. (1977a) have reported on activity of NAD^+ analogs in which analogs of adenine or ade-

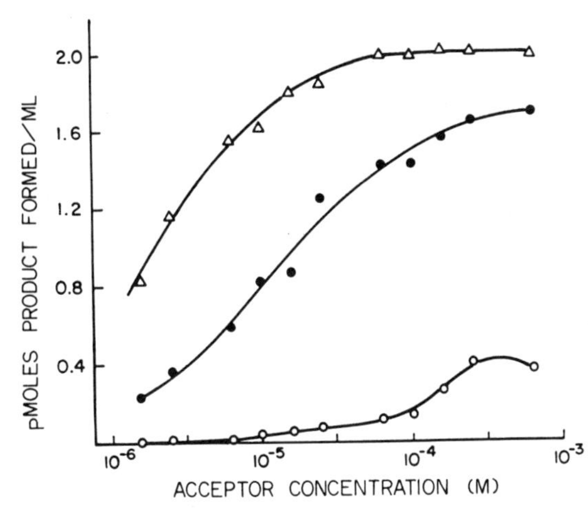

Figure 3.30 The amount of peptide formed from *anti*-compound **I** and *syn*-compound **II** was determined by ethyl acetate extraction. *N*-Ac-Phe-tRNA was used as a peptide donor. *anti*-Compound I (●); *syn*-Compound II (○); puromycin (reference compound) (△). (Bhuta and Zemlicka, unpublished results.)

nosine have replaced the adenine or the adenine-ribose of NAD^+. The kinetic and protein fluorescence quenching data show that replacement of the adenosine moiety of NAD^+ with formycin (NFD^+) does not alter the K_m or V_{max} for either horse liver alcohol, yeast alcohol, or glyceraldehyde-3-phosphate dehydrogenase; however, with lactic acid dehydrogenase there is a 10- and 27-fold decrease in the K_m and V_{max} of the NFD^+. Because the dehydrogenases follow compulsory ordered kinetics, the decrease in the V_{max} decreases the "k_{1on}," that is, the addition of NAD^+ to the enzyme, and "k_{9off}," that is the dissociation of the enzyme·NADH complex. Apparently, the hydrophobic coenzyme domain that accomodates the adenine portion of NAD^+ of lactic acid dehydrogenase is very specific and cannot form a productive complex with NFD^+ as effectively as does NAD^+.

Cytokinin Activity of Adenosine Analogs. N^6-substituted adenine and adenosine analogs are cytokinins that promote cell division and growth (Skoog and Armstrong, 1970; Skoog et al., 1965). To further explore the involvement of exogenously added cytokinins in cAMP metabolism, Hecht and coworkers studied the interaction of certain cytokinins and related compounds with cAMP phosphodiesterase activity (Gallo et al., 1972). In their report, Hecht et al. (1974b) described an excellent synthesis of formycin cyclic 3′:5′-monophosphate. Twenty-three compounds were tested for the inhibition of cAMP conversion to 5′-AMP. Only formycin 3′:5′-monophosphate failed to inhibit this enzyme.

Effect of Formycin on Blood Platelet Aggregation. Formycin has been used in studies on blood platelet aggregation. Blood platelets from humans contain higher concentrations of adenine nucleotides than do erythrocytes or leukocytes (Parks et al., 1975; Scholar et al., 1973). Platelet aggregation can be initiated by low concentrations of ADP plus collagen, epinephrine, or thrombin. Adenosine inhibits ADP-induced aggregation. The effect of 2-fluoroadenosine, cordycepin, ara-A, and formycin with coformycin on the platelet nucleotide pool, aggregation, and release of nucleotides from storage granules in response to thrombin treatment has been studied by Agarwal and coworkers (Agarwal et al., 1975). Although human platelets convert 2-fluoroadenosine to its 5′-triphosphate, very little 2-fluoroadenosine-containing nucleotides are released during aggregation. This is in contrast to the release of ADP and ATP. After thrombin treatment, most of the 2-fluoroadenosine 5′-triphosphate is found in the platelet pellet as the 5′-mono- and 5′-diphosphates. Apparently, 2-fluoroadenosine 5′-triphosphate can replace ATP as the energy source for the aggregation and release phenomenon. Formycin and cordycepin are rapidly deaminated in platelets by adenosine deaminase. When the adenosine deaminase inhibitor, coformycin, is added, the nucleotides of formycin and cordycepin are formed, but they weakly inhibit the ADP-induced aggregation.

Effect on Bovine Adrenal Estrogen Sulfotransferase. Sulfuric acid esterification (sulfurylation) proceeds by way of the appropriate sulfotransferase as do the hydroxylation and methylation reactions. The variegated esterification uses 3′-phosphoadenosine 5′-phosphosulfate as the active sulfate. In an attempt to determine the substrate specificity for estrogen sulfotransferase, Horwitz et al. (1978) reported that the 6-amino group of the adenosine is essential for optimal enzyme activity. The seven-carbon isosteric analog (tubercidin) has a K_i one order of magnitude lower than adenosine-3′,5′-diphosphate. By contrast, the formycin analog, formycin 3′,5′-diphosphate, shows a six-fold decline in the affinity for estrogen sulfotransferase.

Biochemical Properties of Formycin B

Formycin B, an analog of inosine, has been used as an inhibitor of tumors and viruses. It has also been used to study nucleotide metabolism in human erythrocytes, nuclear ADP ribosylation, and fruiting body transformation. Müller et al. (1975) reported that 16 μM formycin B causes a 50% inhibition of the growth of L5178Y mouse tumor cells. Incubation of formycin B with Ehrlich ascites tumor cells from mice yielded no formycin B nucleotides, whereas formycin was phosphorylated. Therefore, the inability of cellular kinases to phosphorylate formycin B has led investigators to study the cellular processes that are affected by this inosine analog at the level of the nucleoside.

Effect on ADP Ribosylation of Chromatin Proteins, Proinsulin Synthesis, and Development. The ADP-ribose moiety of NAD^+ is used for a number of reactions involving mono- and poly(ADP ribosylation) in the prokaryotes and eukaryotes (Hayaishi and Ueda, 1977). Müller et al. (1975) demonstrated that formycin B inhibited the growth of L5178Y mouse tumor cells. Furthermore, they showed that formycin B is a competitive inhibitor of NAD^+ in reactions catalyzed by purified chromatin-bound and soluble poly(ADP-ribose) polymerase isolated from quail oviduct. Müller and Zahn (1975) also reported that formycin B and showdomycin inhibit poly(ADP-ribose) polymerase, while cordycepin, tubercidin, and ara-A do not. With isolated nuclei from rat liver, Suhadolnik et al. (1977b) showed that NFD^+ was equally affective in replacing NAD^+ as an inhibitor of DNA synthesis; 2′-deoxyNAD^+ and 3′-deoxyNAD^+ were 10 times as potent as inhibitors of DNA synthesis than NAD^+ (Fig. 3.6) Jain and Logothetopoulos (1977) have used formycin B as a biochemical probe to study the mechanism of proinsulin synthesis.

Because formycin B is a very potent inhibitor of purine nucleoside phosphorylase in human erythrocytes, the two nucleosides, inosine and guanosine, cannot be hydrolyzed to yield D-ribose, so proinsulin biosynthesis is inhibited. This suggests that intracellular nucleoside metabolic products contribute to the primary signal that controls proinsulin biosynthesis.

Sussman et al. (1975) have used formycin B as a probe to study the effect of interference with guanosine metabolism in *Dictyostelium discoideum*. Formycin B inhibits normal cellular deposition of guanosine in *D. discoideum* by inhibiting purine nucleoside phosphorylase. The result is an inhibition in the construction of fruiting bodies (Cohen and Sussman, 1975).

Interferon Induction. The most effective interferon inducers are the synthetic polynucleotides. To further elicit an interferon response, Witkop and coworkers (Torrence et al., 1975) studied the effect of polynucleotides from purine analogs. The homopolymer of poly(formycin B), synthesized from polynucleotide phosphorylase and FDP (Ikehara and Tezuka, 1974), forms a double-stranded complex in a 1:1 ratio with the octanucleotide of 8,2′-anhydro-8-mercapto-9-β-D-arabinofuranosyladenine; poly(F) does not.

Synthesis of a C-Analog of AICAR from Formycin B. Townsend and coworkers described an excellent method for the synthesis of the 4-amino derivative of pyrazofurin, 4-amino-3-(β-D-ribofuranosyl)pyrazole-5-carboxamide, from formycin B and from the N^6-oxide of formycin. This 4-amino nucleoside can be viewed as a C-analog of 5-amino-carbamideimidazole riboside (AICAR) and may be a key intermediate in the biosynthesis or degradation of formycin (Lewis et al., 1976).

Oxoformycin B

Mouse and rabbit liver aldehyde oxidase can oxidize formycin B to oxoformycin B (Sheen et al., 1970; Johns, 1974). The K_m for formycin B is 0.2 mM, which is in close agreement with its K_i. Parks and coworkers (Sheen et al., 1970) reasoned that formycin B is oxidized at the same catalytic site of aldehyde oxidase as is N^1-methylnicotinamide. The aglycon of formycin B, 7-hydroxypyrazolo-[4,3-*d*]pyrimidine, is also a substrate (K_m = 4 mM). An explanation of these findings has been proposed by Johns (1974). He postulates the necessity of an unsubstituted N-9 atom of a purine for binding to the enzyme prior to the oxidation step; the rapid oxidation of C-8 of azathiopurine, but not that of 9-butylazathiopurine, illustrates this point. Formycin and formycin B, but not inosine, are competitive inhibitors, but not substrates, of xanthine for milk xanthine oxidase (Sheen et al., 1970).

Oxoformycin B does not inhibit the growth of *X. oryzae*, Yoshida sarcoma, or influenza virus.

The synthesis of the 3-methyl derivative of oxoformycin B has been described by Long et al. (1970). The glycosyl torsion angle of crystalline oxoformycin B is $x \simeq 164.1°$, which places this nucleoside in the *syn* form (Koyama et al., 1976). An intramolecular hydrogen bond between N-4 and O-5′ in oxoformycin B stabilizes the *syn* conformation. Because of the widespread occurrence of 3′-terminal poly(A) tracts in many mRNAs, Davies (1976) studied the interaction of poly(A) with potentially complementary monomer molecules. Whereas xanthosine forms a 1:2 complex with poly(A), surprisingly, oxoformycin B complexes only in a 1:1 ratio. This difference in complex formation may be a result of steric hindrances due to the 5-oxo group on oxoformycin B (Davies, 1973).

3.10 THURINGIENSIN

The naturally occurring adenine nucleotide analog (thuringiensin*, Fig. 3.31), is found in the culture medium of certain strains of *Bacillus thuringiensis*. It is highly toxic to insects and mammals (Šebesta et al., 1968). Thuringiensin has been isolated and purified (Šebesta et al., 1969a; Bond et al., 1969a; Benz, 1961; de Barjac and Dedonder, 1968). It contains adenine, ribose, glucose, allaric acid, and phosphoric acid (Bond et al., 1969a; Farkaš et al., 1969; deBarjac and Dedonder, 1968). This unusual nucleotide

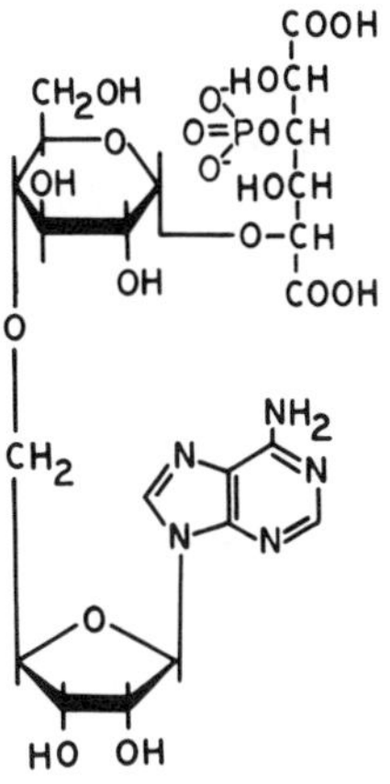

COOH, HOCH, O=POCH, HOCH, O-CH, COOH, CH_2OH, OH, OH, O, CH_2, NH_2, N, N, N, N, O, HO OH

Figure 3.31 Structure of thuringiensin.

* Thuringiensin is the trivial name proposed by several laboratories (Kim and Huang, 1970; Pais and de Barjac, 1975; Farkas et al., 1977). Other names found in the literature for this nucleotide are β-exotoxin, heat-stable exotoxin, thuringiensin B, and thuringiensin A.

is an adenosylglucose with a unique ether bond between D-ribose and D-glucose; the ribosylglucose has been characterized as methyl-4-*O*-(5-deoxy-D-ribofuranosid-5-yl)-D-glucopyranoside; there is an α-glucosidic bond between the anomeric carbon of glucose and C-2 of allaric acid (Prystaš and Šorm, 1971a). A phosphate is attached to C-4 of allaric acid. Prystaš et al. (1976) have reported two excellent chemical syntheses of thuringiensin.

Thuringiensin is toxic to insects, animals, and pathogenic nematodes (Bond et al., 1969a; Šebesta et al., 1969b). The antagonism is manifest mostly in larvae during molting and metamorphosis. Dephosphorylated thuringiensin is not toxic to insects or mice; therefore, it is not rephosphorylated *in vivo*. Thuringiensin inhibits prokaryotic and eukaryotic DNA-dependent RNA polymerases (Šebesta and Horská, 1968, 1970; Beebee and Bond, 1973a). It competes exclusively with ATP (Šebesta and Horská, 1968). Mitotic spindle formation, adenylate cyclase, bacterial membranes, maturation of nuclear RNA, and segregation of nucleolar components are also affected by thuringiensin (Sharma et al., 1976; Grahame-Smith et al., 1975; Johnson, 1976; Smetana et al., 1974). Thuringiensin, like tubercidin, is an effective chemosterilant when ingested orally by the house fly (Kohls et al., 1966). It was reviewed by Bond et al. (1971) as well as by Šebesta et al. (1979). Finally, much of our knowledge on the production, physical, and chemical properties, chemical intermediates, chemical syntheses, and biochemical properties of thuringiensin is due to the in-depth studies by Šebesta and coworkers at the Czechoslovak Academy of Science.

Discovery, Production, and Isolation

B. thuringiensis produces a protein endotoxin that is toxic to insects (Angus, 1954; Hannay, 1953; Heimpel, 1955). McConnell and Richards (1959) reported that the autoclaved supernatant from a culture medium of *B. thuringiensis* was toxic when injected into the larvae of Lepidoptera. This effect is due to a heat-stable exotoxin (thuringiensin). The strains of *B. thuringiensis* that produce this nucleotide have been reviewed in detail (Šebesta et al., 1979).

B. thuringiensis is grown on either Connor's or Cantwell's medium (Connor and Hansen, 1967; Cantwell et al., 1964). Thuringiensin is detected in the culture medium during maximal growth and is complete at sporulation (Šebesta et al., 1973). The growth and production of thuringiensin are shown in Fig. 3.32. The assays of thuringiensin are based on the use of *Galleria mellonella*, *Sarcina flava*, larvae of the house fly, isotope dilution assay (^{32}P), or inhibition of DNA-dependent RNA polymerase (Šebesta et al., 1969a; Rosenberg et al., 1971; Šebesta et al., 1973; de Barjac and Lecadet, 1976). The yields vary from 50 mg to 300 mg/liter (Bond et al., 1969a; Šebesta et al., 1969a, 1973; de Barjac and Lecadet, 1976).

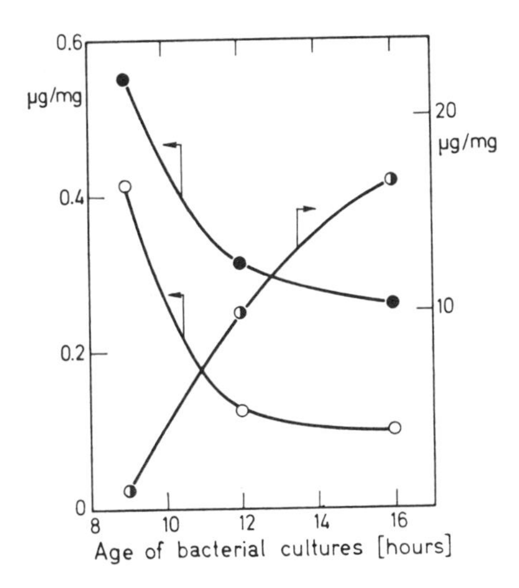

Figure 3.32 Concentration of thuringiensin and ATP in bacterial cells and of thuringiensin in culture medium in micrograms per milligram of dry weight of *B. thuringiensis*. ATP (●); thuringiensin in bacterial cells (○); thuringiensin in culture medium (◑). Reprinted with permission from Horská et al., 1975.

Thuringiensin has been isolated by de Barjac, Šebesta, Bond, Kim, Huang, and their coworkers (de Barjac and Dedonder, 1968; Bond et al., 1969a; Šebesta et al., 1969a; Kim and Huang, 1970). Thuringiensin lactone can be converted to thuringiensin by ion exchange (Dowex 1-OH^-) or by ammonium hydroxide (Šebesta et al., 1969a).

Physical and Chemical Properties

The molecular formula for hydrated thuringiensin is $C_{22}H_{32}N_5O_{19}P \cdot 3\ H_2O$; mol wt 809; $\lambda_{max}^{pH\ 7.0}$ 259 nm ($E_{1\,cm}^{1\,percent}$ 168); $\lambda_{max}^{0.01\,M\,HCl}$ 260 nm (ϵ 13,700); $\lambda_{max}^{0.01\,N\,NaOH}$ 259 nm ($E_{1\,cm}^{1\,percent}$ 170); $[\alpha]_D^{25}$ +30.9° (*c* 0.5, water); ORD spectrum (0.01 *N* HCl): peak $[\Phi]_{236}$ + 194.0°, trough $[\Phi]_{263}$ − 560° (Bond et al., 1969a; deBarjac and Lecadet, 1976; Farkas et al., 1969). Thuringiensin can be dephosphorylated with *E. coli* alkaline phosphatase, calf intestinal alkaline phosphatase, or wheat germ acid phosphatase (Bond et al., 1969a; Šebesta et al., 1969a). The ir spectrum of the dephosphorylated thuringiensin showed an absorption at 1775 cm^{-1} which is characteristic of a five-membered lactone.

Structural Elucidation

On the basis of a stepwise degradation and isolation of adenine, a mixture of anomeric glycosides and allaric acid, Farkas et al. (1969) assigned the structure as shown for thuringiensin (Fig. 3.31). Bond and coworkers (Bond, 1969; Bond et al., 1969a), using nmr, ^{31}P-nmr spectral data, and optical rotatory dispersion, concluded that the anomeric linkage of ribose to adenine was β. Convincing evidence of the α-configuration of the anomeric carbon of the glucosidic bond to allaric acid was established by Prystaš and Šorm (1971a). The 2-*R*-configuration of allaric acid was determined (Kalvoda et al., 1973; Prystaš et al., 1975a) and the position of the phosphate bond was unambiguously determined by ^{13}C-nmr (Pais and de Barjac, 1975). The most complete degradation of thuringiensin and the isolation of structural components is described in a recent report of Farkas et al. (1977). The unequivocal synthesis of the sugar fragment (ribose and glucose) with the β-linkage as described by Prystaš and Šorm (1971a) confirms *inter alia* the structure proposed. Finally, Prystaš and Šorm (1971b) reported on the synthesis of the adenosine-glucosyl nucleoside moiety of thuringiensin as the methyl glycoside.

Chemical Synthesis of Thuringiensin

Two independent total chemical syntheses of thuringiensin were done by Kalvoda et al. (1976a, b) and Prystaš et al., (1975a, b). The methods for these syntheses lend themselves to the general syntheses of thuringiensin analogs. One consists of three major steps: (i) formation of the trisaccharide moiety, (ii) nucleosidation, and (iii) phosphorylation (Scheme 3.3). The ethereal bond of the ribofuranosylglucopyranose disaccharide is formed by transdiaxial epoxide ring opening of compound **8** followed by condensation with **7** to form **10**. The α-glucosidic bond with allaric acid was formed by a stereoselective reaction of the diacetate of **11** and condensation of the product with allaric acid lactone ester (**12**) to give **13**, which was converted to the nucleoside, **17**. The lactone ring of the allaric acid of the nucleoside was opened by alkaline methanolysis to afford **18**. Compound **18** was phosphorylated at C-4 and treated with alkali to hydrolyze the phosphodichloridate (**19**) to obtain thuringiensin. The key intermediate, allaric acid lactone ester (**12**), was prepared from D-glucofuranuronic acid lactone.

Biochemical studies prompted the preparation of several thuringiensin analogs. Those altered in the base moiety of the molecule were prepared mainly by reacting native thuringiensin (Šebesta and Horská, 1970). Analogs containing acids other than allaric acid were synthesized by the method described above (Horská et al., 1976).

Inhibition of Growth

Thuringiensin is a potent inhibitor of mammalian cells in culture, insects, mammals, and bacteria. The insecticidal effects are restricted to species of orders *Diptera*, *Lepidoptera*, *Coleoptera*, *Hymenoptera*, *Isoptera*, and *Orthoptera* (Bond et al., 1971; Krieg, 1968; Šebesta et al., 1979). The larvae are most sensitive at physiologically critical stages of development (i.e.,

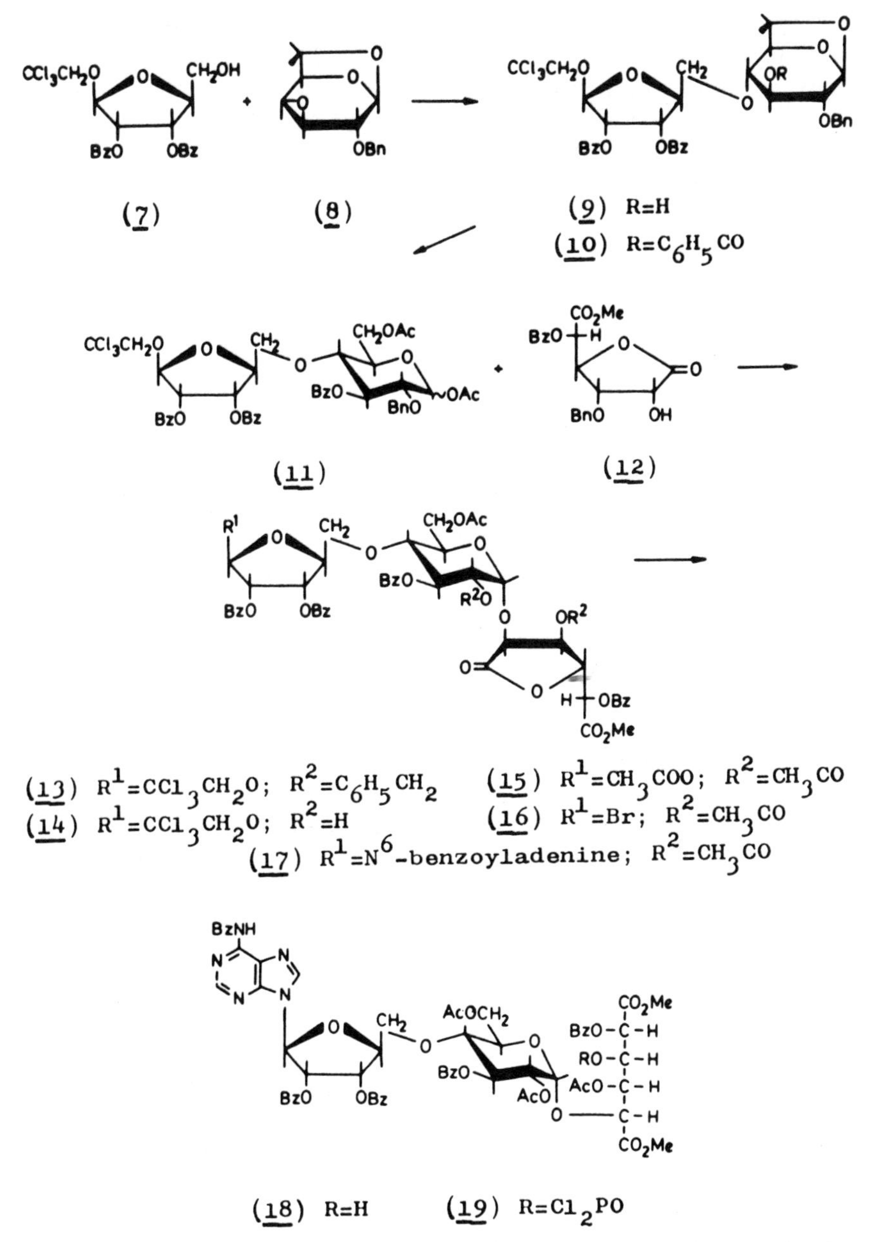

Scheme 3.3 Total chemical synthesis of thuringiensin. Reprinted with permission from Prystaš et al., 1975b.

molting, pupation, and metamorphosis). Purified thuringiensin does not cause anomalies of pupae or imagoes (Šebesta et al., 1969a). Fleas that survive exposure to thuringiensin are sterile (Prokop'ev et al., 1976). Application of thuringiensin to rice effectively controls sheath blight and bacterial leaf blight (Ito, 1973). With *Lepidoptera*, the teratologic symptoms are observed on the head appendages and buccal organs; the proboscis of pupae and adults is atrophied and mouth parts, thorax, and legs are also affected (Burgerjon and Biache, 1967; Kim et al., 1972; Wolfenbarger et al., 1972).

Toxicity

Originally it was reported that thuringiensin was only toxic to insects; however, subsequent studies showed that it is also toxic to mammals (de Barjac and Riou, 1969; Bond et al., 1969b; Šebesta et al., 1968, 1969b). Mice die 3 days after administration (i.p.); the LD_{50} is 18 $\mu g/g$ (Šebesta et al., 1969b). The LD_{50} for caterpillars (*Galleria mellonella*) is 0.5 $\mu g/g$ (Šebesta et al., 1969b). The toxicity of thuringiensin to larvae is 100 times as great as that of DDT (Shieh et al., 1968). The decrease of toxicity of thuringiensin administered per orally to *G. mellonella* caterpillars is due to the rapid release of ^{32}P from [^{32}P]thuringiensin (Vaňková et al., 1975). The hydrolysis is due to an active alkaline phosphatase in the gut (Vaňková and Horská 1975).

Biosynthesis of Thuringiensin

To determine if the biosynthesis of thuringiensin is dependent on adenine, Myslovataya (1974) studied the effect of 8-azaguanine on the growth of *B. thuringiensis*. Azaguanine inhibited the growth of *B. thuringiensis;* adenine or guanine reversed the inhibition. Mutants resistant to 8-azaguanine had lower production of thuringiensin. Therefore, the biosynthesis of thuringiensin may be related to purine biosynthesis.

Biochemical Properties

Thuringiensin and Its Analogs: Effect on RNA, DNA, and Protein Synthesis. In 1961 Benz proposed that thuringiensin might be an antimetabolite of nucleic acids. Šebesta et al. (1968, 1969b) first demonstrated that the administration (i.p.) of thuringiensin to mice inhibited RNA synthesis by 60%; polynucleotide phosphorylase and protein and DNA synthesis were not inhibited. It can cross the mammalian cell wall intact. Given orally, thuringiensin is dephosphorylated to the inactive nucleoside (Beebee et al., 1972; Šebesta and Horská, 1973).

Šebesta and Horská (1968) reported that thuringiensin is a competitive inhibitor of ATP with *E. coli* DNA-dependent RNA polymerase; the adenine moiety of thuringiensin provides for the base pairing with the complementary base of the template, and the allaric acid moiety mimics the triphosphate grouping of ATP (Šebesta and Sternbach, 1970; Horská et al., 1976). Deamination of thuringiensin produces the inosine analog, which is a competitive inhibitor of GTP, but not ATP, for RNA synthesis (Šebesta and Horská, 1970); therefore, the inosine analog enters the GTP binding site. The competition of the inosine analog agrees with the requirements for base pairing, where either guanosine or inosine can hydrogen bond with cytosine. The N^1-oxide is less inhibitory. The inhibition of RNA polymerase markedly decreases by changing either the hydroxyl at C-3 of allaric acid or the carboxyl group at C-5 of allaric acid or shortening the allaric acid by one carbon atom. These results show that there is a rigid structural requirement for thuringiensin and its analogs in order for them to compete with ATP for the ATP binding site of RNA polymerase.

Effect of DNA-Dependent RNA Polymerase in Nuclei Isolated from Rat Liver Pretreated with Thuringiensin. Assays of rat and mouse liver RNA polymerases in animals pretreated with thuringiensin (i.p.) gave more reproducible results than those studies with untreated animals (Čihák et al., 1975; Smuckler and Hadjiolov, 1972).

Essentially they agree that both polymerases are inhibited by thuringiensin to the same extent. On the other hand, results obtained with isolated polymerases from pretreated animals are controversial. While Beebee et al. (1972) found a more pronounced inhibition of the nucleolar polymerase in comparison with the nucleoplasmic enzyme, Smuckler and Hadjiolov (1972) reported a preferential inhibition of the nucleoplasmic (α-amanitin sensitive) RNA polymerase. Experiments with highly purified DNA-dependent RNA polymerases I and II (Horská, unpublished results) indicate that it is the nucleolar enzyme (I) that is more sensitive to thuringiensin inhibition. This result, which confirms the finding of Beebee et al. (1972), also agrees with the results obtained with nuclei from untreated animals (see page 190, Fig. 3.33) as well as results from studies on the inhibition of biosynthesis of the single RNA species by thuringiensin (Mackedonski et al., 1972a, 1972b; Mackedonski, 1975).

Inhibition of DNA-Dependent RNA Polymerases in Nuclei Isolated from Rat Liver of Untreated Animals. Although the experiments on the effect of thuringiensin on nuclei isolated from rat liver from untreated animals show variations, there is one feature that is common to all the results. Nucleolar RNA polymerase I (synthesis of rRNA, α-amanitin resistant) is inhibited more strongly by thuringiensin than is nucleoplasmic RNA polymerase II (synthesis of mRNA, α-amanitin sensitive) (Fig. 3.33) (Beebee et al., 1972;

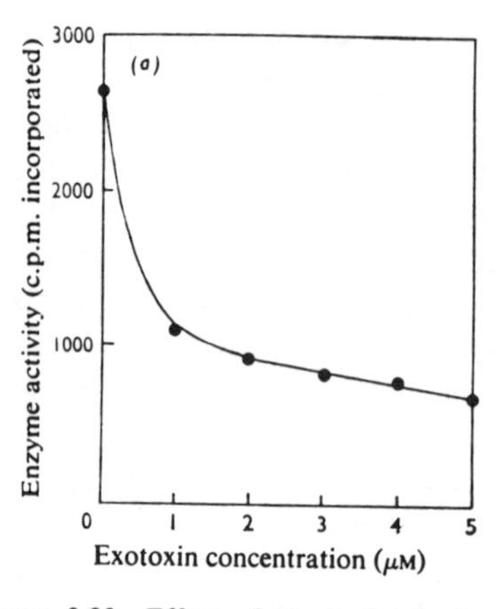

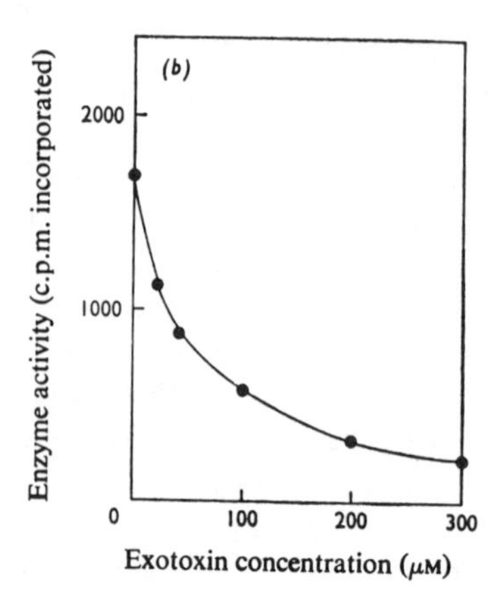

Figure 3.33 Effect of thuringiensin (exotoxin) concentration on the (*a*) nucleolar RNA polymerase I (rRNA synthesis) and (*b*) nucleoplasmic RNA polymerase II (mRNA synthesis) activities in untreated intact nuclei. Fifty percent inhibition of nucleolar RNA polymerase at 0.5 μM; for nucleoplasmic RNA polymerase at 75 μM. Reprinted from Beebee et al., 1972.

Čihák et al., 1975; Beebee and Bond, 1973a, b; Smuckler and Hadjiolov, 1972).

Effect on Maturation of Mouse Nuclear RNA Synthesis. Mackedonski et al. (1972a, b) and Mackedonski (1975) reported that administration (i.p.) of thuringiensin to mice preferentially inhibits the synthesis of nuclear rRNA. There is a preferential inhibition of synthesis of 28, 18 and 5 S rRNA, including the 45 S precursor rRNA. The synthesis of the nuclear "DNA-like" RNA is less affected by the labeling of HnRNA; 4 S RNA was only affected about 50%. Although the *in vivo* administration of thuringiensin inhibits RNA polymerase activities, the *in vivo* action of this nucleotide cannot be explained by the inhibition of RNA polymerases (Mackedonski and Hadjiolov, 1974). Mackedonski (1975) subsequently demonstrated that thuringiensin (25 μg/mouse) does not inhibit the synthesis of 45 S pre-rRNA, but causes a breakdown of these molecules (Fig. 3.34). The conversion of 38 S pre-rRNA into 32 and 21 S pre-rRNA is also inhibited. A summary of the findings described in Figs 3.33 and 3.34 is given below.

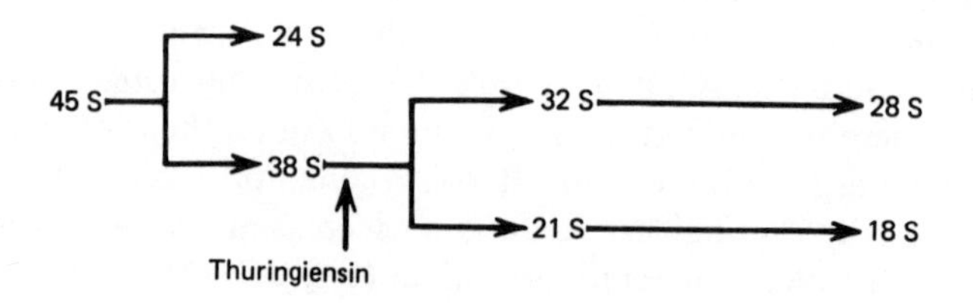

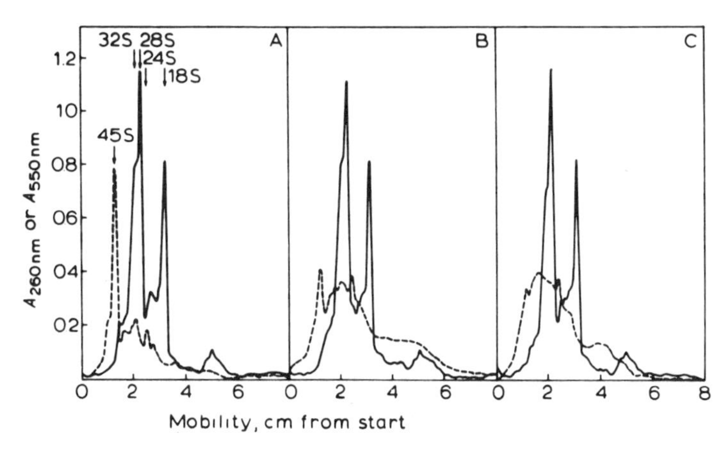

Figure 3.34 Agar gel electrophoresis of mouse liver nuclear rRNA: (*A*) control and (*B* and *C*) thuringiensin-treated animals. Mice were pretreated with 0.025 mg of thuringiensin per mouse for (*B*) 30 and (*C*) 120 min; 25 μCi per mouse of [^{14}C]orotic acid was then injected i.p. (15 min labeling time). Curve represents absorbance recorded from the blackening of radioautograms at 550 nm. Reprinted with permission from Mackedonski, 1975.

Because the formation of 24 S RNA is not affected by thuringiensin, it appears that thuringiensin interferes with the conversion of 38 S into 32 and 21 S pre-RNA. Thuringiensin inhibits the labeling of nuclear 5 S RNA but 4.6 S pre-tRNA is not affected. These studies suggest that thuringiensin may affect RNA synthesis in several ways.

Inhibition of RNA Synthesis by Thuringiensin with* E. coli *DNA-Dependent RNA Polymerase. Šebestá, Horská, and coworkers (Šebesta and Horská, 1970; Šebesta and Sternbach, 1970; Horská et al., 1976) studied the effect of thuringiensin with *E. coli* DNA-dependent RNA polymerase. Thuringiensin is a competitive inhibitor of ATP; GTP and CTP do not reverse the inhibition. Rat liver nucleolar RNA polymerases are 2 orders of magnitude more sensitive to thuringiensin than is the *E. coli* RNA polymerase (Beebee et al., 1972). The inhibition by thuringiensin is competitive with ATP, as in the case of the *E. coli* RNA polymerase. Thuringiensin affects the elongation step in RNA synthesis.

***Effect of Thuringiensin on the DNA-Dependent RNA Polymerase Isolated from* B. thuringiensis.** Johnson et al. (1975) and Klier and Lecadet (1974) reported a unique difference in the sensitivity of the RNA polymerase isolated from *B. thuringiensis*. This sensitivity is dependent on the age of the culture. RNA polymerase isolated from culture of *B. thuringiensis* in the stationary phase of growth was two to five times less sensitive than exponential RNA polymerase activity (Johnson, 1978). This may be attributed to the fact that the sigma factor activity is altered or missing in aged cultures (Brevet, 1974; Klier and Lecadet, 1974; Klier et al., 1973; Linn et al., 1973). This change in sensitivity to thuringiensin does not occur with *E. coli* RNA polymerase at different ages of cell growth. However, when the sigma factor is removed from *E. coli*, the sensitivity to thuringiensin is diminished. These findings strongly suggest that sigma factor activity may affect RNA polymerase sensitivity to thuringiensin. The data of Johnson (1978) are in contrast to the studies of Klier et al. (1973). Klier found no change in sensitivity of thuringiensin between vegetative and stationary-phase RNA polymerase from *B. thuringiensis*. The RNA polymerase from *E. coli* and *B. thuringiensis* is inhibited to about the same extent.

***Effect of Thuringiensin on Normal and Hormone-Stimulated RNA Synthesis in Isolated Nuclei from the Larvae and Adult* Sarcophaga bullata.** Because the molting hormone, ecdysone, stimulated mRNA and rRNA synthesis when injected into ligated *Calliphora erythrocephala*, (Sekeris, 1965), Beebee and Bond (1973b) studied the effects of thuringiensin with normal and ecdysone-treated nuclei from the fat bodies of *S. bullata* larvae. Ecdysone stimulated the activity of the α-amanitin-resistant RNA polymerase I. However, normal and ecdysone-stimulated RNA polymerase are both inhibited by thuringiensin. Beebee and Bond (1973a) also reported on the sensitivity of the larvae to thuringiensin, but the insensitivity of the adult insect. Adult RNA polymerase is less sensitive to thuringiensin than is larval RNA polymerase. Apparently, there is no change in the adult RNA polymerase of *S. bullata* that makes it operationally different from the larval form. Thuringiensin may be useful as a biochemical probe into the changes that take place with the RNA polymerases in various stages of metamorphosis.

Effect of Thuringiensin on Polyphenylalanine Formation. The *in vivo* and *in vitro* studies with bacteria, insects, and mammals indicate that thuringiensin exerts its activity either by inhibiting RNA polymerase (Farkaš et al., 1969; Šebesta and Horská, 1968; Beebee and Bond, 1973b; Šebesta et al., 1969b; Bond et al., 1971) or effecting the maturation of RNA (Mackedonski, 1975). Kim et al. (1972) suggest that thuringiensin also inhibits protein synthesis. Somerville and Swain (1975) demonstrated that 8 m*M* thuringiensin inhibits cellfree protein synthesis. The inhibition is temperature sensitive. The inhibition of protein synthesis at such a high concentration may be a common property of nucleosides because high concentrations of ATP are also inhibitory to protein synthesis (Todde and Campbell, 1969).

Inhibition of Mitotic Spindle Formation by Thuringiensin. Vincristine and vinblastine, the indole alkaloids isolated from *Vinca rosa*, inhibit the mitotic process (Sharma, 1971). Sharma et al. (1976) reported that thuringiensin (100 ppm, 3 h) inhibits mitotic spindle and condenses and scatters chromosomes. Because thuringiensin inhibits RNA polymerase, the inhibition of mitotic spindle probably results from an interference with the synthesis of proteins required in the spindle microtubular systems. This inhibition of spindle formation, which is similar to that of colchicine, prompted Sharma et al. (1976) to caution against the use of thuringiensin for crop plants.

Inhibition of Adenyl Cyclase by Thuringiensin. Because adenyl cyclase belongs to the functional group of enzymes in which a nucleophilic attack by an alcohol group on the α-phosphorus of ribonucleoside triphosphates occurs, Grahame-Smith et al. (1975) studied the effect of thuringiensin on adenyl cyclase. Thuringiensin competitively inhibits adenyl cyclase, as well as the fluoride and hormonal stimulation of this enzyme, and blocks the ACTH-induced increases in adrenal cAMP and adrenal steroidogenesis. When thuringiensin is added to hemolyzed erythrocyte preparations in which adenyl cyclase activity is increased by fluoride, it completely inhibits adenyl cyclase (K_i = 0.5 mM; K_m for ATP = 1.1 mM). The competition of thuringiensin for ATP with adenyl cyclase suggests that thuringiensin is similar in structure to ATP.

Bacterial Membrane Transport of Thuringiensin. Thuringiensin is more toxic to mammalian and other eukaryotic systems than cellfree prokaryotic extracts. While thuringiensin can cross cell and nuclear membranes in mammalian systems (Smucker and Hadjiolov, 1972; Mackedonski et al., 1972a, b; Beebee et al., 1972), it does not inhibit bacteria, since nucleotides are not permeable to bacterial walls. However, thuringiensin does inhibit *E. coli* (Fig. 3.35). To determine the mode of action of thuringiensin with *E. coli*,

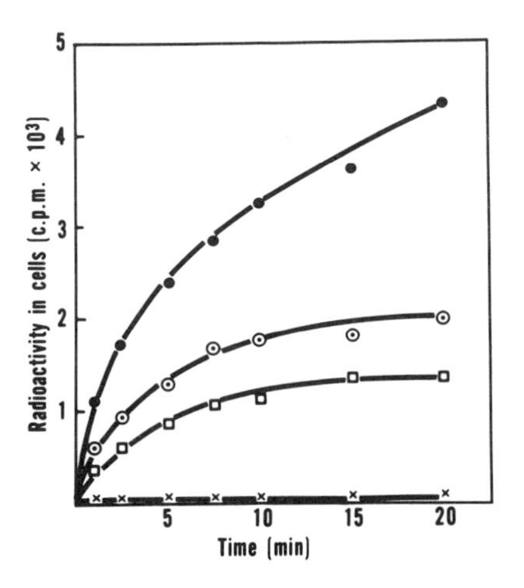

Figure 3.35 Uptake of [^{3}H]uridine and [^{3}H]thuringiensin in cells of *E. coli* treated with EDTA–Tris. [^{3}H]Uridine uptake without rifampin (●); with rifampin (O) (50 μg/ml); with rifampin and unlabeled thuringiensin (100 μg/ml), (□); [^{3}H]thuringiensin uptake (x). Rifampin was added (50 μg/ml) to prevent *in vivo* [^{3}H]uridine incorporation into RNA. Reprinted with permission from Johnson, 1976.

Johnson (1976) reported on the intracellular inhibition of RNA synthesis by thuringiensin. Thuringiensin is totally excluded by the bacterial membranes even in cells rendered permeable by pretreatment with EDTA–Tris. These data suggest that the thuringiensin blocks the uptake of uridine in *E. coli*. *E. coli*, treated with EDTA–Tris which results in increased permeability, do not take up thuringiensin. However, RNA synthesis is inhibited, which suggests that thuringiensin affects membrane transport.

Effect of Thuringiensin on Nucleolar Morphology. Smetana et al. (1974) studied the nucleoli isolated from hepatocytes of mice treated with thuringiensin. Electron micrographs showed changes in the nucleolar morphology, such as the formation of compact to ring-shaped nucleoli with segregated micronucleoli and segregation of nucleolar components. These changes were attributed to an inhibition of nucleolar RNA.

Differentiation of* E. coli *RNA Polymerase from Phage T3 RNA Polymerase by Thuringiensin and α-Amanitin. The fact that two different DNA-dependent RNA polymerases are involved in the development of phage T3 (Dunn et al., 1971; Chamberlain et al., 1970) plus the observation that the phage RNA polymerases have properties that markedly differ from those of the bacterial enzyme led Kupper et al. (1973) to compare the effect of thuringiensin and α-amanitin on the sensitivity of T3 and *E. coli* RNA polymerases. Whereas the *E. coli* RNA polymerase is completely inhibited by thuringiensin, the phage polymerase is not (Fig. 3.36). There is no inhibition of either the *E. coli* or phage RNA polymerase by α-amanitin. An

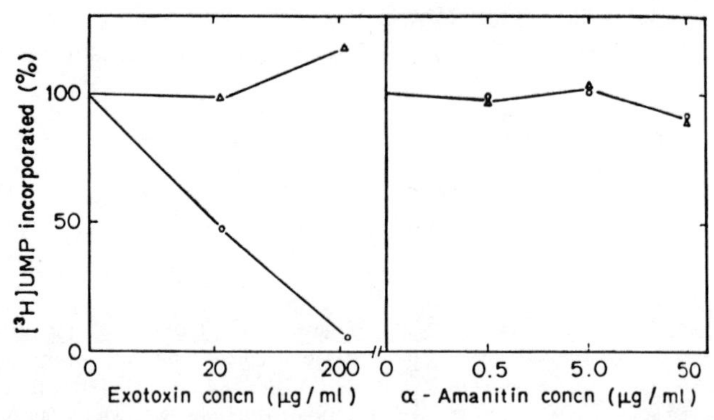

Figure 3.36 Sensitivities of *E. coli* and T3 RNA polymerases to thuringiensin (exotoxin) and α-aminitin. Standard reaction mixtures containing either *E. coli* RNA polymerase (O) or T3 RNA polymerase (Δ) and antibiotic as indicated were incubated at 37°C. After 10 min, the incorporation of [^{3}H]UMP into acid-insoluble material was determined. Reprinted with permission from Kupper et al., 1976.

explanation of the differences in Fig. 3.36 must take into account the facts that phage T3 has a lower molecular weight, has only one polypeptide chain, and has template requirements that differ from *E. coli* RNA polymerase.

Application of Thuringiensin in Insect Control. Formulations of thuringiensin are in active use for the control of pests and insects in the U.S.S.R. (Klassen, 1975). Part of the plan to manage pest populations is accomplished by enriching the biocenoses with beneficial organisms (Shchepetil'nikova et al., 1971). In 1973 *B. thuringiensis*, produced in multiton quantities, was applied to 250,000 hectares (Fedorinchik, 1973). This technique showed that microbial infections decrease or eliminate the immunity of certain insect pests to parasitic fungi or that parasitic insects promote epizootics by vectoring the diseases of insect pests.

3.11 PUROMYCIN AMINONUCLEOSIDE

The aminonucleoside, N^6-dimethyl-9-(3′-amino-3′-deoxy-β-D-ribofuranosyl)adenine (PAN) (Fig. 3.37), is obtained from puromycin by hydrolytic removal of the *p*-methoxyphenylalanyl group. Although puromycin is a broad spectrum antibiotic, it has no antibacterial activity; however, it inhibits cultured hamster embryo cells (Taylor and Stanners, 1968; Lewin and Moscarello, 1968). Although PAN is not a substrate for adenosine kinase, the monodemethylated compound is monophosphorylated. This may explain its nephrotic effect in animals (Wilson et al., 1962; Michaels et al., 1962; Karnofsky and Clarkson, 1963; Kmetec and Tirpack, 1970). The formation of the nucleotide may also explain the inhibition of RNA synthesis in mammalian cells (Taylor and Stanners, 1968; Lewin and Moscarello, 1968). Nair and Emmanuel (1977) reported on the stereospecific synthesis of the "reversed" puromycin (page 96), which is not a substrate for adenosine deaminase. Vince et al. (1976) reported that 5′-deoxy-PAN and the demethylated derivative show no nephrotoxicity.

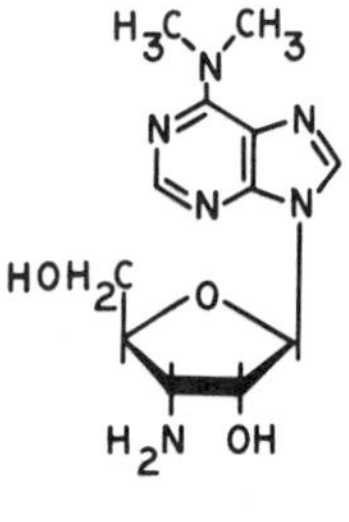

Figure 3.37 Structure of puromycin aminonucleoside.

SUMMARY

The primary action of the 15 nucleosides/nucleotides reviewed in this chapter is concerned with the inhibition of RNA, poly(A) synthesis and immunosuppresive activity. Cordycepin, a cytostatic analog of adenosine, is readily converted to its 5′-triphosphate and incorporated into 28, 10, 5, and 4 S RNA and poly(A), but not into DNA. In regenerating rat livers, cordycepin is equally effective in its inhibition of the major species of nuclear RNA. Cordycepin treated cells inhibit nuclear rRNA as shown by the decrease of 28 and 18 S RNA. The drug preferentially interferes with the maturation of mRNA by blocking the posttranscriptional addition of poly(A) to the 3′-terminus of pre-formed mRNA in nuclei, mitochondria, and viruses. It also interferes with the 2′-*O*-methylation of rRNA and the 5′-cap. 3′-deoxyATP is a moderate inhibitor of DNA-dependent RNA polymerases I, II, and III, but strongly inhibits nuclear poly(A) polymerase or cytoplasmic terminal riboadenylate transferases. The therapeutic efficiency of cordycepin is greatly increased by adenosine deaminase inhibitors.

The replacement of the terminal adenylylate in tRNA by 3′dAMP and 3′-amino-dAMP has greatly aided our understanding of the chemical proofreading needed to minimize miscoding of amino acids for protein synthesis.

In toluene-treated *E. coli*, 3′dATP inhibits RNA and DNA synthesis. However, the inhibition of DNA synthesis occurs by way of the inhibition of the RNA primer. In studies with the dehydrogenases and the ADP ribosylation of nuclear histone and nonhistone proteins, 3′-deoxyNAD$^+$ has been used to replace NAD$^+$.

3′-Amino-3′-deoxyadenosine also inhibits RNA synthesis. The biosynthesis of these two adenosine analogs utilizes the C–N skeleton of adenosine.

The guanosine analog, 2′-aminoguanosine, inhibits protein and RNA synthesis. Although the ^{3}H analog is taken up by *E. coli* very little is incorporated into RNA. The inhibition of protein synthesis may be due to its 5′-triphosphate, which competes for GTP in the protein synthesis steps.

Aristeromycin, an adenosine carbocyclic analog, inhibits RNA synthesis. *S*-Aristeromycinyl-L-homocysteine, an analog of SAH, has been used to study the specificity of the methylases. Aristeromycin also regulates cell division and elongation in plants.

Although there are nine pentopyranines, only A–D have been shown to inhibit RNA synthesis.

Bredinin is a new AICA riboside nucleoside analog. It inhibits the growth of *C. albicans*, vaccinia virus, and tumor cells in culture, but not bacteria. Bredinin inhibits the conversion of IMP to XMP to GMP. Whereas bredinin is cytostatic, its 5′-monophosphate is cytotoxic. It inhibits nucleic acid synthesis without being incorporated and induces chromosomal aberrations in late S-phase of the cycle of L5178Y mouse leukemic cells. Bredinin is an immunosuppressive agent and has antiarthritic activity.

The pyrrolopyrimidine nucleosides, tubercidin, toyocamycin, and sangivamycin, are adenosine analogs that inhibit RNA synthesis. Tubercidin inhibits the initiation step in RNA synthesis. *S*-Tubercidinyl homocysteine is an excellent SAH analog for studies of the SAM-dependent methylations. Finally, tubercidin is effective against basal cell carcinoma, actinic keratoses, and mycosis fungoides. Toyocamycin is an excellent inhibitor of the processing of pre-rRNA. It blocks the processing of ribosomal 45 S precursor and conversion of 27, 25, and 20 S pre-rRNA to mature 28 and 18 S rRNA. HnRNA synthesis is not effected. Poly(A) synthesis is inhibited by toyocamycin in adenovirus-infected HeLa cells. Toyocamycin-treated Rous sarcoma virus infected embryonic chick fibroblast multiplication is inhibited.

The pyrazolopyrimidine nucleosides, formycin, formycin B, and oxoformycin B, are analogs of adenosine, inosine, and xanthosine, respectively. Formycin is readily phosphorylated and incorporated into RNA. It inhibits the processing of the tRNA precursor. It replaces the terminal adenylylate in the tRNA and has been used to study polypeptide synthesis. When formycin replaces adenosine in NAD^+, it is equally effective in the inhibition of DNA synthesis in isolated rat liver nuclei. FTP reacts 17 times slower than ATP at the initiation site of RNA polymerase. The deamination of formycin in erythrocytes, when inhibited by adenosine deaminase inhibitors, results in high concentrations of FTP. This finding makes formycin a useful analog to detect ADA deficiency in humans.

Thuringiensin, an unusual analog of ATP, is elaborated by *B. thuringiensis*. It is toxic to insects, animals, and bacteria. *In vivo* administration of thuringiensin to animals inhibits RNA synthesis. DNA and protein synthesis are not inhibited. Polynucleotide phosphorylase is not inhibited by thuringiensin. RNA polymerases isolated from rat liver and *E. coli* are inhibited by thuringiensin. Nucleoplasmic RNA polymerase (RNA polymerase II) is less sensitive to thuringiensin than is nucleolar RNA polymerase. This makes thuringiensin an excellent compliment for α-amanitin. Dephosphorylated thuringiensin is not inhibitory. Although larvae are very sensitive to thuringiensin, the adult insect is not. This difference in the sensitivity to thuringiensin appears to be due to operational differences in the RNA polymerases. Thuringiensin inhibits polyphenylalanine synthesis, adenylate cyclase, and mitotic spindle formation. Although thuringiensin can cross mammalian cell walls intact, it cannot cross bacterial cell walls even after treatment with EDTA–Tris. However, it does inhibit RNA synthesis in *E. coli* and *B. subtilis*.

REFERENCES

Acs, G., E. Reich, and M. Mori, *Proc. Natl. Acad. Sci. U.S.*, **52,** 493 (1964).

Agarwal, K. C., and R. E. Parks, Jr., *Biochem. Parmacol.*, **24,** 2239 (1975).

Agarwal, R. P., S. M. Sager, and R. E. Parks, Jr., *Biochem. Pharmacol.*, **24,** 693 (1975).

Agarwal, R. P., G. W. Crabtree, R. E. Parks, Jr., J. A. Nelson, R. Keightley, R. Parkman, F. S. Rosen, R. C. Stern, and S. H. Polmar, *J. Clin. Invest.*, **57,** 1025 (1976).

Aizawa, S., T. Hidaka, N. Ōtake, H. Yonehara, K. Isono, N. Igarashi, and S. Suzuki, *Agr. Biol. Chem.*, **29,** 375 (1965).

Alford, B., and S. M. Hecht, *J. Biol. Chem.*, **253,** 4844 (1978).

Altman, S., *Cell*, **4,** 21 (1975).

Angus, T. A., *Nature*, **173,** 545 (1954).

Asano, S., Y. Kurashina, Y. Anraku, and D. Mizuno, *J. Biochem.*, **70,** 9 (1971a).

Asano, S., Y. Anraku, and D. Mizuno, *J. Biochem.*, **70,** 21 (1971b).

August, J. T., P. J. Ortiz, and J. J. Hurwitz, *J. Biol. Chem.*, **237,** 3786 (1962).

Bablanian, R., *Prog. Med. Virol.*, **19,** 40 (1975).

Baksht, E., N. de Groot, M. Sprinzl, and F. Cramer, *FEBS Lett.*, **55,** 105 (1975).

Ballet, J. J., R. Insel, E. Merler, and F. M. Rosen, *J. Exp. Med.*, **143,** 1271 (1977).

Bassleer, R., A. Lepoint, F. De Paermentier, and G. Goessens, *Microsc. Biol. Cell*, **25,** 33 (1976).

Bateman, J. R., in Minutes of the New Drug Liaison Meeting, National Cancer Institute, Bethesda, 1975, p. 32.

Baxter, C., and P. Byvoet, *Cancer Res.*, **32,** 1418 (1974).

Beach, L. R., and J. Ross, *J. Biol. Chem.*, **253,** 2628 (1978).

Beebee, T. J. C., and R. P. M. Bond, *Biochem. J.*, **136,** 9 (1973a).

Beebee, T., A. Korner, and R. P. M. Bond, *Biochem. J.*, **127,** 619 (1972).

Beebee, T. J. C., and R. P. M. Bond, *Biochem. J.*, **136,** 1 (1973b).

Beisler, J. A., M. M. Abbasi, and J. S. Driscoll, *Cancer Treat. Rep.*, **60,** 1671 (1976).

Beisler, J. A., D. L. Hepp, F. R. Quinn, and J. S. Driscoll, *Proc. Am. Assoc. Cancer Res.*, **18,** 171 (1977a).

Beisler, J. A., M. M. Abbasi, J. A. Kelley, and J. S. Driscoll, *J. Carbohyd. Nucleosides Nucleotides*, **4,** 281 (1977b).

Bellet, R. E., P. F. Engstrom, R. P. Custer, J. G. Strawitz, and J. W. Yarbro, *Proc. Am. Assoc. Cancer Res.*, **13,** 87 (1972a).

Bellet, R. E., M. J. Mastrangelo, P. F. Engstrom, R. S. Bornstein, J. G. Strawitz, J. W. Yarbro, and A. J. Weiss, *Clin. Res.*, **20,** 563 (1972b).

Bellet, R. E., M. J. Mastrangelo, P. F. Engstrom, and R. P. Custer, *Neoplasma*, **20,** 303 (1973).

Bellet, R. E., M. J. Mastrangelo, P. F. Engstrom, J. G. Strawitz, A. J. Weiss, and J. W. Yarbro, *Cancer Chemother. Rep.*, **58,** 217 (1974).

Bennett, L. L., Jr., P. W. Allan, and D. L. Hill, *Mol. Pharmacol.*, **4,** 208 (1968).

Benz, G., *Experientia*, **22,** 81 (1961).

Bergy, M. E., and R. R. Herr, *Antimicrob. Agents Chemother.*, **1966,** 625.

Bhalla, R. B., M. K. Schwartz, and M. J. Modak, *Biochem. Biophys. Res. Commun.*, **76,** 1056 (1977).

Bhuyan, B. K., L. G. Scheidt, and T. J. Fraser, *Cancer Res.*, **32,** 398 (1972).

Bisel, H. F., F. J. Ansfield, J. H. Mason, and W. L. Wilson, *Cancer Res.*, **30,** 76 (1970).

Blecher, M., J. T. Ro'Ane, and P. D. Flynn, *Biochem. Pharmacol.*, **20,** 249 (1971).

Bloch, A., *Ann. N.Y. Acad. Sci.*, **255,** 576 (1975).

Bloch, A., R. J. Leonard, and C. A. Nichol. *Biochem. Biophys, Acta*, **138,** 10 (1967).

Bollum, F. J., *J. Biol. Chem.*, **237,** 1945 (1962).

Bond, R. P. M., *J. Chem. Soc. Chem. Commun.*, **1969,** 338.

Bond, R. P. M., C. B. C. Boyce, V. K. Brown, and J. D. Tipton, *Biochem. J.*, **114,** 1p (1969a).

Bond, R. P. M., C. B. C. Boyce, and S. J. French, *Biochem. J.*, **114,** 477 (1969b).

Bond, R. P. M., C. B. C. Boyce, M. H. Rogoff, and T. R. Shieh, in *Microbial Control of Insects and Mites*, H. D. Burgess, and N. W. Hussey, Eds., Academic Press, London, 1971, p. 275.

Borchardt, R. T., and Y-S. Wv, *J. Med. Chem.*, **19,** 197 (1976).

Both, G. W., A. K. Banerjee, and A. J. Shatkin, *Proc. Natl. Acad. Sci., U.S.*, **72,** 1189 (1975).

Brawerman, G., *Prog. Nucleic Acid Res. Mol. Biol.*, **17,** 117 (1976).

Brdar, B., and E. Reich, *Period. Biol.*, **78,** 51 (1976).

Brdar, B., D. B. Rifkin, and E. Reich, *J. Biol. Chem.*, **248,** 2397 (1973).

Brevet, J., *Mol. Gen. Genet.*, **128,** 223 (1974).

Brodniewicz-Proba, T., and J. Buchowicz, *FEBS Lett.*, **65,** 183 (1976).

Bruzel, A., R. J. Suhadolnik, and J. K. Hoober, *Fed. Proc.*, **36,** 909 (1977).

Bruzel, A., R. J. Suhadolnik, and R. G. Wilson, *Fed. Proc.*, **37,** 1636 (1978).

Burchenal, J. H., M. Cole, D. Pomeroy, and H. J. Krakoff, *Proc. Am. Assoc. Cancer Res.*, **13,** 105 (1972).

Burgerjon, A., and G. Biache, *Ann. Soc. Entomol., Fr.*, **3,** 929 (1967).

Burgess, G. H., A. Bloch, H. Stoll, H. Milgram, F. Helm, and E. Klein, *Cancer*, **34,** 250 (1974).

Burridge, P. W., W. Paetkau, and J. F. Henderson, *J. Immunol.*, **119,** 675 (1977).

Caboche, M., and J. P. Bachellerie, *Eur. J. Biochem.*, **74,** 19 (1977).

Cairns, J. A., T. C. Hall, K. B. Olson, C. L. Khuang, J. Horton, J. Colsky, and R. K. Shadduck, *Cancer Chemother. Rep.*, **51,** 197 (1967).

Camiener, G. W., *Biochem. Pharmacol.*, **17,** 1981 (1968).

Cantoni, G. L., *Annu. Rev. Biochem.*, **44,** 435 (1975).

Cantwell, G. E., A. M. Heimpel, and M. J. Thomson, *J. Insect Pathol.*, **6,** 466 (1964).

Carson, D. A. and J. E. Seegmiller, *J. Clin. Invest.*, **57,** 274 (1976).

Chabner, B. A., and R. C. Young, *J. Clin. Invest.*, **52,** 922 (1973).

Chabner, B. A., J. C. Drake, and D. G. Johns, *Biochem. Pharmacol.*, **22,** 2763 (1973).

Chamberlain, M., J. McGrath, and L. Waskell, *Nature*, **228,** 227 (1970).

Chambon, P., J. D. Weill, J. Doly, M. T. Strosser, and P. Mandel, *Biochem. Biophys. Res. Commun.*, **75,** 638 (1966).

Chang, L. M. S., and F. J. Bollum, *J. Biol. Chem.*, **246,** 909 (1971).

Chang, C., and J. K. Coward, *Mol. Pharmacol.*, **11,** 701 (1975).

Chassy, B. M., and R. J. Suhadolnik, *J. Biol. Chem.*, **243,** 3538 (1968).
Cheng, C. S., B. C. Hinshaw, R. P. Panzica, and L. B. Townsend, *J. Am. Chem. Soc.*, **98,** 7870 (1976).
Chenon, M. T., R. P. Panzica, J. C. Smith, R. J. Pugmire, D. M. Grant, L. B. Townsend, *J. Am. Chem. Soc.*, **98,** 4736 (1976).
Chiu, T. M. K., H. Ohrui, K. A. Watanabe, and J. J. Fox, *J. Org. Chem.*, **38,** 3622 (1973).
Chung, H. L. and J. Žemlička, *J. Heterocycl. Chem.*, **14,** 135 (1977).
Čihák, A., and H. M. Rabes, *Neoplasma*, **21,** 497 (1974).
Čihák, A., *Collect. Czech. Chem. Commun.*, **39,** 3782 (1974).
Čihák, A., *Eur. J. Cancer*, **14,** 117 (1978).
Čihák, A., and F. Šorm, *Collect. Czech. Chem. Commun.*, **30,** 2091 (1965).
Čihák, A., and J. Veselý, *Collect. Czech. Chem. Commun.*, **34,** 910 (1969).
Čihák, A., and J. Veselý, *Biochem. Pharmacol.*, **21,** 3257 (1972).
Čihák, A., and J. Veselý, *FEBS Lett.*, **78,** 244 (1977).
Čihák, A., and J. Veselý, *Neoplasma*, in press (1979).
Čihak, A., J. Skoda, and F. Šorm, *Collect. Czech. Chem. Commun.*, **29,** 300 (1964).
Čihák, A., R. Tykva, and F. Šorm, *Collect. Czech. Chem. Commun.*, **31,** 3015 (1966).
Čihák, A., J. Veselý, and F. Šorm, *Biochim. Biophys. Acta*, **134,** 486 (1967a).
Čihák, A., J. Veselý, and F. Šorm, *Collect. Czech. Chem. Commun.*, **32,** 3427 (1967b).
Čihák, A., H. Veselá, and F. Šorm, *Biochim. Biophys. Acta*, **166,** 277 (1968).
Čihák, A., J. Veselý, and F. Šorm, *Collect. Czech. Chem. Commun.*, **34,** 1060 (1969).
Čihák, A., M. Seifertová, J. Veselý, and F. Šorm, *Int. J. Cancer*, **10,** 20 (1972).
Čihák, A., L. M. Narurkar, and H. C. Pitot, *Collect. Czech. Chem. Commun.*, **38,** 948 (1973a).
Čihák, A., C. Lamar, and H. C. Pitot, *Arch. Biochem. Biophys.*, **156,** 176 (1973b).
Čihák, A., J. W. Weiss, and H. C. Pitot, *Cancer Res.*, **34,** 3003 (1974).
Čihák, A., K. Horská, and K. Šebesta, *Collect. Czech. Chem. Commun.*, **40,** 2912 (1975).
Čihák, A., M. Seifertová, and P. Riches, *Cancer Res.*, **36,** 37 (1976).
Cohen, A., and M. Sussman, *Proc. Natl. Acad. Sci. U.S.*, **72,** 4479 (1975).
Cole, F. X. and P. R. Schimmel, *J. Am. Chem. Soc.*, **100,** 3957 (1978).
Coleman, C. N., R. G. Stroller, J. C. Drake, and B. A. Chabner, *Blood*, **46,** 791 (1975).
Coleman, M. S., J. J. Hutton, P. De Simone, and F. J. Bollum, *Proc. Natl. Acad. Sci. U.S.*, **71,** 4404 (1974).
Coleman, M. S., M. F. Greenwood, J. J. Hutton, F. J. Bollum, B. Lampkin, and P. Holland, *Cancer Res.*, **36,** 120 (1976).
Coles, E., P. S. Thayer, V. Reinhold, and L. Gaudio, *Proc. Am. Assoc. Cancer Res.*, **15,** 72 (1974).
Connor, R. M., and P. A. Hansen, *J. Invert. Pathol.*, **9,** 12 (1967).
Cory, J. G., R. J. Suhadolnik, B. Resnick, and M. A. Rich, *Biochim. Biophys. Acta*, **103,** 646 (1965).
Coward, J. K., and E. P. Slisz, *J. Med. Chem.*, **16,** 460 (1973).
Coward, J. K., and P. A. Crooks, Conference on Transmethylation, NIH, Oct 16–19, 1978, E.

Udsin, C. R. Creveling, and R. T. Borchardt, Eds. Elsevier-North Holland, Amsterdam, 1979 in press.

Coward, J. K., D. L. Bussolotti, and C. D. Chang, *J. Med. Chem.*, **17,** 1286 (1974).

Coward, J. K., N. C. Motola, and J. D. Moyer, *J. Med. Chem.*, **20,** 500 (1977).

Crabtree, G. W., R. P. Agarwal, R. E. Parks, Jr., A. F. Lewis, L. L. Wotring, and L. B. Townsend, *Biochem. Pharmacol.*, in press (1979).

Cunningham, K. G., S. A. Hutchinson, W. Manson, and F. S. Spring, *J. Chem. Soc.*, **1951,** 2299.

Cunningham, T. J., T. Nemoto, D. Rosner, E. Knight, S. Taylor, C. Rosenbaum, J. Horton, and T. Dao, *Cancer Chemother. Rep.*, **58,** 677 (1974).

Dabeva, M. D., K. P. Dudov., A. A. Hadjiolov, I. Emanuilov, and B. N. Todorov, *Biochem. J.*, **160,** 495 (1976).

Daluge, S., and R. Vince, *Tetrahedron Lett.*, **35,** 3005 (1976).

Daluge, S., and R. Vince, *J. Org. Chem.*, **43,** 2311 (1978).

Darlix, J. L., P. Fromageot, and E. Reich, *Biochemistry*, **10,** 1525 (1971).

Darnell, J. E., R. Wall, and R. J. Tushinski, *Proc. Natl. Acad. Sci. U.S.*, **68,** 1321 (1971a).

Darnell, J. E., L. Philipson, R. Wall, and M. Adesnik, *Science*, **174,** 507 (1971b).

Darnell, J. E., W. R. Jelinek, and G. R. Molloy, *Science*, **181,** 1215 (1973).

Darnell, J. E., *Science*, **202,** 1257 (1978).

Darnell, J. E., in W. J. Whelan and J. Schultz, Eds., *Miami Winter Symposia*, Vol. 16, Academic Press, New York, in press, 1979.

Daves, G. D. and C. C. Cheng, *Prog. Med. Chem.*, **13,** 303 (1976).

Davies, R. J. H., *J. Mol. Biol.*, **73,** 317 (1973).

Davies, R. J. H., *Eur. J. Biochem.*, **61,** 225 (1976).

deBarjac, H., and R. Dedonder, *Bull. Soc. Chim. Biol.*, **50,** 941 (1968).

deBarjac, H., and M. M. Lecadet, *C. R. Acad. Sci.*, **282,** 2119 (1976).

deBarjac, H., and J. V. Riou, *Rev. Pathol. Comp. Med. Exp.*, **6,** 367 (1969).

DeClercq, E., P. F. Torrence, and B. Witkop, *Proc. Natl. Acad. Sci. U.S.*, **71,** 182 (1974).

Delseny, M., M. T. Peralta, and Y. Guitton, *Biochem. Biophys. Res. Commun.*, **64,** 1278 (1975).

Dicioccio, R. A., and B. I. S. Srivastava, *Eur. J. Biochem.*, **79,** 411 (1977).

Diez, J., and G. Brawerman, *Proc. Natl. Acad. Sci. U.S.*, **71,** 4091 (1974).

Doskočil, J., *Biochem. Biophys. Acta*, **282,** 393 (1972).

Doskočil, J., *Biochem. Biophys. Res. Commun.*, **56,** 997 (1974).

Doskočil, J., and A. Holý, *Nucleic Acids Res.*, **1,** 491 (1974).

Doskočil, J., and F. Šorm, *Biochem. Biophys. Res. Commun.*, **38,** 569 (1970a).

Doskočil, J., and F. Šorm, *Collect. Czech. Chem. Commun.*, **35,** 1880 (1970b).

Doskočil, J., and F. Šorm, *Eur. J. Biochem.*, **13,** 180 (1970c).

Doskočil, J., and F. Šorm, *FEBS Lett.*, **19,** 30 (1971).

Dreyer, C., and P. Hausen, *Nucleic Acids Res.*, **5,** 3325 (1978).

Drysdale, J. W., and H. N. Munro, *J. Biol. Chem.*, **241,** 3630 (1966).

Dunn, J. J., F. A. Bautz, and E. K. F. Bautz, *Nature, New Biol.*, **230,** 94 (1971).

Edmonds, M., and R. Abrams, *J. Biol. Chem.*, **235,** 1142 (1960).

Edmonds, M., and R. Abrams, *J. Biol. Chem.*, **238,** 1186 (1963).

Edmonds, M., M. H. Vaughan, Jr., and H. Nakazato, *Proc. Natl. Acad. Sci., U.S.*, **68,** 1336 (1971).

Elstner, E. F., and R. J. Suhadolnik, *J. Biol. Chem.*, **246,** 6973 (1971).

Evans, F. E., and R. H. Sarma, *Cancer Res.*, **35,** 1458 (1975).

Farkaš, J., Šebesta, K., K. Horská, Z. Samek, L. Dolejs, and F. Šorm, *Collec. Czech. Chem. Commun.*, **34,** 1118 (1969).

Farkaš, J., K. Šebesta, K. Horská, Z. Samek, L. Dolejs, and F. Šorm, *Collec. Czech. Chem. Commun.*, **42,** 909 (1977).

Fersht, A. R., and C. Dingwall, *Biochemistry*, **18,** 1238, 1245, 1250 (1979).

Fedorinchik, N. A., in Papers Read by the Soviet Specialists, The Soviet-American Conference on the Integrated Pest Control, Ukranian Scientific Research Institute for Agriculture, Kiev, U.S.S.R., 1973, p. 38.

Fouquet, H., R. Wick, R. Böhme, H. W. Sauer, and K. Scheller, *Arch. Biochem. Biophys.*, **168,** 273 (1975).

Fox, J. J., and I. Goodman, *J. Am. Chem. Soc.*, **73,** 3256 (1951).

Fox, I. H., E. C. Keystone, D. D. Gladman, M. Moore, and D. Cane, *Immunol. Commun.*, **4,** 419 (1975).

Fraser, T. H., and A. Rich, *Proc. Natl. Acad. Sci. U.S.*, **70,** 2671 (1973).

Frederiksen, S., and H. Klenow, *Biochem. Biophys. Res. Commun.*, **17,** 165 (1964).

Fučik, V., A. Michaelis, and R. Rieger, *Mutat. Res.*, **9,** 599 (1970).

Furuichi, Y., A. La Fiandra, and A. J. Shatkin, *Nature*, **266,** 235 (1977).

Gallo, R. C., S. M. Hecht, J. Whang-Peng, and S. O'Hopp, *Biochim. Biophys. Acta*, **281,** 488 (1972).

Georgiev, G. P., and V. L. Mantieva, *Biochim. Biophys. Acta*, **61,** 153 (1962).

Giziewicz, J., and D. Shugar, *Acta Biochim. Polon.*, **24,** 231 (1977).

Giziewicz, J., E. De Clereg, M. Luczak, and D. Shugar, *Biochem. Pharmacol.*, **24,** 1813 (1975).

Glazer, R. I., *Biochim. Biophys. Acta*, **418,** 160 (1975).

Glazer, R. I., *Toxicol. Appl. Pharmacol.*, **46,** 191 (1978).

Glazer, R. I., and J. F. Kuo, *Biochem. Pharmacol.*, **26,** 1287 (1977).

Glazer, R. I., and A. L. Peale, *Biochem. Biophys. Res. Commun.*, **81,** 521 (1978).

Glazer, R. I., T. J. Lott, and A. L. Peale, *Cancer Res.*, **38,** 2233 (1978).

Gotoh, S., N. Nikolaev, E. Battaner, C. H. Birge, and D. Schlessinger, *Biochem. Biophys. Res. Commun.*, **59,** 972 (1974).

Gottesman, M. E., Z. N. Canellakis, and E. S. Canellakis, *Biochim. Biophys. Acta*, **61,** 34 (1962).

Grage, T. B., F. B. Rochlin, A. J. Weiss, and W. L. Wilson, *Cancer Res.*, **30,** 79 (1970).

Grahame-Smith, D. G., P. Isaac, and D. J. Heal, *Nature*, **253,** 58 (1975).

Gumport, R. I., E. B. Edelheit, T. Uematsu, and R. J. Suhadolnik, *Biochemistry*, **15,** 2804 (1976).

Hadjivassiliou, A., and G. Brawerman, *J. Mol. Biol.*, **20,** 1 (1966).

Hadjivassiliou, A., and G. Brawerman, *Biochemistry*, **6,** 1934 (1967).

Hagimoto, H., H. Yoshikawa, and H. Tamura, 4th Annual Meeting of the Society for Chemical Regulation of Plants, Shiobara, Japan, 1969.

Hammett, J. R., and F. R. Katterman, *Biochemistry*, **14,** 4375 (1975).

Hampton, A., and T. Sasaki, *Biochemistry*, **12,** 2188 (1973).

Hañka, L. J., and J. J. Clark, *Proc. Am. Assoc. Cancer Res.*, **16,** 113 (1975).

Hañka, L. J., J. S. Evans, D. J. Mason, and A. Dietz, *Antimicrob. Agents Chemother.*, **1966,** 619.

Hannay, C. L., *Nature*, **172,** 1004 (1953).

Hardesty, C. T., N. A. Chaney, V. S. Waravdekar, and J. A. R. Mead, *Cancer Res.*, **34,** 1005 (1974).

Harris, B., and L. S. Dure, *Biochemistry*, **13,** 5463 (1974).

Hayaishi, O., and K. Ueda, *Annu. Rev. Biochem.*, **46,** 95 (1977).

Hayashi, M., T. Hirano, M. Yaso, K. Mizuno, and T. Ueda, *Chem. Pharm. Bull.*, **23,** 245 (1975).

Heby, O., and D. H. Russell, in *Polyamines in Normal and Neoplastic Growth*, D. H. Russell, Ed., Raven Press, New York, 1973a, p. 221.

Heby, O., and D. H. Russell, *Cancer Res.*, **33,** 159 (1973b).

Hecht, S. M., J. W. Kozarich, and F. J. Schmidt, *Proc. Natl. Acad. Sci., U.S.*, **71,** 4317 (1974a).

Hecht, S. M., R. D. Faulkner, and S. D. Hawrelak, *Proc. Natl. Acad. Sci., U.S.*, **71,** 4670 (1974b).

Heimpel, A. M., *Can. J. Zool.*, **33,** 311 (1955).

Heine, U., *Cancer Res.*, **29,** 1875 (1969).

Henderson, J. F., C. M. Smith, F. F. Snyder, and G. Zombor, *Ann. N.Y. Acad. Sci.*, **255,** 489 (1975).

Henshaw, E., M. Revel, and H. Hiatt, *J. Mol. Biol.*, **14,** 241 (1965).

Hershfield, M. S., F. F. Snyder, and J. E. Seegmiller, *Science*, **197,** 1284 (1977).

Hilbert, G. E., and E. F. Jansen, *J. Am. Chem. Soc.*, **58,** 60 (1936).

Hirasawa, K., and K. Isono, *J. Antibiot.*, **31,** 628 (1978).

Hirsch, J., and O. J. Martelo, *Life Sci.*, **19,** 85 (1976).

Hobbs, J. B., and F. Eckstein, *J. Org. Chem.*, **42,** 714 (1977).

Hori, M., E. Ito, T. Tokida, G. Koyama, T. Takeuchi, and H. Umezawa, *J. Antibiot.*, **17A,** 96 (1964).

Horwitz, J. P., R. S. Misra, J. Rozhin, J. P. Neenan, A. H. Viriane, C. Godefroi, K. D. Philips, H. L. Chung, G. Butke, and S. C. Brooks, *Biochim. Biophys. Acta*, **525,** 364 (1978).

Horská, K., J. Vaňková, and K. Šebesta, *Z. Naturforsch.*, **30C,** 120 (1975).

Horská, K., L. Kalvoda, and K. Šebesta, *Collect. Czech. Chem. Commun.*, **41,** 3837 (176).

Hovi, T., J. F. Smyth, A. C. Allison, and S. C. Williams, *Clin. Exp. Immunol.*, **23,** 395 (1976).

Hoyer, B. H., B. J. McCarthy, and E. T. Bolton, *Science*, **140,** 1408 (1963).

Hrodek, O., and J. Veselý, *Neoplasma*, **18,** 493 (1971).

Iapalucci-Espinoza, S., S. Cereghini, and M. T. Franze-Fernandez, *Biochemistry*, **16,** 2885 (1977).

Ikehara, M., and T. Fukui, *J. Biochem.*, **73,** 945 (1973).

Ikehara, M., and T. Fukui, *Biochim. Biophys. Acta*, **338,** 512 (1974).

Ikehara, M., and T. Tezuka, *Nucleic Acids Res.*, **1,** 907 (1974).

Ikehara, M., I. Tazawa, and T. Fukui, *Biochemistry*, **8,** 736 (1969).

Ingoglia, N. A., *J. Neurochem.*, **30,** 1029 (1978).

Israili, Z. H., W. R. Vogler, E. S. Mingioli, J. L. Pirkle, R. W. Smithwick, and J. H. Goldstein, *Cancer Res.*, **36,** 1453 (1976).

Ito, H., Japanese Patent No. 32, 648 (1973).

Iwata, H., H. Iwaki, T. Masukawa, S. Kasmatsu, and H. Okamoto, *Experientia*, **33,** 502 (1977).

Jaffe, J. J., E. Meymarian, and H. M. Doremus, *Nature*, **230,** 408 (1971).

Jaffe, J. J., H. M. Doremus, H. A. Dunsford, and E. Meymarian, *Am. J. Trop. Med. Hyg.*, **24,** 835 (1975).

Jain, K., and J. Logothetopoulos, *Endocrinology*, **100,** 923 (1977).

Jelinek, W., M. Adesnik, M. Salditt, D. Sheiness, R. Wall, G. Molloy, L. Philipson, and J. E. Darnell, *J. Mol. Biol.*, **75,** 515 (1973).

Johns, D. G., and R. H. Adamson, *Biochem. Pharmacol.*, **25,** 1441 (1976).

Johns, D. G., A. C. Sartorelli, and D. G. Johns, Eds., in *Antineoplastic and Immuno-suppressive Agents, Part I*, Springer-Verlag, New York, 1974, p. 277.

Johnson, D. E., *Nature*, **260,** 333 (1976).

Johnson, D. E., *Can. J. Microbiol.*, **24,** 537 (1978).

Johnson, D. E., L. A. Bulla, Jr., and K. W. Nickerson, in *Spores*, Vol. VI, P. Gerhardt, R. N. Castilow, and H. L. Sadoff, Eds., American Society of Microbiology, Washington, D.C., 1975, p. 248.

Jurovčík, M., K. Raška, F. Sorm, and Z. Sormová, *Collect. Czech. Chem. Commun.*, **30,** 3370 (1965).

Kaehler, M., J. Coward, and F. Rottman, *Biochemistry*, **16,** 5770 (1977).

Kalousek, F., K. Rašks, M. Jurovčík, and F. Sorm, *Collect. Czech. Chem. Commun.*, **31,** 1421 (1966).

Kalvoda, L., *J. Carbohyd. Nucleosides Nucleotides*, **3,** 47 (1976).

Kalvoda, L., *Collect. Czech. Chem. Commun.*, **43,** 1431 (1978).

Kalvoda, L., M. Prystaš, and F. Sorm, *Collect. Czech. Chem. Commun.*, **38,** 2529 (1973).

Kalvoda, L., M. Prystaš, and F. Šorm, *Collect Czech. Chem. Commun.*, **41,** 788 (1976a).

Kalvoda, L., M. Prystaš, and F. Šorm, *Collect. Czech. Chem. Commun.*, **41,** 800 (1976b).

Kann, H. E., Jr., and K. W. Kohn, *Mol. Pharmacol.*, **8,** 551 (1972).

Karnofsky, D. A., and B. D. Clarkson, *Annu. Rev. Pharmacol.*, **3,** 357 (1963).

Karon, M., and W. F. Benedict, *Science*, **178,** 62 (1972).

Karon, M., L. Siegel, S. Leimbrock, J. Z. Finkelstein, M. Nesbitt, and J. J. Swaney, *Blood*, **42,** 359 (1973).

Kasai, H., Y. Kuchino, K. Nihei, and S. Nishimura, *Nucleic Acids Res.*, **2,** 1931 (1975a).

Kasai, H., Z. Ohashi, F. Harada, S. Nishimura, N. J. Oppenheimer, P. F. Crain, J. G. Liehr, D. L. von Minden, and J. A. McCloskey, *Biochemistry*, **14,** 4198 (1975b).

Kates, J., *Cold Spring Harbor Symp. Quant. Biol.*, **35,** 743 (1970).

Keefer, R. D., D. J. McNamara, D. E. Schumm, D. F. Billmire, and T. E. Webb, *Biochem. Pharmacol.*, **24,** 1287 (1975).

Kim, Y. T., and H. T. Huang, *J. Invert. Pathol.*, **15,** 100 (1970).

Kim, Y. T., B. G. Gregory, and C. M. Ignoffe, *J. Invert. Pathol.*, **20,** 46 (1972).

Kisselev, L. L., and O. O. Favorova, *Adv. Enzymol.*, **40,** 141 (1974).

Klassen, W., *Impressions of Applied Insect Pathology in the U.S.S.R.* U.S. Department of Agriculture, Agriculture Research Service, Dec. 1975.

Klein, E., G. A. Burgess, A. Bloch, H. Milgram, and O. A. Halterman, *Ann. N.Y. Acad. Sci.*, **255,** 216 (1975).

Klenow, H., *Biochem. Biophys. Res. Commun.*, **5,** 156 (1961).

Klenow, H., and K. Overgaard-Hansen, *Biochim. Biophys. Acta*, **80,** 500 (1964).

Klier, A. F., M. M. Lecadet, and R. Dedonder, *Eur. J. Biochem.*, **361,** 317 (1973).

Klier, A., and M. M. Lecadet, *Eur. J. Biochim.*, **47,** 111 (1974).

Kmetec, E., and A. Tirpack, *Biochem. Pharmacol.*, **19,** 1493 (1970).

Kohls, R. E., A. J. Lemin, and P. W. O'Connell, *J. Econ. Entomol.*, **59,** 745 (1966).

Koyama, G., H. Nakamura, H. Umezawa, and Y. Iitaka, *Acta Cryst.*, **B32,** 813 (1976).

Krieg, A., *J. Invert. Pathol.*, **12,** 478 (1968).

Kuchino., Y., H. Kasai, K. Nihei, and S. Nishimura, *Nucleic Acids Res.*, **3,** 393 (1976).

Kumar, S. A., J. S. Krakow, and D. C. Ward, *Biochim. Biophys. Acta*, **477,** 112 (1977).

Kunimoto, T., T. Sawa, T. Wakashiro, M. Hori, and J. Umezawa, *J. Antibiotics*, **24,** 253 (1971).

Kupper, H. A., W. T. McAllister, and E. K. F. Bautz, *Eur. J. Biochem.*, **38,** 581 (1973).

Kuroyanagi, T., in Y. Ōtake and T. Matsuhashi, Eds., *Therapy, Immuno-suppressive Therapy*, Igaku Shoin, Tokyo, 1971, p. 29.

Kusaka, T., *J. Antibiot.*, **24,** 756 (1971).

Kusaka, T., *J. Takeda Res. Lab.*, **31,** 85 (1972).

Kusaka, T. H., H. Yamamoto, M. Shibata, M. Muroi, T. Kishi, and K. Mizuno, *J. Chem. Soc. Chem. Commun.*, **1967,** 852.

Kusaka, T., H. Yamamoto, M. Shibata, M. Muroi, T. Kishi, and K. Mizuno, *J. Antibiot.*, **21,** 255 (1968).

Lavers, G. C., J. H. Chen, and A. Spector, *J. Mol. Biol.*, **82,** 15 (1974).

Lawrence, F., D. J. Shire, and J. P. Waller, *Eur. J. Biochem.*, **41,** 73 (1974).

Lee, T., *Proc. Am. Assoc. Cancer Res.*, **14,** 94 (1973).

Lee, T. T., and R. L. Momparler, *Anal. Biochem.*, **71,** 60 (1976).

Lee, T. T., and R. L. Momparler, *Biochem. Pharmacol.*, **26,** 403 (1977).

Lee S. Y., J. Mendecki, and G. Brawerman, *Proc. Natl. Acad. Sci. U.S.*, **68,** 1331 (1971).

Lee, T., M. Karon, and R. L. Momparler, *Cancer Res.*, **34,** 3482 (1974).

Lee, T., and M. R. Karon, *Biochem. Pharmacol.*, **25,** 1737 (1976).

Legraverend, M., R. I. Glazer, and D. G. Johns, *Proc. 19th Annu. Meet. Am. Assoc. Cancer Res.*, **19,** Abstr. 437 (1978).

Leinwand, L., and F. H. Ruddle, *Science*, **197,** 381 (1977).

Lennon, M. B., and R. J. Suhadolnik, *Biochim. Biophys. Acta*, **425,** 532 (1976).

Lennon, M. B., and R. J. Suhadolnik, *Biochim. Biophys. Acta*, in press (1979).

Lennon, M. B., J. Wu, and R. J. Suhadolnik, *Biochem. Biophys. Res. Commun.*, **72,** 530 (1976).

Levey, I. L., and R. L. Brinster, *Exp. Cell. Res.*, **109,** 397 (1977).

Levi, J. A., and P. H. Wiernik, *Cancer Chemother. Rep.*, **59,** 1043 (1975).

Levitan, I. B., and T. E. Webb, *Biochim. Biophys. Acta*, **182,** 491 (1969).

Levitan, I. B., H. P. Morris, and T. E. Webb, *Biochim. Biophys. Acta*, **240,** 287 (1971).

Lewin, P. K., and M. A. Moscarello, *Lab. Invest.*, **19,** 265 (1968).

Lewis, A. F., and L. B. Townsend, *J. Heterocyc. Chem.*, **11,** 71 (1974).

Lewis, A. F., and L. B. Townsend, Abstracts Medi 42, Meeting of the 175th American Chemical Society, March 13, 1978.

Lewis, A. F., R. A. Long, L. W. Roti-Roti, and L. B. Townsend, *J. Heterocyc. Chem.*, **13,** 1359 (1976).

Li, L. H., E. J. Olin, H. H. Buskirk, and L. M. Reineke, *Cancer Res.*, **30,** 2760 (1970a).

Li, L. H., E. J. Olin, T. J. Fraser, and B. K. Bhuyan, *Cancer Res.*, **30,** 2770 (1970b).

Lim, L., and E. S. Canellakis, *Nature*, **227,** 710 (1970).

Linder-Horowitz, M., R. T. Ruettinger, and H. N. Munro, *Biochim. Biophys. Acta.*, **200,** 442 (1970).

Linn, T. G., A. L. Greenleaf, R. G., Shorestein, and R. Losick, *Proc. Natl. Acad. Sci. U.S.*, **70,** 1865 (1973).

Loftfield, R. B., *Prog. Nucleic Acid Res. Mol. Biol.*, **12,** 87 (1972).

Lomen, P. L., L. H. Baker, G. L. Neil, and N. K. Samson, *Cancer Chemother. Rep.*, **59,** 1123 (1975).

Lon, U., *J. Mol. Biol.*, **112,** 661 (1977).

Long, R. A., J. F. Gerster, and L. B. Townsend, *J. Heterocyc. Chem.*, **7,** 863 (1970).

Long, R. A., A. F. Lewis, R. K. Robins, and L. B. Townsend, *J. Chem. Soc., Sect. C*, **1971,** 2443.

Maale, G., G. Stein, and R. Mans, *Nature*, **255,** 80 (1975).

McConnell, E., and A. G. Richards, *Can. J. Microbiol.*, **5,** 161 (1959).

McCredie, K. B., G. P. Bodey, M. A. Burgess, J. U. Gulterman, V. Rodriguez, M. P. Sullivan, and E. J. Freireich, *Cancer Chemother. Rep.*, **57,** 319 (1973).

McCredie, K. B., in Minutes of New Drug Liaison Meeting, National Cancer Institute, Bethesda, 1975, p. 31.

McGuire, W., K. Grotzinger, and R. Young, *Biochem. Pharmacol.*, **27,** 745 (1978).

Mackedonski, V. V., *Biochim. Biophys. Acta*, **390,** 319 (1975).

Mackedonski, V. V., and A. A. Hadjiolov, *Compt. Rend. Bulg. Acad. Sci.*, **27,** 1117 (1974).

Mackedonski, V. V., A. A. Hadjiolov, and K. Šebesta, *FEBS Lett.*, **21,** 211 (1972a).

Mackedonski, V. V., N. Nikolaev, K. Šebesta, and A. A. Hadjiolov, *Biochim. Biophys. Acta*, **272,** 56 (1972b).

McNamara, D. J., and T. E. Webb, *Biochim. Biophys. Acta*, **313,** 356 (1973).

McNamara, D. J., and T. E. Webb, *Arch. Biochem. Biophys.*, **163,** 777 (1974).

Maelicke, A, M. Sprinzl, F. von der Haar, T. A. Khwaja, and F. Cramer, *Eur. J. Biochem.*, **43,** 617 (1974).

Mahy, B. W. J., N. J. Cox, S. J. Armstrong, and R. D. Barry, *Nature, New Biol.*, **234,** 172 (1973).

Majima, R., K. Tsutsumi, H. Suda, and K. Shimura, *J. Biochem.*, **82,** 1161 (1977).

Makabe, O., M. Nakamura, and S. Umezawa, *J. Antibiot.*, **28,** 492 (1975).

Marumoto, R., T. Nishimura, and M. Honjo, *Chem. Pharm. Bull.*, **23,** 2295 (1975).

Marumoto, R., Y. Yoshioka, Y. Furukawa, and M. Honjo, *Chem. Pharm. Bull.*, **24,** 2624 (1976).

Marumoto, R., Y. Yoshioka, T. Naka, S. Shima, O. Miyashita, Y. Maki, T. Suzuki, and M. Honjo, *Chem. Pharm. Bull.*, in press (1979).

Mendecki, J., S. Y. Lee and G. Brawerman, *Biochemistry*, **11,** 792 (1972).

Michaels, A. F., Jr., H. D. Venters, H. B. Worthen, and R. A. Good, *Lab. Invest.*, **11,** 1266 (1962).

Michelot, R. J., N. Lesko, R. W. Stout and J. K. Coward, *Mol. Pharmacol.*, **13,** 368 (1977).

Miko, M., and L. Drobnica, *Experientia*, **31,** 832 (1975).

Milne, G. H. and L. B. Townsend, *J. Chem. Soc. Perkin Trans.* **1972,** 2677.

Mintz, B., *J. Exp. Zool.*, **157,** 85 (1964).

Mizuno, K. and T. Miyazaki, *Chem. Pharm. Bull.*, **24,** 2248 (1976).

Mizuno, K., M. Tsujino, M. Takada, M. Hayashi, K. Otsumi, K. Asano, and T. Matsuda, *J. Antibiot.*, **27,** 775 (1974).

Mizuno, K., S. Yaginuma, M. Hayashi, M. Takada, and N. Muto, *J. Ferment, Technol.*, **53,** 609 (1975).

Modak, M. J., Biochemistry, **17,** 3116 (1978).

Moertel, C. G., A. J. Schutt, R. J. Reitemeir, and R. G. Hahn, *Cancer Chemother. Rep.*, **56,** 649 (1972).

Momparler, R. L., J. Goodman, and M. Karon, *Cancer Res.*, **35,** 2853 (1975).

Momparler, R. L., S. Siegel, F. Airla, T. Lee, and M. Karon, *Biochem. Pharmacol.*, **25,** 389 (1976).

Monesi, V., M. Molinaro, E. Spalletta, and C. Davoli, *Exp. Cell Res.*, **59,** 197 (1970).

Moss, B., in *Comprehensive Virology*, Vol. 3, H. Fraenkel-Conrat and R. R. Wagner, Eds., Plenum Press, New York, 1974, p. 405.

Much, R. P., A. W. Senft, and D. G. Senft, *Biochem. Pharmacol.*, **24,** 407 (1975).

Müller, W. E. G., and R. K. Zahn, *Experientia*, **31,** 1014 (1975).

Müller, W. E. G., H. J. Rohde, R. Steffan, A. Maidhof, M. Lachman, R. K. Zahn, and H. Umezawa, *Cancer Res.*, **35,** 3673 (1975).

Müller, W. E. G., G. Seibert, R. Beyer, H. J. Breyer, A. Maidhof, and R. K. Zahn, *Cancer Res.*, **37,** 3824 (1977).

Müller, W. E. G., R. K. Zahn, and J. Arendes, *FEBS Lett.*, **94,** 47 (1978).

Muthukrishnan, S., W. Filipowicz, J. M. Sierra, G. W. Both, A. J. Shatkin, and S. Ochoa, *J. Biol. Chem.*, **250,** 9336 (1975a).

Muthukrishnan, S., G. W. Both, Y. Furuichi, and A. J. Shatkin, *Nature*, **255,** 33 (1975b).

Myslovataya, M. L., *Genetika*, **10,** 151 (1974).

Nair, V., and P. J. Emmanuel, *J. Am. Chem. Soc.*, **99,** 1571 (1977).

Nair, C. N., and M. J. Owens, *J. Virol.*, **13,** 535 (1973).

Nair, C. N., and D. L. Panicali, *J. Virol.*, **20,** 170 (1976).

Nakanishi, T., F. Tomita, and Furuya, A., *J. Antibiot.*, **30,** 743 (1977).

Nakazato, H., M. Edmonds, and P. W. Kopp, *Proc. Natl. Acad. Sci., U.S.*, **71,** 200 (1974).

Neil, G. L., T. E. Moxley, S. L. Kuentzel, R. C. Manak., and L. J. Hanka, *Cancer Chemother. Rep.*, **59,** 459 (1975).

Neil, G. L., A. E. Barger, B. K. Bhuyan, and D. C. De Sante, *Cancer Res.*, **36,** 1114 (1976).

Nevins, J., and J. E. Darnell, *Cell*, **15,** 1477 (1978).

Nevins, J. R., and W. K. Joklik, *Virology*, **63,** 1 (1975).

Niessing, J., Eur. *J. Biochem.*, **59,** 127 (1975).

Nitschke, R., *Proc. Am. Assoc. Cancer Res.*, **15,** 127 (1974).

O'Brien, T. G., *Cancer Res.*, **36,** 2644 (1976).

O'Brien, S. J., and C. W. Boone, *J. Gen. Virol.*, **35,** 511 (1977).

Ochi, K., S. Iwamoto, E. Hayase, S. Yashima, and Y. Okami, *J. Antibiotics*, **27,** 909 (1974).

Ochi, K., S. Yashima, and Y. Eguchi, *J. Antibiotics*, **28,** 965 (1975).

Ochi, K., S. Kikuchi, S. Yashima, and Y. Eguchi, *J. Antibiotics*, **29,** 638 (1976).

Ofengand, J., and C. M. Chen, *J. Biol. Chem.*, **247,** 2049 (1972).

Okada, N., S. Noguchi, H. Kasai, N. Shinado-Okada, T. Ohgi, and S. Nishimura, *J. Biol. Chem.*, **254,** 3067 (1979).

Okada, N., and S. Nishimura, *J. Biol. Chem.*, **254,** 3061 (1979).

Omura, G. A., *Cancer Treat. Rep.*, **61,** 915 (1977a).

Omura, G. A., *Proc. Am. Assoc. Cancer Res.*, **18,** 25 (1977b).

Pačes, V., J. Doskočil, and F. Šorm, *Biochim. Biophys. Acta*, **161.** 352 (1968).

Paik, W. K., in *Conference on Transmethylation* (NIH, Oct. 16–19, 1978), E. Udsin, C. R. Creveling, and R. T. Borchardt, Eds., Elsevier-North Holland, Amsterdam, 1979, in press.

Païs, M., and H. de Barjac, *J. Carbohydr. Nucleosides Nucleotides*, **1,** 213 (1975).

Palm, P. E., and C. J. Kensler, U.S. Clearing House Fed. Sci. Tech. Inform. P. B. Rep. 194791, 1970, p. 191.

Palm, P. E., E. P. Arnold, P. C. Rachwall, and M. S. Nick, *Toxicol. Appl. Pharmacol*, **25,** 492 (1973).

Panicali, D. L., and C. N. Nair, *J. Virol.*, **25,** 124 (1978).

Parkman, R., E. W. Gelfand, F. S. Rosen, A. Sanderson, and R. Hirschhorn, *N. Engl. J. Med.*, **292,** 714 (1975).

Parks, R. E., and P. R. Brown, *Biochem.*, **12,** 3294 (1973).

Parks, R. E., G. W. Crabtree, C. M. Kong, R. P. Agarwal, K. C. Agarwal, and E. M. Scholar, *Ann. N. Y. Acad. Sci., U.S.* **255,** 412 (1975).

Penman, S., M. Rosbash, and M. Penman, *Proc. Natl. Acad. Sci. U.S.*, **67,** 1878 (1970).

Périès, J., M. Canivet, M. Olivié, and M. Tavitian, *C. R. Acad. Sci., Ser. D.* **278,** 2079 (1974).

Perry, R. P., *Annu. Rev. Biochem*, **45,** 605 (1976).

Perry, R. P., and D. E. Kelley, *J. Mol. Biol.*, **70,** 265 (1972).

Person, A., and G. Beaud, *J. Virol.*, **25,** 11 (1978).

Philipson, L., R. Wall, G. Glickman, and J. E. Darnell, *Proc. Natl. Acad. Sci., U.S.*, **68,** 2806 (1971).

Phillips, S. G., and D. M. Phillips, *J. Cell. Biol.*, **49,** 785 (1971).

Pískala, A., and F. Šorm, *Collect. Czech. Chem. Commun.*, **29,** 2060 (1964).

Pitha, J., R. N. Jones, and P. Pithová, *Can. J. Chem.*, **44,** 1045 (1966).

Plunkett, W., and S. S. Cohen, *Cancer Res.*, **35,** 1547 (1975).

Presant, C. A., T. Vietti, and F. Valeriote, *Cancer Res.*, **35,** 1926 (1975).

Presant, C. A., T. Vietti, and F. Valeriote, *Proc. Assoc. Cancer Res.*, **18,** 73 (1977).

Prokop'ev, V. N., B. M. Yakunin, V. D. Chervyakov, and A. M. Dybitskii, *Parazitologiya*, **10,** 222 (1976).

Prusiner, P., T. Brennan, and M. Sundaralingam, *Biochemistry*, **12,** 1196 (1973).

Prystaš, M., and F. Šorm, *Collect. Czech. Chem. Commun.*, **36,** 1448 (1971a).

Prystaš, M., and F. Šorm, *Collect. Czech. Chem. Commun.*, **36,** 1472 (1971b).

Prystaš, M., L. Kalvoda, and F. Šorm, *Collect. Czech. Chem. Commun.*, **40,** 1775 (1975a).

Prystaš, M., L. Kalvoda, and F. Šorm, *Nucleic Acids Res.*, Spec. Publ., **1,** S77 (1975b).

Prystaš, M., L. Kalvoda, and F. Šorm, *Collect. Czech. Chem. Commun.*, **41,** 1426 (1976).

Pugh, C. S. G., R. T. Borchardt, and H. O. Stone, *Biochemistry*, **16,** 3928 (1977).

Quagliana, J. M., J. Costanzi, and R. O'Bryan, *Proc. Am. Assoc. Cancer Res.*, **15,** 121 (1974).

Rao, M. S., B. C. Wu, J. Waxman, and H. Busch, *Biochem. Biophys. Res. Commun.*, **66,** 1186 (1975).

Raška, K., M. Jurovčík, V. Fučík, R. Tykva, Z. Sormová, and F. Šorm, *Collect Czech. Chem. Commun.*, **31,** 2809 (1966).

Reichenbach, N. L., R. J. Suhadolnik, A. Bruzel, T. Uematsu, and J. Monahan, *J. Biol. Chem.*, in press (1980).

Reichman, M., and S. Penman, *Biochim. Biophys. Acta*, **324,** 282 (1973).

Reichman, M., D. Karlan, and S. Penman, *Biochim. Biophys. Acta*, **299,** 173 (1973).

Rich, M. A., P. Meyers, G. Weinbaum, J. G. Cory, and R. J. Suhadolnik, *Biochim. Biophys. Acta*, **95,** 194 (1965).

Richardson, L. S., R. C. Ting, R. C. Gallo, and A. M. Wu, *Int. J. Cancer*, **15,** 451 (1975).

Riquelme, P. T., L. O. Burzio, and S. S. Koide, *Fed. Proc.*, **36,** 785 (1977).

Rizzo, A. J., and T. E. Webb, *Biochim. Biophys. Acta*, **169,** 163 (1968).

Rizzo, A. J., and T. E. Webb, *Biochim. Biophys. Acta*, **195,** 109 (1969).

Rizzo, A. J., P. Heilpern, and T. E. Webb, *Cancer Res.*, **31,** 876 (1971).

Rizzo, A. J., C. Kelly, and T. E. Webb, *Can. J. Biochem.*, **50,** 1010 (1972).

Rose, K. M., and S. T. Jacob, *Biochemistry*, **15,** 5046 (1976).

Rose, K. M., L. E. Bell, and S. T. Jacob, *Nature*, **267,** 178 (1977a).

Rose, K. M., L. E. Bell, and J. T. Jacob, *Biochim. Biophys. Acta*, **475,** 548 (1977b).

Rose, K. M., F. J. Roe, and S. T. Jacob, *Biochim. Biophys. Acta*, **478,** 180 (1977c).

Rosenberg, G., G. Carlberg, and H. G. Gyllenberg, *J. Appl. Bacteriol.* **34,** 417 (1971).

Ross, A. F., and J. J. Jaffee, *Biochim. Pharmacol.*, **21,** 3059 (1977).

Rottman, F., and A. J. Guarino, *Biochim. Biophys. Acta*, **89,** 465 (1964).

Rowinski, J., D. Salter, and H. Koprowski, *J. Exp. Zool.*, **192,** 133 (1975).

Saiki, J. H., K. McCredie, T. Vietti, J. Hewlett, and F. Morrison, *Am. Assoc. Cancer Res.*, **18,** 339 (1977).

Sakaguchi, K., M. Tsujino, M. Yoshizawa, K. Mizuno, and K. Hayano, *Cancer Res.*, **35,** 1643 (1975a).

Sakaguchi, K., M. Tsujino, K. Mizuno, K. Hayano, and N. Ishida, *J. Antibiot.*, **28,** 798 (1975b).

Sakaguchi, K., M. Tsujino, M. Hayashi, K. Kawai, K. Mizuno, and K. Hayano, *J. Antibiot.*, **29,** 1320 (1976a).

Sakaguchi, K., M. Tsujino, K. Mizuno, and K. Hayano, *Jap. J. Genet.*, **51,** 61 (1976b).

Salvatore, F., E. Borek, H. G. Williams-Ashman, and F. Schlenk, *The Biochemistry of Adenosylmethionine*, Columbia University Press, New York, 1977.

Salzman, N. P., A. J. Shatkin, and E. D. Sebring, *J. Molec. Biol.*, **8,** 405 (1964).

Sawa, T., Y. Fukagawa, I. Homma, T. Takeuchi, and H. Umezawa, *J. Antibiot.*, **20A,** 227 (1967).

Sawicki, S. G., W. Jelinek, and J. E. Darnell, *J. Mol. Biol.*, **113,** 219 (1977).

Scholar, E. M., P. R. Brown, R. E. Parks, Jr., and P. Calabresi, *Blood*, **41,** 927 (1973).

Schramm, K. H., and L. B. Townsend, *Tetrahedron Lett.*, **14,** 1345 (1974).

Schramm, K. H., S. J. Manning, and L. B. Townsend, *J. Heterocyc. Chem.*, **12,** 1021 (1975).

Schramm, V. L., W. E. DeWolf, and F. A. Fullin, *Fed. Proc.*, **37,** Abstr. 1742 (1978).

Šebesta, K., J. Farkaš, K. Horská, and J. Vaňková, in *Microbial Control of Insects, Mites, and Plant Diseases*, H. D. Burgess, Ed., Academic Press, London 1979, in press.

Šebesta, K., and K. Horská, *Biochim. Biophys. Acta*, **169,** 281 (1968).

Šebesta, K., and K. Horská, *Biochim. Biophys. Acta*, **209,** 357 (1970).

Šebesta, K., K. Horská, and J. Vaňková, *Collect. Czech. Chem. Commun.*, **38,** 298 (1973).

Šebesta, K., K. Horshá, and J. Vaňková, Abstracts, 5th Meeting FEBS, Prague, Czechoslovakian Biochemical Society, Prague, 1968, p. 250.

Šebesta, K., K. Horská, and J. Vaňková, *Collect. Czech. Chem. Commun.*, **34,** 891 (1969a).

Šebesta, K., K. Horská, and J. Vaňková, *Collect. Czech. Chem. Commun.*, **34,** 1786 (1969b).

Šebesta, K., and K. Horská, *Collect. Czech. Chem. Commun.*, **38,** 2533 (1973).

Šebesta, K., and H. Sternbach, *FEBS Lett.*, **8,** 233 (1970).

Sehgal, P. B., E. Derman, J. E. Darnell, and I. Tamm, *J. Cell Biol*, **70,** 244a (1976a).

Sehgal, P. B., E. Derman, G. R. Molloy, I. Tamm, and J. E. Darnell, *Science*, **194,** 431 (1976b).

Sekeris, C. E., in *Mechanism of Hormone Action*, P. Karlson, Ed., Academic Press, New York, 1965, p. 165.

Senft, A. W., R. P. Much, P. R. Brown, and D. G. Senft, *Int. J. Parasitol.*, **2,** 1 (1972).

Senft, A. W., G. W. Crabtree, K. C. Agarwal, E. M. Scholar, and R. E. Parks, *Biochem. Pharmacol.* **22,** 449 (1973).

Seto, H., *Agr. Biol. Chem.*, **37,** 2415 (1973).

Seto, H., and H. Yonehara, Annual Meeting of the Agricultural Chemical Society of Japan, Sapporo, Japan, 1975, p. 87.

Seto, H., I. Yamaguchi, N. Ōtake, and H. Yonehara, *Agr. Biol. Chem.*, **32,** 1292 (1968).

Seto, H., N. Ōtake, and H. Yonehara, *Tetrahedron Lett.*, **1972,** 3991.

Seto, H., N. Ōtake, and H. Yonehara, *Agr. Biol. Chem.*, **37,** 2421 (1973).

Seto, H., K. Furihata and H. Yonehara, Annual Meeting of the Agricultural Chemical Society of Japan, Kyoto, Japan, 1976a, p. 258.

Seto, H., K. Furihata, and H. Yonehara, *J. Antibiot.*, **29,** 595 (1976b).

Shapiro, S. K., and F. Schlenk, Eds., *Transmethylation and Methionine Biosynthesis*, The University of Chicago Press, Chicago, 1965.

Sharma, C. B. S. R., *J. Sci. Ind. Res.*, **30,** 571 (1971).

Sharma, C. B. S. R., S. S. V. Prasad, S. B. Pai, and S. Sharma, *Experientia*, **32,** 1465 (1976).

Shatkin, A. J., *Prog. Nucleic Acids Res. Mol. Biol.*, **19,** 3 (1976).

Shchepetil'nikova, V. A., N. S. Fedorinchik, and G. V. Gusev, in *Biological Methods of Protecting Fruit and Vegetable Crops from Pests, Diseases, and Weeds as a Basis for Integrated Systems* (Summaries of Reports from Symposium at Kishinev, U.S.S.R., 1971, English translation by R. L. Busbey and W. Klassen, 1972), 1972, p. 88.

Shealy, Y. F., and J. D. Clayton, *J. Am. Chem. Soc.*, **88,** 3885 (1966); **91,** 3075 (1969).

Sheen, M. R., H. F. Martin, and R. E. Parks, Jr., *Mol. Pharmacol.*, **6,** 255 (1970).

Shieh, T. R., R. F. Anderson, and M. H. Rogoff, *Bacteriol. Proc.*, **1968,** 6.

Shigeura, H. T., and G. E. Boxer, *Biochem. Biophys. Res. Commun.*, **17,** 758 (1964).

Shimotohno, K., Y. Kodama, J. Hashimoto, and K.-I. Miura. *Proc. Natl. Acad. Sci., U.S.*, **74,** 2734 (1977).

Shutt, R., and R. G. Krueger, *J. Immunol.*, **108,** 819 (1972).

Sibatani, A., S. R. deKloet, V. G. Allfrey, and A. E. Mirsky, *Proc. Natl. Acad. Sci. U.S.*, **48,** 471 (1962).

Siev, M., R. Weinberg, and S. Penman, *J. Cell Biol.*, **41,** 510 (1969).

Shalko, R. G., and J. M. D. Morse, *Teratology*, **2,** 47 (1969).

Skoog, F., and D. J. Armstrong, *Annu. Rev. Plant Physiol.*, **21,** 359 (1970).

Skoog, F., F. M. Strong, and C. O. Miller, *Science*, **148,** 532 (1965).

Smetana, K., I. Raška, and K. Šebesta, *Exp. Cell Res.*, **87,** 351 (1974).

Smith, J. D., *Prog. Nucleic Acids Res. Mol. Biol.*, **16,** 25 (1976).

Smith, C. G., G. D. Gray, R. G. Carlson and A. R. Hanze, *Advan. Enz. Regu.*, **5,** 121 (1973).

Smith, C. G., L. M. Reineke, M. R. Burch, A. M. Shefner, and E. E. Muirhead, *Cancer Res.*, **30,** 69 (1970).

Smuckler, E. A., and A. A. Hadjiolov, *Biochem. J.*, **129,** 153 (1972).

Smulson, M., and R. J. Suhadolnik, *J. Biol. Chem.*, **242,** 2872 (1967).

Snyder, F. F., J. Mendelsohn, and J. E. Seegmiller, *J. Clin. Invest.*, **58,** 654 (1977).

Söll, D., and P. R. Schimmel, *The Enzymes*, **10,** 3rd Ed., 489 (1974).

Somerville, H. J., and H. M. Swain, *FEBS Lett.*, **54,** 330 (1975).

Šorm, F., and Veselý, J., *Neoplasma*, **15,** 339 (1968).

Šorm, F., A. Piskala, A. Čihǎk, and J. Veselý, *Experientia*, **20,** 202 (1964).

Spiegel, S., and A. Marcus, *Nature*, **256,** 228 (1975).

Sprinzl, M., and F. Cramer, *Nature, New Biol.*, **245,** 3 (1973).

Sprinzl, M., and F. Cramer, *Eur. J. Biochem.*, **81,** 579 (1977).

Sprinzl, M., and F. Cramer, *Prog. Nucleic Acid Res. Mol. Biol.*, **22,** 2 (1978).

Srivastava, B. I. S., *Cancer Res.*, **34,** 1015 (1974).

Stegman, R. J., A. W. Senft, P. R. Brown, and R. E. Parks, *Biochem Pharmacol.*, **22,** 459 (1973).

Streeter, D. G., J. T. Witkowski, G. P. Khare, R. W. Sidwell, R. J. Bauer, R. K. Robins, and L. N. Simon, *Proc. Natl. Acad. Sci., U.S.*, **70,** 1174 (1973).

Suhadolnik, R. J., *Nucleoside Antibiotics*, Wiley, New York, 1970.

Suhadolnik, R. J., and T. Uematsu, *J. Biol. Chem.*, **245,** 4365 (1970).

Suhadolnik, R. J., and T. Uematsu, *Carbohydr. Res.*, **61,** 545 (1978).

Suhadolnik, R. J., G. Weinbaum, and H. P. Meloche, *J. Am. Chem. Soc.*, **86,** 948 (1964).

Suhadolnik, R. J., T. Uematsu, and H. Uematsu, *Biochim. Biophys. Acta*, **149,** 41 (1967).

Suhadolnik, R. J., S. I. Finkel, and B. M. Chassy, *J. Biol. Chem.*, **243,** 3532 (1968a).

Suhadolnik, R. J., T. Uematsu, H. Uematsu, and R. G. Wilson, *J. Biol. Chem.*, **243,** 2761 (1968b).

Suhadolnik, R. J., M. B. Lennon, T. Uematsu, J. E. Monahan, and R. Baur, *J. Biol. Chem.*, **252,** 4125 (1977a).

Suhadolnik, R. J., R. Baur, D. M. Lichtenwalner, T. Uematsu, J. H. Roberts, S. Sudhakar, and M. Smulson, *J. Biol. Chem.*, **252,** 4134 (1977b).

Sussman, M., S. Alexander, C. Boschwitz, R. Brackenbury, A. Cohen, and J. Schindler, *Developmental Biology, Pattern Formation, and Gene Regulation*, Vol. 2, S. McMahon and C. F. Fox, Eds. Benjamin, Menlo Park, 1975, p. 89.

Swart, C., and L. D. Hodge, *Virology*, **84,** 374 (1978).

Tabor, C. W., and H. Tabor, *Annu. Rev. Biochem.*, **45,** 285 (1976).

Tabor, H., S. M. Rosenthal, and C. W. Tabor, *J. Biol. Chem.*, **233,** 907 (1958).

Tan, C., J. R. Burchenal, B. Clarkson, M. Feinstein, E. Garcia, J. Sidhu, and I. H. Krakoff, *Proc. Am. Assoc. Cancer Res.*, **14,** 97 (1973).

Tasca, R. J., and N. Hillman, *Nature*, **225,** 1022 (1970).

Tavitian, A., S. C. Uretsky, and G. Acs, *Biochim. Biophys. Acta*, **157,** 33 (1968).

Tavitian, A., S. C. Uretsky, and G. Acs, *Biochim. Biophys. Acta*, **179,** 50 (1969).

Taylor, J. M., and C. P. Stanners, *Biochim. Biophys. Acta*, **155,** 424 (1968).

Thomson, J. L., and J. D. Biggers, *Exp. Cell Res.*, **41,** 411 (1966).

Tilghman, S. M., R. W. Hanson, L. Reshef, M. F. Hopgood, and F. J. Ballard, *Proc. Natl. Acad. Sci. U.S.*, **71,** 1304 (1974).

Tobey, R. A., *Cancer Res.*, **32,** 2720 (1972).

Todde, P. S., and P. N. Campbell, in *Techniques in Protein Biosynthesis, Vol. 2*, P. N. Campbell and J. R. Sargent, Eds., Academic Press, New York, 1969, p. 251.

Torrence, P. F., E. DeClercq, J. A. Waters, and B. Witkop, *Biochemistry*, **13,** 4400 (1974).

Torrence, P. F., E. DeClercq, J. A. Waters, and B. Witkop, *Biochem. Biophys. Res. Commun.*, **62,** 658 (1975).

Townsend, L. B., in *Handbook of Biochemistry and Molecular Biology, Nucleic Acids*, Vol. I, 3rd ed. G. D. Fasman, Ed., CRC Press, Cleveland, 1975, p. 271.

Townsend, L. B., and G. H. Milne, *Ann. N. Y. Acad. Sci.*, **255,** 91 (1975).

Townsend, L. B., R. A. Long, J. P. McGraw, D. W. Miles, R. K. Robins, and H. Eyring, *J. Org. Chem.*, **39,** 2023 (1974).

Troetel, W. M., A. J. Weiss, J. E. Stambaugh, J. F. Laucius, and R. W. Manthel, *Cancer Chemother. Rep.*, **56,** 405 (1972).

Truman, J. T., and S. Frederiksen, *Biochim. Biophys. Acta*, **182,** 36 (1969).

Truman, J. T., and H. Klenow, *Mol. Pharmacol.*, **4,** 77 (1968).

Tsutsumi, K., R. Majima-Tsutumi, and K. Shimura, *J. Biochem.*, **84,** 169 (1978).

Uesugi, S., T. Tezuka, and M. Ikehara, *Biochemistry*, **14,** 2903 (1975).

Uretsky, S. C., G. Acs, E. Reich, M. Mori, and L. Altwerger, *J. Biol. Chem.*, **243,** 306 (1968).

Vadlamudi, S., J. N. Choudry, V. S. Waravdekar, I. Kline, and A. Goldin, *Cancer Res.*, **30,** 362 (1970).

Vaňková, J., and K. Horská, *Acta Entomol. Bohemoslov.*, **72,** 7 (1975).

van Oortmerssen, E. A. C., B. A. Regènsburg, J. G. Tasseron-de Jong, and L. Bosch, *Virology*, **65,** 238 (1975).

Venkov, P. V., L. I. Stateva, and A. A. Hadjiolov, *Biochim. Biophys. Acta*, **474,** 245 (1977).

Veselý, J., and A. Čihák, *Cas. lek. Cesk.*, **114,** 605 (1975).

Veselý, J., and A. Čihák, *Cancer Res.*, **37,** 3684 (1977).

Veselý, J., and A. Čihák, *Pharmac. Ther.*, **2,** 813 (1978).

Veselý, J., J. Seifert, A. Čihák, and F. Šorm, *Int. J. Cancer*, **1,** 31 (1966).

Veselý, J., A. Čihák, and F. Šorm, *Cancer Res.*, **28,** 1995 (1968).

Vince, R., R. G. Almquist, C. L. Ritter, F. N. Shirota, and H. T. Nagasawa, *Life Sci.*, **18,** 345 (1976).

Vogler, W. R., D. S. Miller, and J. W. Keller, *Blood*, **48,** 331 (1976).

Von der Haar, F., and E. Gaertner, *Proc. Natl. Acad. Sci. U.S.*, **72,** 1378 (1975).

Von der Haar, F., and F. Cramer, *Biochemistry*, **15,** 4131 (1976).

Von der Haar, F., and F. Cramer, *Biochemistry*, **17,** 3139 (1978a).

Von der Haar, F., and F. Cramer, *Biochemistry*, **17,** 4509 (1978b).

Von Hoff, D. D., H. Handelsman, and M. Slavík, Clinical Brochure, National Cancer Institute, 1975.

Von Hoff, D. D., M. Slavík, and F. M. Muggia, *Ann. Intern. Med.*, **85,** 237 (1976).

Voytek, P., B. Futterman, and M. M. Abbasi, *Proc. Am. Assoc. Cancer Res.*, **18,** 126 (1977).

Wainfan, E., and B. Landsberg, *Biochem. Pharmacol.*, **22,** 493 (1973).

Wainfan, E., J. Chu, and G. B. Chheda, *Biochem. Pharmacol.*, **24,** 83 (1973).

Walter, R. D., *Biochim. Biophys. Acta*, **429,** 137 (1976).

Walter, R. D., and F. Ebert, *Hoppe-Seyler's Z. Physiol. Chem.*, **358,** 23 (1977).

Ward, D. C., and E. Reich, *Biochemistry*, **61,** 1491 (1968).

Watanabe, K. A., I. M. Wempen, and J. J. Fox, *Carbohydr. Res.*, **21,** 148 (1972).

Watanabe, K. A., T. M. K. Chiu, D. H. Hollenberg, and J. J. Fox, *J. Org. Chem.*, **29,** 2482 (1974).

Watanabe, K. A., T. M. K. Chiu, U. Reichman, C. K. Chiu, and J. J. Fox, *Tetrahedron*, **32,** 1493 (1976a).

Watanabe, K. A., D. H. Hollenberg, and J. J. Fox, *J. Antibiot.* (*Tokyo*), **29,** 597 (1976b).

Weinberg, R. A., *Arch. Biochem. Biophys.*, **42,** 329 (1973).

Weinberg, R. A., and S. Penman, *J. Mol. Biol.*, **47,** 169 (1970).

Weiss, S. R., and M. A. Bratt, *J. Virology*, **16,** 1575 (1975).

Weiss, J. W., and H. C. Pitot, *Arch. Biochem. Biophys.*, **160,** 119 (1974a).

Weiss, J. W., and H. C. Pitot, *Cancer Res.*, **34,** 581 (1974b).

Weiss, J. W., and H. C. Pitot, *Biochemistry*, **14,** 316 (1975).

Weiss, A. J., E. E. Stambaugh, M. J. Mastrangelo, J. F. Laucius, and R. E. Bellet, *Cancer Chemother. Rep.*, **56,** 413 (1972).

Weiss, A. J., G. E. Metter, T. E. Nealon, J. P. Keanan, G. Ramirez, A. Swaiminatha, W. S. Fletcher, S. E. Moss, and R. W. Manthei, *Cancer Treat. Rep.*, **61,** 55 (1977).

Wolberg, G., T. P. Zimmerman, K. Hiemstra, M. Winston, and L. C. Chu, *Science*, **187,** 957 (1975).

Wolf, S. F., and D. Schlessinger, *Biochemistry*, **16,** 2783 (1977).

Wu, A. M., R. C. Ting, M. Paran, and R. C. Gallo, *Proc. Natl. Acad. Sci. U.S.*, **69,** 3820 (1972).

Wilson, S. G. F., W. Heyman, and D. Goldthwait, *Pediatrics*, **25,** 228 (1962).

Winicov, I., *Biochemistry*, **18,** 1575 (1979).

Wolfenbarger, D. A., A. A. Guerra, H. T. Dulmage, and R. D. Garcia, *J. Econ. Entomol.*, **65,** 1245 (1972).

Yoshioka, H., K. Nakatsu, M. Hayashi, and K. Mizuno, *Tetrahedron Lett.*, No. **46,** 4031 (1975).

Zadražil, S., V. Fučík, P. Barth, Z. Sormová, and F. Šorm, *Biochim. Biophys. Acta*, **108,** 701 (1965).

Zähringer, J., A. M. Konijn, B. S. Baliga, and H. N. Munro, *Biochem. Biophys. Res. Commun.*, **65,** 583 (1975).

Zähringer, J., B. S. Baliga, and H. N. Munro, *Proc. Natl. Acad. Sci. U.S.*, **73,** 857 (1976).

Zain, B. S., R. L. P. Adams, and R. C. Imrie, *Cancer Res.*, **33,** 40 (1973).

Žemlička, J., *J. Am. Chem. Soc.*, **97,** 5896 (1975).

Žemlička, J., S. Chladek, D. Ringer, and K. Quiggle, *Biochemistry*, **14,** 5239 (1975).

Chapter 4 Inhibition of DNA Synthesis, Viruses, and Neoplastic Tissue

4.1 9-β-D-ARABINOFURANOSYLADENINE 217

Chemical Synthesis of [5′-^{2}H]- and [5′-^{3}H]Ara-A, Ara-A 5′-Monophosphate (Ara-AMP), and Ara-A 219
Derivatives of Ara-A 219
Biosynthesis of Ara-A 221
Biochemical Properties 221
Mechanism of Action of Ara-A 221
Incorporation of Ara-ATP into RNA and the 3′-Terminus of tRNA 224
Effect of β-Ara-ATP and α-Ara-ATP on Terminal Deoxynucleotidyl Transferase (TdT) 225
Inhibition of Adenosine Deaminase 225
Suicide Inactivation of S-Adenosylhomocysteine (AdoHcy) Hydrolase by Ara-A and Cordycepin 227
Effect on Protein Synthesis 227
Antiviral Activity Against DNA Viruses in Cell Culture 227
Antagonism of Deoxyadenosine to Ara-A 229
Antiviral Activity Against RNA Viruses in Cell Culture 229
Effect of Ara-A on Viral Infections in Humans 230
Antiviral Activity of Ara-A in Animals 231
Cancer Chemotherapy in Humans 231
Human Pharmacology Trials of Ara-A 232
Effect of Ara-A on Postnatal Growth and Development of the Rat 232
Toxicology of Ara-AMP 232

4.2 1-β-D-ARABINOFURANOSYLTHYMINE 233

Biochemical Properties 233
Antibacterial Activity 233
Antiviral Activity 233
Inhibition of Mammalian DNA Polymerase-α 236

4.3 OXAZINOMYCIN 236

Discovery, Production, and Isolation 237
Physical and Chemical Properties 237
Structural Elucidation 237
Chemical Synthesis of Minimycin 238
Biosynthesis of Oxazinomycin 238
Biochemical Properties 240

4.4 1-METHYLPSEUDOURIDINE 241

Discovery, Production, and Isolation 242
Physical and Chemical Properties 242
Chemical Synthesis 243
Inhibition of Growth 243
Biochemical Properties 243

4.5 NEBULARINE 244

Biochemical Properties 244
Protection of Mice Against Nebularine Toxicity by Nitrobenzylthioinosine (MBMPR) 245
Pharmacology 245

4.6 5,6-DIHYDRO-5-AZATHYMIDINE 245

Discovery, Production, and Isolation 246
Physical and Chemical Properties 247
Inhibition of Growth 247
Biosynthesis of DHAdT 247
Biochemical Properties 249
Antiviral Activity 249
Immunosuppressive Activity 249
Resistance to 5,6-Dihydro-5-azathymidine 249

SUMMARY 250

REFERENCES 251

This chapter reviews the naturally occurring nucleoside analogs that inhibit DNA synthesis in eukaryotes and prokaryotes, have antiviral activity, and are antineoplastic. These analogs are 9-β-D-arabinofuranosyladenine (ara-A, Vidarabine), 1-β-D-arabinofuranosylthymine (ara-T), 5,6-dihydro-5-azathymidine (DHAdT), oxazinomycin (minimycin), 1-methylpseudouridine, and nebularine.

4.1 9-β-D-ARABINOFURANOSYLADENINE

Bergmann and coworkers isolated three purine and pyrimidine nucleosides from the Caribbean sponge, *Tethya crypta* (Fig. 4.1) (Bergmann and Feeney, 1950, 1951; Bergmann and Stempien, 1957). Ara-A has been isolated from the culture filtrates of *Streptomyces antibioticus* by the Warner-Lambert/Parke, Davis research group (1967). The chemical synthesis of ara-A was first reported by Goodman and coworkers (Lee et al., 1960; Reist et al., 1962).

Historically, the initial, outstanding contributions by Cohen and his coworkers on the biochemical properties of ara-A in prokaryotes and eukaryotes set the stage for the numerous studies on ara-A and its degradation products that were to follow. Cohen and coworkers were the first to demonstrate that ara-A inhibited *E. coli* as well as mouse fibroblasts in culture (Hubert-Habart and Cohen, 1962; Doering et al., 1966; Leung et al., 1966; for more details, see the first review of ara-A biochemistry by Cohen, 1966).

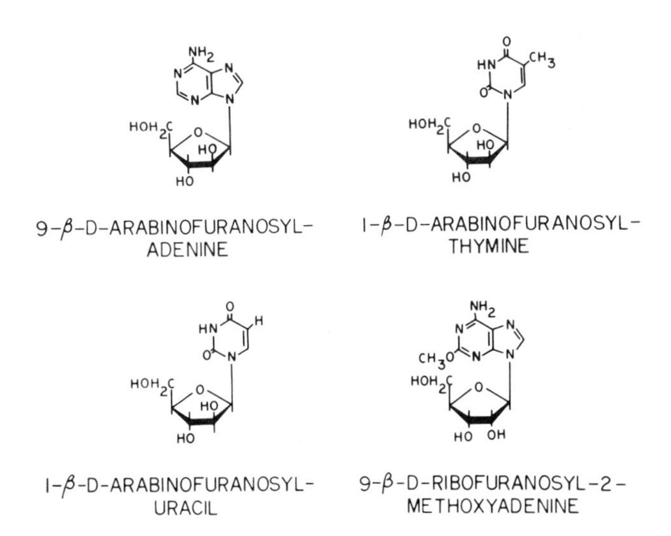

Figure 4.1 Structures for ara-A, ara-T, ara-U, and spongosine.

Hubert-Habart and Cohen (1962) and Leung et al. (1966) reported on the lethality of ara-A to their polyauxotropic strain of *E. coli*. The inhibition in *E. coli* and mouse fibroblasts is due to the inhibition of DNA synthesis (Cohen, 1966; Hubert-Habart and Cohen, 1966; Doering et al., 1966; Plunkett and Cohen, 1975a). Adenine in a 1:1 ratio with ara-A reversed the inhibition of a purine-deficient strain of *E. coli* (Hubert-Habart and Cohen, 1962).

Cohen (1966) was also the first to report on the deamination of ara-A to ara-Hx, as well as the isolation of ara-Hx. Although ara-Hx is nontoxic, Cohen (1966) and Leung et al. (1966) demonstrated that ara-Hx is lethal to *E. coli* mutant $T^-A^-U^-$ad. Subsequently, Furth and Cohen (1968) synthesized the 5′-mono-, 5′-di-, and 5′-triphosphates of ara-A and demonstrated that ara-ATP is a competitive inhibitor of dATP with the partially purified calf thymus DNA-dependent DNA polymerase. Moore and Cohen (1968) subsequently demonstrated that ara-ADP and ara-ATP inhibit rat tumor ribonucleotide reductase. York and LePage (1966) had reported earlier that ara-ATP is a noncompetitive inhibitor of dATP with DNA-dependent DNA polymerase from TA3 ascites tumor cells. Nichols (1964) reported that ara-A induced chromosome breaks in human leukocytes. The reports by deGarilhe and deRudder (1964), by Parke, Davis (1967), and Brink and LePage (1965) on the antiviral activity and toxicity of ara-A toward herpes simplex virus and mammalian cells in culture stimulated research into the biochemical and pharmacological properties of ara-A.

In terms of the effect of ara-A on tumor-bearing animals, Brink and LePage (1964a, b) and LePage (1970) demonstrated that ara-A increased the survival time of the host animals. Subsequent studies showed that ara-ATP inhibits DNA polymerases, bacteria, mammalian cells in culture, DNA viruses, and RNA viruses. The DNA viruses inhibited by ara-A are types I and II herpes simplex, cytomegalo, varicella zoster, herpes B, Saimiri, vaccinia, pseudorabies, myxoma, monkey pox, and fish lymphomcytis virus. Ara-A shows little activity against adenovirus and no activity against polyoma viruses. Although ara-A does not inhibit RNA viruses such as poliovirus or influenza virus, it does inhibit vesicular stomatitis virus, rabies virus, and infectious pancreatic necrosis virus replication in cell culture. Ara-A also inhibits some RNA tumor viruses, which have the ability to produce virus-specific DNA on an RNA template. Because DNA viruses and RNA tumor viruses can code for DNA polymerases, a possible molecular basis for the antiviral activity of ara-A might be the inhibition of viral DNA polymerase or virus-specific enzymes that are essential for DNA synthesis.

The report of double-blind and open studies by Pavan-Langston and Buchanan (1976) on the use of ara-A in humans suffering from herpes simplex keratitis is the main publication that has resulted in ara-A becoming the first drug to be approved by the FDA for treatment of a life-threatening viral infection. Whitley et al. (1977) subsequently used ara-A to treat type 1 herpes simplex encephalitis. Administration of ara-A in the course of infection, before the advent of coma, was reported to have beneficial effects. These studies have also contributed to FDA approval for use of ara-A in the treatment of herpes simplex encephalitis. Other human viral infections (i.e., neonatal herpes, herpes zoster, and mucocutaneous herpes) have since been included in these studies (Whitley, private communication). In addition, ara-AMP has been successfully administered to humans to treat type 1 herpes simplex encephalitis. The successful use of ara-AMP again is attributed to the work of Cohen and coworkers (Ortiz et al., 1972; Cohen, 1975; Cohen and Plunkett, 1975; Plunkett and Cohen 1975b; Plunkett et al., 1974), who demonstrated that ara-AMP and 3′,5′-c-ara-AMP are taken up intact by mammalian cells. Ara-A is a potentially useful antineoplastic agent. It has been employed in patients with chronic myelogeneous leukemia in acute blast crisis.

The reactions influenced by ara-A have been summarized by Müller (Fig. 4.2). The chemical, biochemical, and antiviral properties of ara-A have been extensively reviewed (Cohen, 1966, 1976; Suhadolnik, 1970; Cozzarelli, 1977; Townsend, 1975; Bloch, 1978; Pavan-Langston et al., 1975a).

Chemical Synthesis of [5′-^{2}H]- and [5′-^{3}H]Ara-A, Ara-A 5′-Monophosphate (Ara-AMP), and Ara-A

Pharmacological studies require the labeling of ara-A or ara-AMP that is metabolically resistant to the removal of ^{2}H or ^{3}H from positions C-2 and C-8 of ara-A. Consequently, Baker and Haskell (1975) described the synthesis of [5′-^{2}H]- and [5′-^{3}H]ara-A and [5′-^{3}H]ara-AMP. The synthesis required the oxidation of C-5′ of ara-A to the aldehyde followed by reduction with $NaBD_4(T_4)$.

Ranganathan (1975) has described a simple and novel approach to the stereospecific synthesis of ara-A.

Derivatives of Ara-A

The syntheses of prodrugs of ara-A have been described (Baker et al., 1978, 1979). These prodrugs are 5′-(*O*-acyl)-2′,3′- and 5′-di-*O*-acyl derivatives of ara-A. These derivatives offer a range of solubilities and lipophilicities for

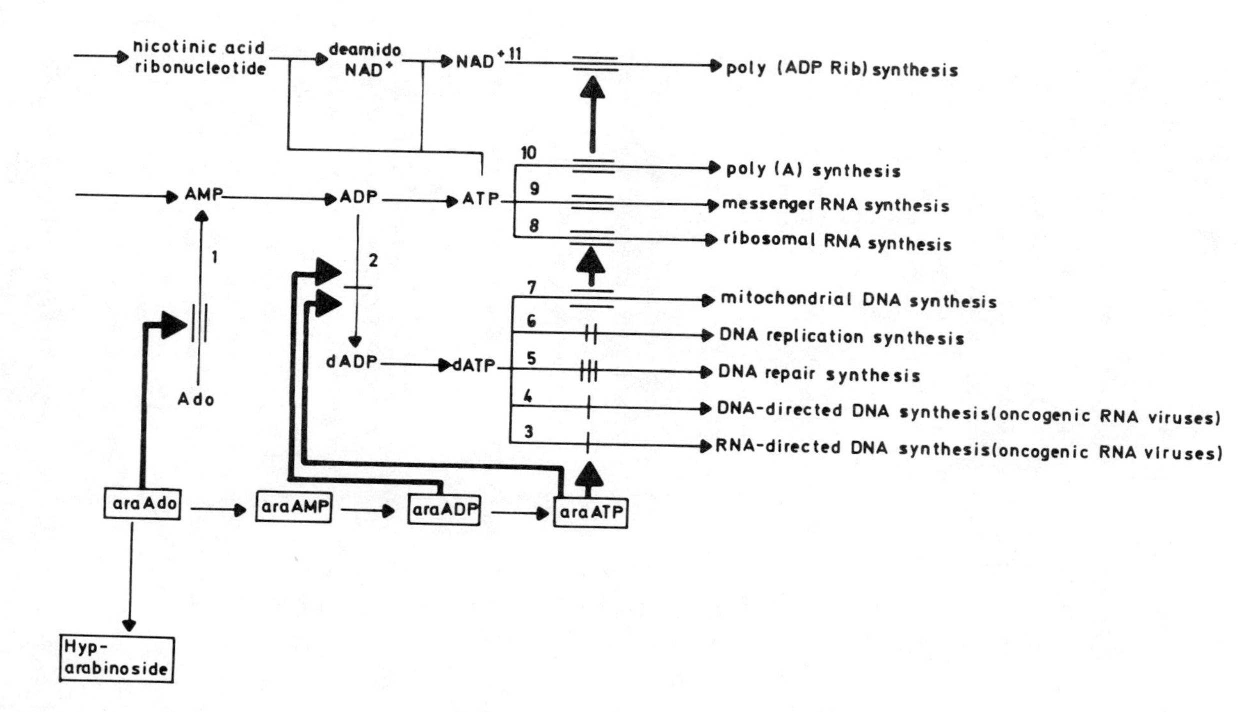

Figure 4.2 Known influences of ara-A or its metabolic derivatives on enzymes from eukaryotic cells and oncogenic RNA viruses that are involved in nucleic acid syntheses. The thin arrows mark the metabolic pathways; the heavy arrows mark known effects of ara-A and its phosphorylated derivatives on specific enzyme systems. Enzyme systems: (*1*) nucleoside kinase; (*2*) adenosine diphosphate reductase; (*3*) RNA-directed DNA polymerase; (*4*) DNA-directed DNA polymerase; (*5*) DNA polymerase-β; (*6*) DNA polymerase-α; (*7*) mitochondrial DNA-dependent DNA polymerase; (*8*) DNA-dependent RNA polymerase I; (*9*) DNA-dependent RNA polymerase II: (*10*) poly(A) polymerase; (*11*) poly(adenosine diphosphate ribose) polymerase. (═══) no inhibition; (—+—) weak inhibition; (—++—) strong inhibition; (—+++—) strongest inhibition. Ado, adenosine; Hyp-arabinoside, ara-Hx. Reprinted with permission from Müller et al. 1977a.

improved membrane transport. All are resistant to deamination. The 5′-(*O*-valeryl) derivative of ara-A shows a marked increase in antiviral activity over ara-A. In addition, ara-A-5′-valerate is a competitive inhibitor of calf intestinal adenosine deaminase (Lipper et al., 1978) (see Chapter 5).

Biosynthesis of Ara-A

Cohen (1966) reported that *Tethya crypta* did not incorporate radioactive glucose, uracil, or D-arabinose into thymine or the arabinonucleosides; no $^{14}CO_2$ was produced. With the discovery of ara-A in the culture filtrates of *S. antibioticus*, it has been possible to study the biosynthesis of ara-A *in vivo* and *in vitro*. As with puromycin, cordycepin, and 3′-aminoadenosine, adenosine is a direct precursor for the biosynthesis of ara-A by *S. antibioticus*. This was demonstrated by the use of ^{3}H, ^{14}C, and ^{15}N labeled adenosine (Farmer and Suhadolnik, 1972; Farmer et al., 1973). In more recent studies, adenosine 2′-epimerase has been isolated and purified tenfold from *S. antibioticus* (Wu and Suhadolnik, manuscript in preparation). Adenosine is converted to ara-A by the partially purified 2′-epimerase. Although tritium from $NaBH_4$ added to enzyme assays is not incorporated into ara-A, tritium from $^{3}H_2O$ is incorporated exclusively into the arabinosyl moiety of ara-A. The addition of [2′-^{3}H]adenosine results in the release of $^{3}H_2O$. These data suggest that C-2′ of adenosine could be oxidized to 2′-keto-adenosine followed by an epimerization to the 2′,3′-enediol and reduction to ara-A. Tubercidin is converted to ara-tubercidin by this enzyme.

Biochemical Properties

Mechanism of Action of Ara-A. Ara-A is a cytostatic analog of 2′-deoxyadenosine. Ara-ATP is a competitive inhibitor of 2′dATP. Ara-A is deaminated to ara-Hx which then forms ara-IMP, which is oxidized to ara-XMP, which can be converted to ara-GMP (Miller and Adamczyk, 1976; Spector, 1975). Neither ara-A nor ara-Hx is a substrate for purified calf spleen nucleoside hydrolase (Sweetman et al., 1975).

Ara-ATP inhibits DNA polymerase-α and DNA-polymerase-β of calf thymus and uninfected rabbit kidney cells (Müller et al., 1977a; Okura and Yoshida, 1978). DNA polymerase-α is more sensitive to ara-ATP than is DNA polymerase-β. With quail oviduct, DNA polymerase-α and DNA polymerase-β are inhibited by ara-ATP (Müller et al., 1975); however, human leukemic T-cell lines (Molt-4) are about 50 times as sensitive to the cytotoxic action of ara-A as are B-cell lines (RPMI-8432) (DiCioccio and Srivastava, 1977). The data of Müller et al. show that ara-ATP inhibits DNA polymerase-α and DNA polymerase-β, but not DNA polymerase-γ when assayed with poly(A)·$(dT)_{12-18}$ as template primer and dTTP as substrate. The selective inhibition of viral DNA synthesis in KB cells infected with HSV prompted Reinke et al. (1978) to examine the effect of ara-ATP on HeLa cell DNA polymerase-α, polymerase-β, and polymerase-γ from uninfected HeLa cells and from HSV-1 infected HeLa cells (Fig. 4.3). The data in Fig. 4.3 show that whereas ara-ATP selectively inhibits the HSV-induced (type 1-induced DNA polymerase, HDP-1) DNA polymerase; the cellular DNA polymerase-β and DNA polymerase-γ are insensitive to ara-ATP. In HSV-infected cells treated with ara-A, the molecular weight of viral DNA is less than 2.6 million. This DNA is not assembled into high molecular weight DNA (Fig. 4.4). The HSV-DNA from the uninfected cells (Fig. 4.4*A*), compared to HSV-infected cells incubated with ara-A, results in an incorporation of ara-AMP into the DNA. The incorporated ara-AMP results in a loss of synthesis of the viral DNA (Fig. 4.4*B*).

Müller et al. (1977a, c) and Waqar et al. (1971) concluded that the DNA-dependent DNA polymerase induced by HSV type 1 incorporates ara-AMP into the 3′-end of DNA (Müller et al., 1977a, c). More recently, Pelling et al. (1978) showed by enzymatic digestion and chromatographic analysis that 92–96% of the [^{3}H]ara-AMP incorporated into HSV-1 DNA occurs by means of internucleotide linkage. Ara-A is a chain terminator in this system (Drach, personal communication).

With Rous sarcoma virus, Müller et al. (1975) reported that the RNA-directed DNA polymerase and the DNA-directed DNA polymerase are less affected by ara-ATP than are the cellular DNA polymerases. While studying the alterations of the activities of the distinct DNA and RNA polymerase species after infection of rabbit kidney cells with HSV, Müller et al.

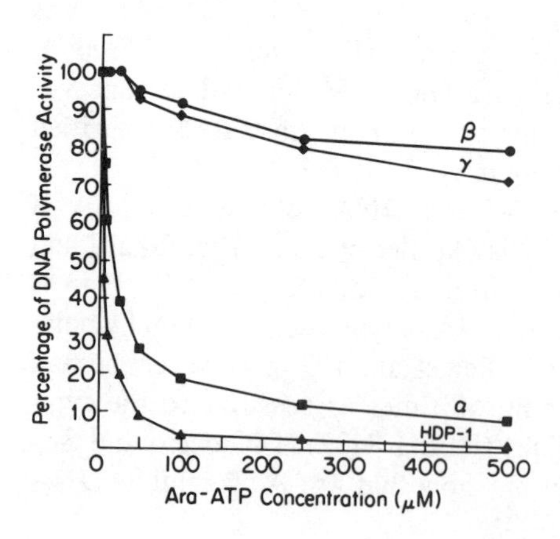

Figure 4.3 Inhibition of HeLa cell DNA polymerases by ara-ATP. Polymerase-α, polymerase-β, and polymerase-γ were isolated from uninfected cells. HDP-1 was isolated from cells infected with HSV-1. Reprinted with permission from Reinke et al., 1978.

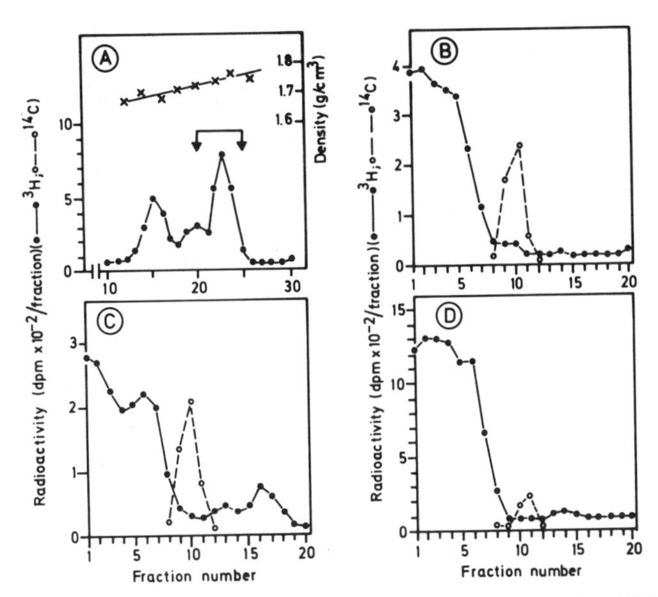

Figure 4.4 Analysis of the location of ara-AMP moieties incorporated into HSV-DNA. The HSV-infected cells were incubated 7 h postinfection with [³H]ara-A as follows: experiment 1, incubation for 1 h followed by harvesting; experiment 2, incubation for 1 h, washing the cells free of unincorporated ara-A, and a subsequent incubation for an additional 5 h; experiment 3, incubation for 6 h. The cells were harvested and the material from five parallel assays was pooled. DNA was extracted and HSV-DNA was separated from cellular DNA by isopycnic centrifugation in CsCl gradients, as shown for experiment 1 in *A*. Viral DNA (density 1.721–1.750 g/cm³; marked by arrows in *A*) from the CsCl gradients was dialyzed and analyzed on alkaline sucrose gradients. Alkaline sucrose sedimentation of HSV-DNA from experiment 1 (*B*), from experiment 2 (*C*), and experiment 3 (*D*). Direction of sedimentation is from left to right. [³H]Ara-A-labeled HSV-DNA (●); T7 phage-marked DNA (○). Reprinted with permission from Müller et al., 1977c.

(1978a) determined the sensitivity of three DNA-dependent DNA polymerases with α- and β-ara-ATP. The K_i's (β-ara-ATP) for DNA polymerase-α, DNA polymerase-β, and HSV DNA polymerase were 7.4, 5.6, and 0.14 μM, respectively. As an indication of the relative affinities of the polymerase for inhibition and substrate in competitive inhibition (i.e., K_m/K_i), the HSV-DNA polymerase has the highest affinity for β-ara-ATP. α-Ara-ATP is a potent inhibitor of mammalian DNA polymerase-α, but does not inhibit either DNA polymerase-β or HSV-induced DNA polymerase. With mouse lymphoma cells in culture, α-ara-A is cytostatic. Deoxyadenosine, but not adenosine, reverses the inhibition of ara-A. DNA, but not RNA or protein, synthesis is inhibited by α-ara-A. α-Ara-A is converted to α-ara-ATP and incorporated into DNA (Müller, 1978b, 1979). In HSV-infected L5178Y cells, α-ara-A is intracellularly phosphorylated to α-ara-ATP. α-Ara-ATP has no effect on HSV-DNA polymerase, but is a potent inhibitor of host cell DNA polymerase-α (Müller, 1979).

Schwartz et al. (1976) and Shipman et al. (1976) noted that ara-A exerted a selective inhibition of viral DNA. Most DNA synthesis was not as sensitive to ara-A. Drach and Shipman (1977) have subsequently developed an evaluation procedure to quantitate the selective activity of ara-A and other potentially useful inhibitors.

Ara-ATP does not inhibit *E. coli* DNA polymerase or *E. coli* RNA polymerase (Cardeilhac and Cohen, 1964; Furth and Cohen, 1967). In addition, eukaryotic or prokaryotic DNA-dependent RNA polymerase, nuclear poly(A) polymerase, and poly(ADP-ribose) polymerase are not inhibited by ara-ATP (Müller et al., 1975, 1977a; Müller, 1976). The incorporation of ara-AMP into DNA may explain the damaging effect of ara-A on chromosomes as reported by Nichols (1964). The portion of the cell cycle most sensitive to ara-A is the S-phase; there is a larger effect on viruses that have a DNA phase in their cycle (Cozzarelli, 1977; Müller et al., 1977b). Ara-A preferentially inhibits the replication of the RNA tumor virus, murine leukemia virus, which has reverse transcriptase activity (Shannon et al., 1974). Rose and Jacob (1978) and Leonard and Jacob (1979) have reported that 3′dATP inhibits RNA synthesis while ara-ATP selectively inhibits the polyadenylation reaction *in vitro*.

Incorporation of Ara-ATP into RNA and the 3′-Terminus of tRNA. Although ara-ATP is incorporated into RNA, most of the radioactivity in RNA from growing mouse fibroblasts treated with [^{3}H]ara-A was associated with the 2′(3′)AMP after alkaline hydrolysis (Plunkett and Cohen, 1975a; Müller et al., 1975). Most of the radioactivity associated with the nucleoside portion of these alkaline hydrolysates was associated with ara-A, which suggested the possibility that ara-AMP may be the 3′-terminus. Müller et al. (1975) reported similar findings. Recent evidence for this possibility is the report of Sprinzl et al. (1977) who showed that ara-ATP acts as a substrate for ATP(CTP):tRNA nucleotidyl transferase and thus forms the 3′-terminus of tRNA *in vitro*. These findings are consistent with the findings of Hubert-Habart and Cohen (1962), who reported on the isolation of ara-A from the 3′-RNA termini in *E. coli* treated with ara-A. Sprinzl et al. (1977) used 26 ATP analogs to study the structural requirements of the terminal adenosine of tRNA. They observed that positions 1, 2, 6, and 8 of the adenine ring of ATP cannot be modified, whereas the free hydroxyls can be replaced with an amino group or hydrogen atom without loss of substrate properties.

Effect of β-Ara-ATP and α-Ara-ATP on Terminal Deoxynucleotidyl Transferase (TdT). Müller et al. (1977d, 1978c) studied the sensitivity of purified TdT towards β-ara-ATP and α-ara-ATP. α-Ara-ATP has no influence on TdT activity. However, β-ara-ATP is incorporated into oligo[d(pA)$_3$], but is not a chain terminator. ATP and 3′dATP inhibit the TdT reaction by reducing the initiation capacity of this enzyme by acting as chain terminators (see page 123 for action of 3′dATP on TdT).

Inhibition of Adenosine Deaminase. One of the most exciting developments in the protection of the amino group of adenosine analogs is the use of adenosine deaminase inhibitors. The specificity of adenosine deaminase was first studied by Brink and LePage (1964b), Cory and Suhadolnik (1965), and Bloch et al. (1967). Ara-A is rapidly deaminated by adenosine deaminase. Subsequently, Cory et al. (1967) reported that fetal calf liver has a very active adenosine deaminase. Therefore, dose–inhibition studies are distorted unless deamination of the adenosine analog is prevented. Subsequently, Chassy and Suhadolnik (1967) reported on the inhibitory properties of adenosine deaminase by N^6-substituted adenosine derivatives; N^6-methyladenosine was one analog of adenosine that inhibited adenosine deaminase. Later experiments by Connor et al. (1974) with viral-infected monkey kidney cells in culture showed that N^6-methyladenosine increased the antiviral activity of ara-A; without N^6-methyladenosine, the half-life of ara-A was 2–3 h; inhibition by N^6-methyladenosine has been attributed by Trewyn and Kerr (1977) to N^6-methyladenosine 5′-triphosphate, which also inhibits DNA synthesis.

There are excellent inhibitors of adenosine deaminase (see Fig. 5.1 and 5.3 for structures). The two most potent inhibitors are the naturally occurring nucleoside analogs, coformycin (CF) and 2′-deoxycoformycin (dCF) (see page 259). Three chemically synthesized inhibitors are 1,6-dihydro-6-hydroxymethylpurine ribofuranoside (DHMPR), *erythro*-9-(2-hydroxy-3-nonyl)adenine (EHNA), and 5′-(*O*-valeryl)ara-A. With ara-A as the substrate, the $K_{m,\mathrm{app}}$ and K_i are 110 μM and 11 μM, respectively. Ara-A-5′-valerate is not a substrate for adenosine deaminase (Lipper et al., 1978).

Because of the rapid deamination of ara-A to ara-Hx by adenosine deaminase, the effectiveness of ara-A is limited. Two approaches have been used to minimize the deamination of adenosine analogs (i.e., ara-A). One approach is to use inhibitors of adenosine deaminase and the second approach is to add the analog as the 5′-monophosphate. Using the first approach, Plunkett and Cohen (1975b, 1977a) and North and Cohen (1978) have shown that the therapeutic effectiveness of ara-A can be increased by the simultaneous administration of *erythro*-9-(2-hydroxy-3-nonyl)adenine (EHNA), an inhibitor of adensosine deaminase. The second approach is to use ara-AMP. The advantage of using ara-AMP on L cells, as stated by Plunkett and Cohen (1975b), is that ara-AMP is (i) inert to adenosine deaminase, (ii) the physiological intermediate on the way to ara-ATP, (iii) more soluble than the nucleoside, and (iv) potentially capable of bypassing resistance due to a lack of nucleoside kinase.

Evidence that nucleotides can penetrate the cell was provided by the report of Chaudry and Gould (1970) in which they demonstrated that externally added ATP did indeed enter soleus muscle. Subsequently, it was demonstrated that ara-AMP enters the cell intact and is converted to ara-ATP as reported by Cohen and coworkers (Plunkett et al., 1974; Cohen and Plunkett, 1975; Plunkett and Cohen, 1975b). Plunkett and Cohen (1977b) have also shown that dAMP and 2′,3′-dideoxy-AMP could penetrate the cell intact. By using nucleotides of ara-A, therefore, Cohen and coworkers have discarded the dictum that nucleotides "do not penetrate," which had been a powerful deterrent to research in the area of nucleotide biochemistry. The antiviral activity of ara-AMP and ara-HxMP has been summarized by Allen et al. (1974) and Sloan (1975).

In carefully detailed experiments, Müller et al. (1978d) have recently reported on the intracellular 5′-triphosphate pools of adenosine, deoxyadenosine, and ara-A in mouse lymphoma L5178Y cells. With the addition of coformycin at noncytostatic concentrations, the growth inhibitory potency of adenosine, deoxyadenosine, and ara-A was increased; the intracellular pool of ATP and dATP remained constant, whereas ara-ATP pools increased drastically. However, at high concentrations of coformycin, the ara-ATP pool decreases and the concentration of ara-Hx increases. The authors suggest that ara-Hx is formed by the ara-AMP to ara-IMP to ara-Hx pathway. Finally, coformycin is not degraded, but is phosphorylated to the 5′-triphosphate (Müller et al., 1978d).

Another approach to increase the antiviral activity of ara-A has been to synthesize the carbocyclic ara-A derivative, C-ara-A, (±)-9-[2α,3β-dihydroxy-4α-(hydroxymethyl)cyclopenl-1α-yl]adenine (Vince and Daluge,

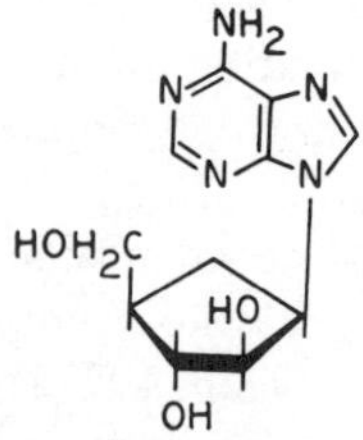

NH_2 N N N N HOH$_2$C HO OH

C–ARA–A

1977). C-Ara-A has significant antiviral activity. It is not a substrate for either adenosine deaminase or the nucleoside hydrolase or phosphorylase.

The complications of inactivating adenosine deaminase by the addition of inhibitors to inactivate the adenosine analogs are compounded by the interactions of these adenosine deaminase inhibitors with other enzymes. For example, North and Cohen (1978) have reported that the adenosine deaminase inhibitor, EHNA, specifically inhibits DNA synthesis in herpes simplex virus. This difficulty has now been overcome by the isolation of a mammalian cell line that has no adenosine deaminase activity (Shipman and Drach, 1978). Incubation of these adenosine deaminase-deficient cells in 5.4 μM ara-A inhibited DNA synthesis by 50% (Drach et al., 1977). When ara-A is removed, the cells recover. In the presence of adenosine deaminase, twice the ara-A concentration was essential for the same inhibition of DNA synthesis.

Suicide Inactivation of S-Adenosylhomocysteine (AdoHcy) Hydrolase by Ara-A and Cordycepin. Although the inhibition of growth by adenosine analogs in bacterial, cultured mammalian cells, and viruses following the formation of intracellular nucleotides is clearly documented by the many studies referred to in this book, there is also a very interesting "nucleotide independent" mechanism of purine toxicity for 2′-deoxyadenosine, ara-A, and cordycepin. Hershfield (1979) has demonstrated that these nucleosides bind tightly and irreversibly to AdoHcy hydrolase from human lymphoblasts. This finding suggests a basis for an unrecognized inactivation of adenosine analogs at the nucleoside level. This binding which occurs with first order kinetics suggests a "suicide" inactivation of the hydrolase.

Effect on Protein Synthesis. Ilan et al. (1970, 1977) reported that the malarial parasite, *Plasmodium berghei*, can phosphorylate ara-A to ara-ATP. They found that, while some protein synthesis was markedly inhibited, ara-A stimulated the synthesis of other proteins that were not normally expressed by the malarial genome.

Antiviral Activity Against DNA Viruses in Cell Culture. Ara-A has a very broad spectrum of activity against DNA viruses in cell culture. The antiviral activity of ara-A against HSV can be increased greatly by the addition of coformycin. The effect of ara-A, ara-A plus CF, and ara-Hx (ara-H) on DNA synthesis in KB cells and HSV-infected cells is described (Fig. 4.5). Ara-A (25 μM) is eight times as active in the presence of CF. A lower concentration of ara-A produces a larger inhibition of DNA synthesis of either viral or cellular DNA when CF is added. Similarly, Cohen, in a continua-

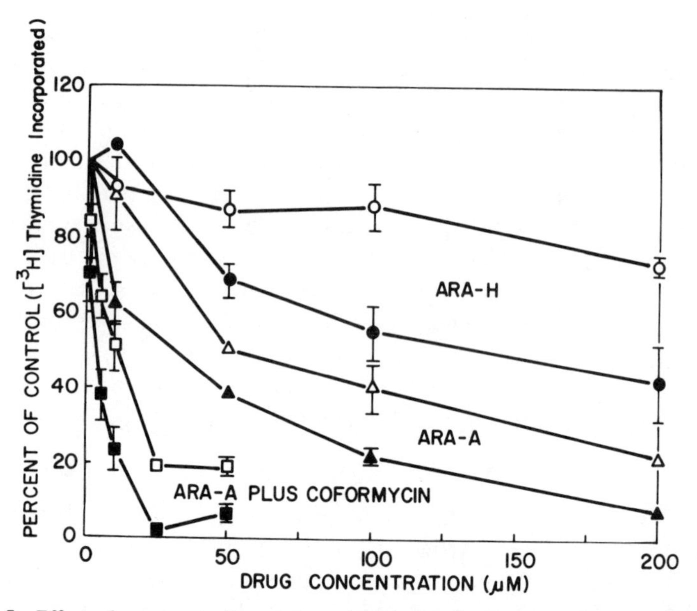

Figure 4.5 Effect of ara-A, ara-Hx, and ara-A plus CF on DNA synthesis in HSV-infected KB cells grown in nonolayer culture. DNA was labeled and viral DNA was separated from cellular DNA by isopycnic centrifugation. The amount of label incorporated into viral DNA (filled symbols) and cellular DNA (open symbols) is expressed as a percentage of the amount incorporated into the respective DNA species in cell culture without drugs. Reprinted with permission from C. M. Reinke, J. C. Drach, C. Shipman, Jr., and A. Weissbach (1978), Differential inhibition of mammalian DNA polymerases α, β, and γ and herpes simplex virus-induced DNA polymerase by the 5′-triphosphates of arabinosyladenine and arabinosylcytosine. In G. deThe, W. Henle, and E. Rapp, eds., *Oncogenesis and Herpesviruses III*, Lyon, International Agency for Research on Cancer (IARC Scientific Publications No. 24), pp. 999–1005.

tion of earlier studies on increasing the effectiveness of analogs of adenosine by adenosine deaminase inhibitors (Plunkett and Cohen, 1975a, b), has reported that EHNA markedly potentiates the anti-HSV effect of ara-A by EHNA (North and Cohen, 1978) (Fig. 4.6). North and Cohen (1978) have reported the surprising result that EHNA alone significantly inhibits HSV production, but has little or no effect on the growth or viability of uninfected HeLa cells. Deoxycoformycin does not inhibit HSV. Therefore, the effect of EHNA cannot be solely on adenosine deaminase. Gradients (CsC1) show that EHNA inhibits HSV-DNA, but not HeLa-DNA synthesis. EHNA may be preferable to dCF in the treatment of HSV infections when used with analogs of adenosine. EHNA (10^{-6} *M*) alone inhibits HSV production by 30%; EHNA in combination with ara-A (10^{-5} *M*) inhibits HSV production by more than 99%.

Antagonism of Deoxyadenosine to Ara-A. Although adenosine reverses the ability of ara-A to inhibit DNA synthesis in mammalian cells and *E. coli*, in viral-infected cells, deoxyadenosine can only exert its antagonism toward the antiviral effect of ara-A in the presence of CF. In the absence of CF, deoxyadenosine is deaminated; consequently, the antagonistic effect of deoxyadenosine cannot be detected. Because adenosine deaminase inhibitors such as CF were not available until recently, it was not possible to observe the antagonistic capacity of deoxyadenosine (Smith et al., 1978).

Antiviral Activity Against RNA Viruses in Cell Culture. Ara-A is inactive against the "classical" nononcogenic RNA viruses (i.e., orthomyxoviruses, paramyxoviruses, arboviruses, poliovirus, and influenza virus) (Shannon, 1975). However, of the rhabdovirus group, vesicular stomatitis virus and rabies are inhibited by ara-A. The inhibition of oncornavirus replication by ara-A *in vitro* was reported by Schabel (1968) and Miller et al. (1968). Rous sarcoma virus, an RNA tumor virus, is sensitive to ara-A. These findings are in sharp contrast to the finding of inactivity of ara-A against most of the nononcogenic RNA viruses. Baltimore (1970) and Temin and Mizutani (1970) reported that RNA tumor viruses replicate through a DNA intermediate because of the RNA-directed DNA polymerase. Therefore, it is logical that ara-A, as an inhibitor of DNA synthesis, would show antiviral activity against the oncornaviruses. Indeed, Shannon (1975) reported that ara-A inhibits gross murine leukemia virus and Rauscher leukemia virus *in vitro*. Similarly, ara-A inhibits infectious pancreatic necrosis virus (IPNV)

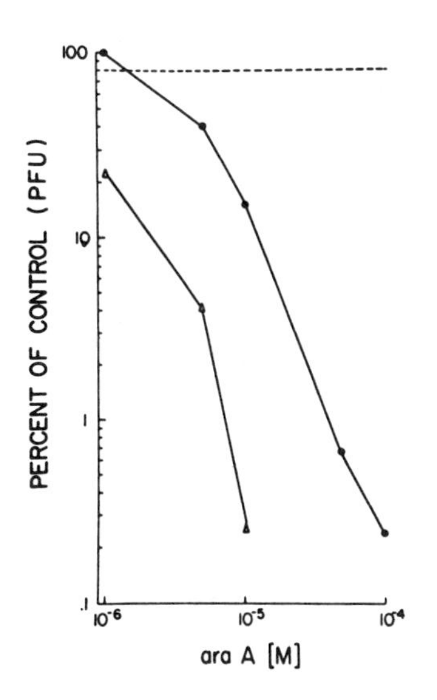

Figure 4.6 Effect of ara-A on HSV production in the presence (▲) and absence (●) of 10^{-6} *M* EHNA. The dashed line represents HSV production in the presence of 10^{-6} *M* EHNA alone (no ara-A). Reprinted with permission from North and Cohen, 1978.

of trout and salmon. Although IPNV has not been classified, it is believed to contain a single stranded RNA genome of about 16 S (Kelly and Loh, 1972). IPNV may represent an entirely new group of viruses.

Effect of Ara-A on Viral Infections in Humans. Because of the limited space, it is difficult to review in detail here the studies that have been done with ara-A on viral infections in humans. Therefore, a very limited review is included. For a more detailed, accurate description of the effect of ara-A on human viral infections, the reader is referred to the excellent comprehensive review by Pavan-Langston et al. (1975a). Ch'ien et al. (1976) have also reported that ara-A may be a useful antiviral drug for systemic treatment of herpes zoster infections; forty-seven immunosuppressed patients with disseminated varicella zoster treated with ara-A showed more rapid healing and significant reduction in pain. This suggests that ara-A may be a useful antiviral drug for treatment of varicella zoster infections. Toxicity of ara-A posed no problem. Ara-A is most effective if administered during the first 6 days of the disease (Whitley et al., 1976).

Ch'ien et al. (1975) studied the effect of ara-A in the treatment of severe HSV type 1 and 2 mucocutaneous infections of immunosuppressed patients. Lesions infected with HSV type 1 appeared to be more responsive than those infected with type 2. Although 5-iodo-2′-deoxyuridine (IUDR) was first used in the successful treatment of herpetic keratitis (Kaufman et al., 1962), evidence of toxicity and intolerance has made it a less-desirable compound. Attempts to use ara-A, trifluoromethylthymidine, and poly(I-C) in place of IUDR have been unsuccessful because of toxicity, expense, or difficulty in synthesis. Ara-A penetrates the cornea only if the cornea is not intact. Ara-A or metabolic products are not found in the aqueous humor with intact cornea. Therefore, treatment of uveitis is only beneficial under these conditions. Humans given topical 3% ara-A 8–10 times in 48 h did not show evidence of corneal disease following routine cataract surgery. In addition, there were no signs of toxicity in regenerating epithelium of eyes treated with ara-A (Pavan-Langston et al., 1975).

Clinical testing has demonstrated that IUDR and ara-A (at equimolar concentrations) have the same therapeutic efficacy in the treatment of established herpes simplex keratoconjunctivitis (Pavan-Langston et al., 1975c; Pavan-Langston and Buchanan, 1976). Ara-A has an advantage over IUDR in that ara-A penetrates the intraocular tissue more rapidly, is less toxic, does not interfere with wound healing, and is effective against viruses resistant to IUDR.

Ara-A (3% ointment) has been investigated with patients having complicated herpes keratitis who were resistant to IUDR or who had toxic reactions to IUDR (O'Day et al., 1975). Of 18 patients treated with ara-A, 16 patients responded. Application of ara-A as a 3% ointment to patients with active stromal keratitis with intercurrent hepatic epithelial keratitis showed a good response. Ara-A inactivated the stromal keratitis upon healing of the epithelium. Ara-A (3% ointment) is an effective substitute for IUDR in HSV epithelial keratitis. Combined corticosteroid–ara-A therapy is beneficial in HSV stromal keratitis (Jones, 1975).

Cytomegalovirus (CMV) infections cause many disease syndromes, including congenital infection, mononucleosis, and disseminated infection in immunosuppressed hosts (Weller, 1971a, b). In a limited study, Ch'ien et al. (1974) found that ara-A had an antiviral effect. Because ara-A did not eradicate viral excretion, it may be that an intracellular reservoir of CMV is capable of replicating once ara-A is removed.

The antiviral properties of ara-A can be summarized as follows: (i) ara-A is relatively nontoxic to the cell, (ii) it does not suppress immune function, (iii) it appears to be free of any tumorigenic activity, and (iv) ara-A is rapidly metabolized in patients who are suffering from serious viral diseases.

Antiviral Activity of Ara-A in Animals. Sloan (1975) has summarized the antiviral activity of ara-A and its derivatives in rabbit or hamster eyes against DNA-containing viruses. Most mice infected with DNA viruses survived when challenged with either ara-A, ara-Hx, ara-AMP, or ara-HxMP. Of interest is the finding that the titers of the herpes virus ranged up to 10^4 mouse infectious units per milligram of brain tissue in the untreated hamster brain, whereas the hamsters treated with ara-A showed no measurable infectious units in the brain samples.

Ara-A has also been successful in treatments of infections of vaccinial keratitis in rabbits, cutaneous vaccinia virus disease, intracerebral viral infections, intraperitoneal virus infections in mice, and herpes virus hominis type 2 infections in mice (Hyndiuk and Kaufman, 1967; Hyndiuk et al., 1976; Klein et al., 1974; Sidwell et al., 1973; Allen and Sidwell, 1972; Schmidt-Ruppin, 1971; Allen et al., 1974; Nahmias and Roizman, 1973; Poste et al., 1972). The antiviral activity of ara-AMP against HSV infections of rabbit cornea is markedly superior to that of ara-A (Falcon and Jones, 1977). However, ara-AMP is very toxic to the cornea.

Cancer Chemotherapy in Humans. Because of the early studies on the antitumor and antiviral activity of ara-A in experimental animals by Lee et al. (1960), York and LePage (1966), and Schabel (1968), patients with advanced hematologic malignancies have received ara-A. It has been speculated that ara-A has some activity against hematologic malignancies and chronic myelogenous leukemia. If ara-A becomes a useful antitumor agent, it will probably be used in combination with other antitumor agents. One advantage of ara-A is that it has low myelosuppressive activity in humans.

Human Pharmacology Trials of Ara-A. The human pharmacologic studies by Ch'ien et al. (1971), LePage and coworkers (LePage et al., 1972, 1973), and Kinkel and Buchanan (1975) showed that ara-A is rapidly distributed in the tissues and is rapidly metabolized to ara-Hx ($t_{1/2}$ 1.5–3.5 h). Although ara-A levels are below detection in the red blood cell or the cerebral spinal fluid, it has the advantage that it does penetrate the central nervous system. This property is important for treatment of central nervous system viral diseases. Occasionally, ara-A causes nausea and loss of weight in humans; 88–97% of the ara-A is excreted as ara-Hx in 12–24 h.

Effect of Ara-A on Postnatal Growth and Development of the Rat. Ara-A and ara-Hx were tested for their effects on the growth and development of newborn rats. Neither ara-A nor ara-Hx had any effect on rat mortality, body growth, cerebellogenesis, or behavior. Ara-C has profound effects on development, causes permanent neurologic damage, and causes high mortality (Fishaut et al., 1975). Ara-A toxicity is 100 times less than that of ara-C.

Toxicology of Ara-AMP. Kurtz et al. (1977) reported on acute i.v. studies in rats with ara-AMP. The LD_{50} for ara-AMP is about 1200 mg/kg. The i.p. LD_{50} in rats is about 2300 mg/kg for ara-A and 1700 mg/kg for the monophosphate. In terms of chronic tolerance, i.v. administration of ara-AMP at 150 mg/kg is tolerated in rats. With rabbits, muscle irritation by ara-AMP (20%) shows no more local reaction than physiological saline. Administration of 100 mg/kg of the monophosphate (i.v.) resulted in a very slight loss of weight. Ara-A was teratogenic in rabbits at 5 mg/kg i.m. when given from 6 to 18 days of pregnancy. There were multiple gross anomalies and skeletal abnormalities in some kits. Rats are less sensitive. Ara-AMP did not produce the teratogenic effect in animals. This difference between ara-AMP and ara-A may be due to the increased solubility of the 5′-monophosphate. Studies with tritium-labeled ara-AMP show that it leaves the injection site very rapidly, which explains the metabolism and rapid excretion. Ara-A suspensions remain at the injection site for many days, resulting in an escalated blood level 3–4 days following the injection. Therefore, with ara-AMP, the fetus has only short bursts of exposure to the nucleotide, but longer exposure to the nucleoside. Ara-A has weak mutagenic properties. The mutagenic assays of ara-AMP were negative by the bacterial plate tests of Ames.

4.2 1-β-D-ARABINOFURANOSYLTHYMINE

Biochemical Properties

Antibacterial Activity. In 1956 Cohen and Barner demonstrated that 1-β-D-arabinofuranosylthymine (ara-T) was moderately toxic to *E. coli*. Thymine was slowly liberated from ara-T.

Antiviral Activity. 1-β-D-Arabinofuranosylthymine (ara-T, spongothymidine) (Fig. 4.1), a nucleoside isolated from the sponge *Tethya crypta* (Bergman and Feeney, 1950, 1951) is a selective inhibitor of the replication of some viruses grown in cells in culture, but does not affect the growth of uninfected control cells (Figs. 4.7 and 4.8) at effective antiviral concentrations (Gentry and Aswell, 1975). Baby hamster kidney (BHK) cells are totally unaffected by ara-T at 2×10^{-4} *M*, whereas replication of HSV-1 and HSV-2 is completely blocked (Fig. 4.8) at this concentration. Aswell et al. (1977) subsequently used CsCl isopycnic centrifugation to demonstrate that in HSV-infected and uninfected cell lysates, ara-T reduced the incorporation of [^{3}H]hypoxanthine into viral DNA by 91% and also inhibited host DNA synthesis of infected cells. DNA synthesis in uninfected cells is not affected by ara-T. The conclusion that viral pyrimidine deoxyribonucleoside kinase (dPyK) is responsible for the selective antiviral activity of ara-T is shown by experiments with $dPyK^+$ and $dPyK^-$ mutants of HSV-1

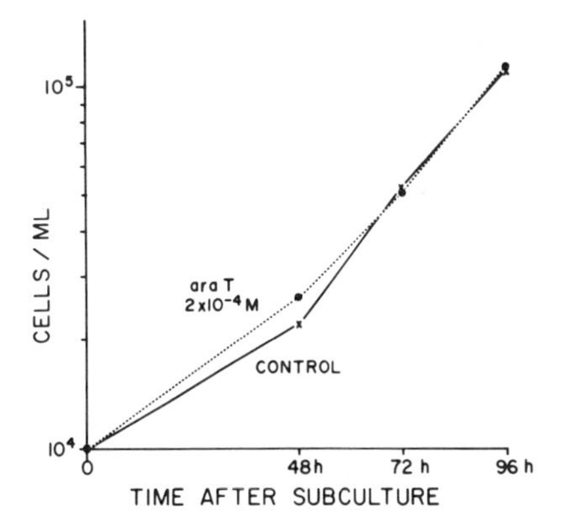

Figure 4.7 Replicate monolayers of BHK cells were prepared and sampled at the indicated intervals. To each well at 0 time was added 1 ml of Eagle's medium–tryptose phosphate broth calf serum with ara-T (2×10^{-4} *M*) and 10^4 cells. Ara-T was omitted from the control wells. Reprinted with permission from Gentry and Aswell, 1975.

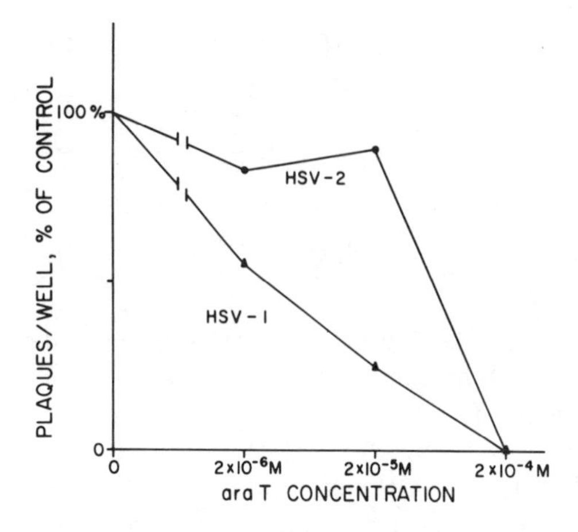

Figure 4.8 Monolayers of BHK cells were infected with HSV-1 or HSV-2. After 2 h of virus fixation, 1 ml of Eagle's medium–tryptose phosphate broth serum containing the desired amount of ara-T was added to each well. Plaques were counted 24 h later. The control well (no ara-T) with HSV-1 produced 40 plaques; that with HSV-2 produced 18. Reprinted with permission from Gentry and Aswell, 1975.

and HSV-2. The wild-type HSV-1 is drastically inhibited by ara-T in culture, while the $dPyK^-$ mutants are unaffected by ara-T (Fig. 4.9). Substrate inhibition studies show that ara-T, in assay mixtures of wild-type HSV-1-induced deoxythymidine kinase or deoxycytidine kinase activities, inhibits the phosphorylation of deoxythymidine and deoxycytidine. The inhibition of DNA synthesis by ara-T in viral-infected cells indicates that the mechanism of the selective inhibition by ara-T is dependent on its initial phosphorylation by viral-induced dPyK (Aswell et al., 1977). The phosphorylated ara-T blocks viral DNA synthesis in infected cells, but not in uninfected cells.

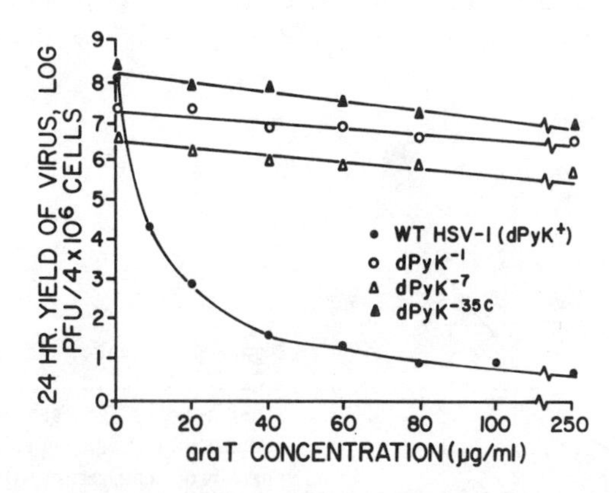

Figure 4.9 Effect of ara-T on the replication of wild-type [WT HSV-1 ($dPyK^+$)] and PyK^- HSV-1. Twenty-four hour yield of virus is plotted against the concentrations of ara-T. Reprinted with permission from Aswell et al., 1977.

Resistance of dPyK$^-$ mutants of HSV further supports this hypothesis. With uninfected mammalian cells in culture, ara-T is not phosphorylated. Thus ara-T is a relatively specific antiviral agent.

In subsequent studies, Miller et al. (1977) demonstrated that ara-T is also effective in inhibition replication of varicella zoster virus (VZV) in human embryo fibroblasts. This inhibition by ara-T is also attributed to the induction of a viral deoxythymidine kinase with a substrate specificity similar to that of HSV. Miller et al. (1977) were also able to show that cytomegalovirus (CMV) replication is relatively resistant to ara-T. They concluded that CMV does not induce the deoxythymidine kinase in infected cells with a substrate specificity similar to those of HSV and VZV. These studies indicate that ara-T may be a useful probe for detecting viral deoxythymidine kinases (Aswell et al., 1977).

As with the lack of inhibition of normal mammalian cells in culture, ara-T has little inhibitory effect on the multiplication of tumor cells in culture (Chu and Fischer, 1962; DeGarilhe and DeRudder, 1970). Deoxythymidine, but not ara-T, is hydrolyzed to thymine and deoxyribose 1-phosphate by the high speed supernatants of homogenized hamster liver (Gentry et al., 1977). Four hours postinjection, 50% of the ara-T appears in the urine. Gentry and coworkers concluded that urinary excretion is the primary limiting factor in the *in vivo* anti-herpes viral activity of ara-T in hamsters.

Ara-T is an attractive pyrimidine analog for studies in animals because it is not toxic to uninfected mammalian cells in culture, but is an effective inhibitor of herpes simplex virus types 1 and 2. Therefore, Aswell and Gentry (1977), appreciating the concept of herpes viral coded enzymes initiated by Kits and Dubbs (1965) with dTK and the broader specificity of viral enzymes (Jamieson et al., 1974; Jamieson and Subak-Sharpe, 1977), expanded their tissue culture studies to equine herpes virus–Syrian hamster as the animal model system. This viral–animal system has several features that make it attractive for studying anti-herpes viral compounds. Although ara-T (one 5 mg injection) prolonged the survival time of viral-infected animals, there was only one survivor. When ara-T is administered in a series of 5 mg doses spaced 4 h apart for 72 h, ara-T effectively promotes survival of animals infected with herpes virus. The toxicity to animals is mild (Gentry et al., 1979). Figure 4.10 summarizes the probable metabolism of ara-T in herpes virus-infected cells. Because the hamster has a high level of dC deaminase, it appeared likely that 4-amino-ara-T or 5-methyl-ara-C would have advantages over ara-T. 5-Methyl-ara-C is inactive in cells lacking deoxycytidine (dC) deaminase, but is active in cells in which the deaminase is present. In the latter case, 5-methyl-ara-C becomes the intracellular donor of ara-T, which is then phosphorylated to ara-TMP and ultimately

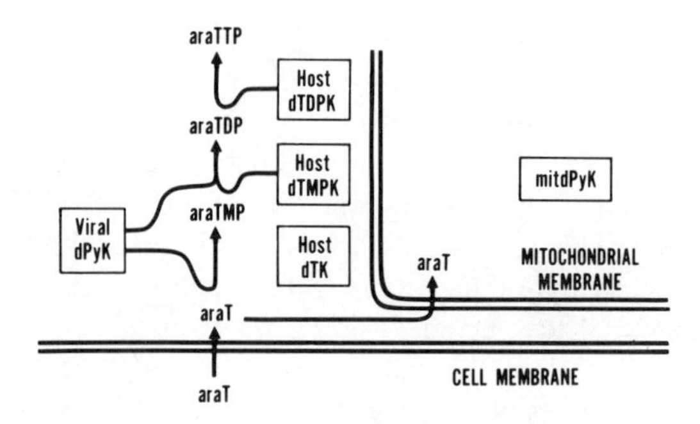

Figure 4.10 Suggested metabolism of ara-T in the herpes virus-infected cell. Abbreviations: ara-T, arabinofuranosylthymine; ara-TMP, ara-T-5′-monophosphate; ara-TDP, ara-T-5′-diphosphate; ara-TTP, ara-T-5′-triphosphate; dTK, dT kinase; dTMPK, dTMP kinase; dTDPK, dTDP kinase; dPyK, pyrimidine deoxyribonucleoside kinase; dPyK, pyrimidine deoxyribonucleoside kinase; mitdPyK, mitochondrial dPyK. Reprinted with permission from Gentry et al., 1979.

inhibits viral replication. The wide distribution of dC deaminase in animal tissues suggests that 5-methyl-ara-C should be included in further studies of the antiviral activity of ara-T in animals and eventually in man (Aswell and Gentry, 1977).

Inhibition of Mammalian DNA Polymerase-α. Matsukage et al. (1978) reported that the incorporation of [^{3}H]dTMP into DNA was inhibited competitively by 1-β-D-arabinofuranosylthymine 5′-triphosphate (ara-TTP) and noncompetitively by ara-CTP. Incorporation of [^{3}H]dCMP was inhibited competitively with ara-CTP and noncompetitively with ara-TTP. Substitution experiments ruled out the possibility that ara-CTP or ara-TTP can replace deoxycytidine 5′-triphosphate or deoxythymidine 5′-triphosphate at reduced efficiency in the elongation of DNA chains. Therefore, ara-TTP or ara-CTP is incorporated into the DNA polymerase-α–DNA complex or into the 3′-end of the growing DNA chains, which results in stopping further elongation.

4.3 OXAZINOMYCIN

Oxazinomycin (minimycin) is a C-riboside analog having the structure, 5-β-D-ribofuranosyl-1,3-oxazine-2,4-dione (Fig. 4.11). This carbon-linked nucleoside analog was first reported by Haneishi et al. (1971) from a soil sample collected at Tanesahi, Aomori Prefecture, Japan. This same nucleoside antibiotic was isolated by Kusakabe et al. (1972) who referred to it as minimycin from the culture medium of *Streptomyces hygroscopicus*. The structural elucidation of minimycin was reported by Sasaki et al. (1972). Minimycin inhibits gram-positive and gram-negative bacteria and shows antitumor activity against transplantable tumors. The biosynthesis of minimycin has been reported by Isono and Suhadolnik (page 238). Oxazinomycin has the same properties as mediocidin, the analog isolated by Okami et al. (1954) from *Streptomyces mediocidicus*.

Discovery, Production, and Isolation

Oxazinomycin was isolated by Haneishi et al. (1971) and Kusakabe et al. (1972) from culture filtrates of *S. tanesashinensis* and *S. hygroscopicus*. The isolation of oxazinomycin by both research groups followed essentially the same techniques, namely adsorption on charcoal followed by elution with aqueous acetone. Further purification was achieved by either silica gel chromatography or Sephadex G-15 chromatography; yield 2 g/100 liters; minimycin is crystallized from methanol, water–ethanol, or acetone.

Physical and Chemical Properties

Molecular formula $C_9H_{11}NO_7$; mol wt 245; mp 164–166°C; $[\alpha]_D^{20}$ +19.7° (c 1.0, H_2O); $\lambda_{231}^{H_2O}$ (ϵ = 4500, 4700); it is soluble in water and methanol; it is stable in acid, but very unstable in alkali. The ultraviolet spectra, nmr spectra, and X-ray analysis have been described [Haneishi et al., 1971; Kusakabe et al., 1972; Sasaki et al., 1972; West German patent 2,043,946 (1971), Kaken Chemical Co., Ltd.; DeBernardo and Weigele, 1977].

Structural Elucidation

Sasaki et al. (1972) reported on the structural elucidation of oxazinomycin by a combination of nmr and mass spectrometry. The nmr data for minimycin are shown in Fig. 4.12 and the proton chemical shifts and coupling con-

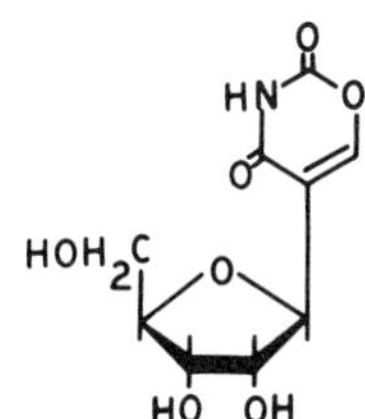

Figure 4.11 Structure of oxazinomycin (minimycin).

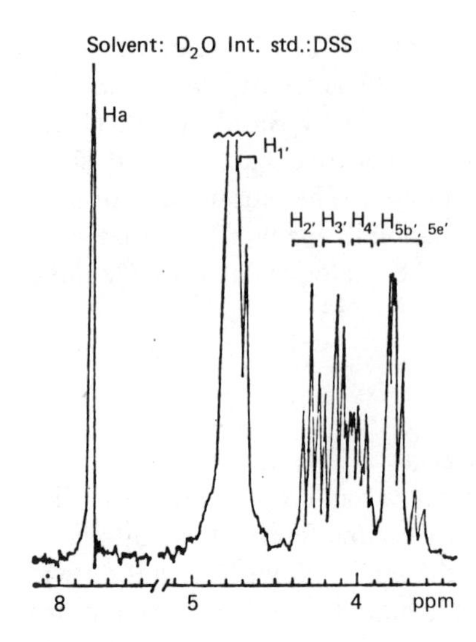

Figure 4.12 The nmr spectrum of minimycin (oxazinomycin). Reprinted with permission from Sasaki et al., 1972.

stants (*J*) for minimycin and pseudouridine are shown in Table 4.1. The isolation of the ribosyl moiety of minimycin was accomplished by treating minimycin in water with hydrazine at 100°C for 5 h. After cooling, benzaldehyde was added and the mixture was extracted with ether. Ribose was identified by paper chromatography.

Chemical Synthesis of Minimycin

DeBernardo and Weigele (1977) reported on the chemical synthesis of minimycin starting with 2′,3′-*O*-isopropylidene-5′-*O*-trityl-D-ribofuranosylacetonitriles (Scheme 4.1). Compounds **3** and **4** were formylated to give compounds **5** and **6**, which by reaction with hydroxylamine were converted to the amino isooxazoles, **8** and **9**. Compounds **8** and **9** were reduced to the acrylamides, **10** and **11**. Hydrolysis of **10** and **11** and subsequent reaction of the resulting **12** with *N*,*N*′-carbonyldiimidazole furnished the 1,3-oxazine-2,4-dione derivatives, **13** and **14**, which after removal of the protective groups gave oxazinomycin (**1**) and its α-anomer, **15**. The physical and chemical properties of the synthetic nucleoside antibiotic were identical with those of the naturally occurring nucleoside.

Biosynthesis of Oxazinomycin

Isono and Suhadolnik (1977) demonstrated that carbons 3, 4, and 5 of glutamate serve as a three-carbon precursor for carbons 4–6 of the aglycon of

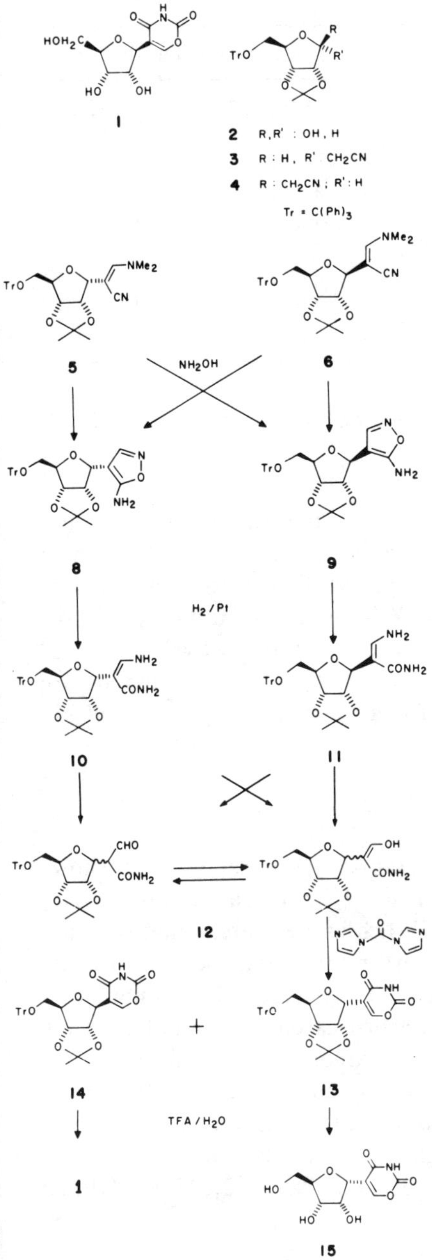

Scheme 4.1 Chemical synthesis of minimycin. Modified from De Bernardo and Weigele (1977), reprinted with permission.

Table 4.1 Proton Chemical Shifts and Coupling Constants (J) of Minimycin and β-Pseudouridine

	Minimycin			β-Pseudouridine		
Proton	δ (ppm)[a]	J (Hz)		δ (ppm)[a]	J (Hz)	
H_a	7.78 (s)			7.63 (s)		
$H_{1'}$	4.65 (d)	$J_{1'2'}$	5.0	4.67 (d)	$J_{1'2'}$	5.0
$H_{2'}$	4.27 (t)	$J_{2'3'}$	5.0	4.27 (t)	$J_{2'3'}$	5.0
$H_{3'}$	4.13 (t)	$J_{3'4'}$	5.2	4.13 (t)	$J_{3'4'}$	5.2
$H_{4'}$	4.00 (ddd)	$J_{4'5b'}$	3.0	4.00 (ddd)	$J_{4'5b'}$	3.5
$H_{5b'}$	3.81 (dd)	$J_{4'5c'}$	4.3	3.82 (dd)	$J_{4'5c'}$	4.5
$H_{5c'}$	3.71 (dd)	$J_{5b'5c'}$	−12.5			

Reprinted with permission from Sasaki et al. 1972.
[a] (s) singlet; (d) doublet; (t) triplet; (dd) double doublet; (ddd) doubling double doublet.

oxazinomycin (Fig. 4.13). The 23.4% retention of the tritium on C-3 of glutamate at C-6 of the aglycon establishes the asymmetric incorporation of carbons 3–5 of glutamate into the 1,3-oxazine-2,4-dione ring. Subsequently, Isono and Uzawa (1977) reported that [1-^{13}C]- and [2-^{13}C]acetate are incorporated into C-4 and C-5 of the 1,3-oxazine aglycon respectively. The ^{13}C-acetate data support the finding that carbons 3, 4, and 5 of glutamate are incorporated into carbons 6, 5, and 4 of oxazinomycin (Fig. 4.13 and Fig 1.34).

Biochemical Properties

Oxazinomycin inhibits gram-positive and gram-negative bacteria, but shows no activity against yeast or fungi (Kusakabe, 1972). The antitumor activity of oxazinomycin was examined against Ehrlich ascites carcinoma, sarcoma 180 (Ascites), and sarcoma 180 (solid form) in mice. Intraperitoneal injection of minimycin at a dose of 2 mg/kg for 10 days resulted in an inhibition of tumor growth of the ascites form of Ehrlich carcinoma. Similar effects were observed on sarcoma 180 (Ascites) and sarcoma 180 (solid form). The LD_{50} for mice was 30 mg/kg intraperitoneally, 20 mg/kg subcutaneously, and 80 mg/kg intravenously. Oxazinomycin is also useful as a growth promoter in children who are underweight for reasons such as eczema, anorexia, and persistent diarrhea [West German Patent 2,043,946 (1971), Kaken Chemical Co., Ltd.,].

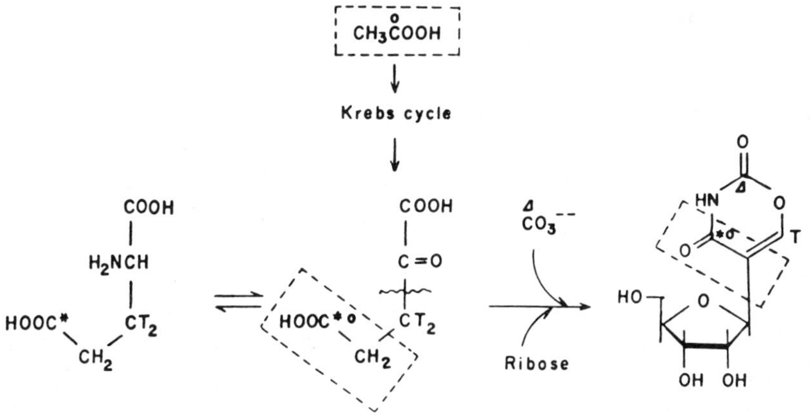

Figure 4.13 Proposed biosynthesis of minimycin. Reprinted with permission from Isono and Suhadolnik, 1977.

4.4 1-METHYLPSEUDOURIDINE

Argoudelis and Mizsak (1976) reported on the isolation of 1-methylpseudouridine (Fig. 4.14) from *Streptomyces platensis* var. *clarensis*. This is the same *Streptomyces* from which DeBoer et al. (1979) and Bannister (manuscript in preparation) have isolated 5,6-dihydro-5-azathymidine, an antibacterial and antiviral agent (Fig. 4.17). Pseudouridine was first discovered as a natural constituent of RNA by Cohn (1961). Suhadolnik (1970) and Isono and Suhadolnik (1975) isolated pseudouridine from the culture filtrates of the *Streptomyces* that produces 5-azacytidine and oxazinomycin (see pages 135 and 236). Argoudelis and Mizsak (1976) reported on the discovery of 1-methylpseudouridine as a naturally occurring nucleoside antibiotic. Cohn (1961) and Scannell et al. (1959) had made reference to 1-methylpseudouridine earlier. 1-Methylpseudouridine and its triacetate are inactive against gram-negative and gram-positive organisms

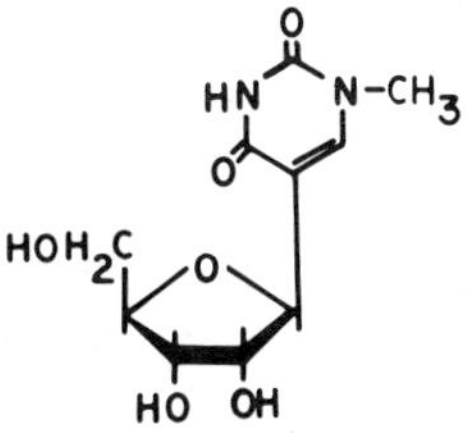

Figure 4.14 Structure of 1-methylpseudouridine.

and have marginal activity against herpes virus replication. L1210 cells are not inhibited by 1-methylpseudouridine.

Discovery, Production, and Isolation

Argoudelis and Mizsak (1976) discovered 1-methylpseudouridine in the culture filtrates of *S. platensis*. The medium is as described by DeBoer et al. (1976). Purification of 1-methylpseudouridine is accomplished by charcoal, countercurrent distribution, and silica gel chromatography. The yields of crystalline uridine, 5,6-dihydro-5-azathymidine, 1-methylpseudouridine, and pseudouridine per 5000 liters of medium are 100, 189, 307, and 70 mg, respectively.

Physical and Chemical Properties

The molecular formula of 1-methylpseudouridine is $C_{10}H_{14}N_2O_6$; mol wt 258; mp 181–184°C; $[\alpha]_D^{25}$ −25° (*c* 1, 59% aqueous ethanol); it is soluble in water and lower alcohols and insoluble in ketones, ether, and acetone; infrared spectra: 3390, 3350 cm^{-1} (OH, NH) and 3060, 3040 cm^{-1} (C=CH); 1672 cm^{-1} (pyrimidine system); ultraviolet spectral properties: $\lambda_{max}^{H_2O}$ 209 nm (ϵ = 7546), 270 nm (ϵ = 9081); $\lambda_{max}^{pH\,1}$ 209 nm (sh) (ϵ = 10,342), 270 nm (ϵ 8923); $\lambda_{max}^{pH\,11}$ 267 nm (ϵ = 6192); nmr spectra (D_2O): δ 3.42 (s, 3H) assigned to an NCH_3; δ 3.85 (2H) due to —CH_2O—; doublet at δ 4.71 (1H, J = 4.0 Hz); singlet at δ 7.81 (1H); anomeric proton, doublet at δ 4.71. The δ doublet 4.71 is similar to carbon–carbon linked nucleosides such as showdomycin and pseudouridine, δ 4.82 and δ 4.71, respectively. This compares with a δ of 6.0 for the anomeric hydrogen of the ribose in uridine. The β-configuration was assigned to the anomeric carbon from the nmr absorption pattern of the C-3′, C-4′, C-5′ proton region in the nmr spectrum, which is identical with that of ψ (Argoudelis and Mizsak, 1976). In addition, Cohn (1961) reported that the nmr spectrum of the α-anomer of pseudouridine shows the anomeric proton 0.33 ppm downfield from that of the β-anomer. The mass spectrum for 1-methylpseudouridine is like those reported (Townsend and Robins, 1969) for formycin and showdomycin. Rice and Dudek (1969) in their studies with pseudouridine reported a fragmentation pattern containing base peak at B + 30 (where B is the aglycon). The mass spectra reported by Argoudelis and Mizsak (1976) showed a peak of m/e at 155, which agrees with a B + 30 and a C-nucleoside; the chemical shifts found in the ^{13}C-nmr spectrum of the triacetate of 1-methylpseudouridine are consistent with the proposed structure.

Chemical Synthesis

Reichman et al. (1977) and Earl and Townsend (1977) reported on the chemical synthesis of 1-methylpseudouridine. The synthesis by Reichman et al. (1977) involved either the silylation and methylation or the synthesis of 4,5′-anhydro-2′,3′-*O*-isopropylidene-1-methylpseudouridine. Earl and Townsend (1977) synthesized 1-methylpseudouridine by reaction of 2′,3′,5′-tri-*O*-acetylpseudouridine with bistrimethylsilylacetamide; yield 62.6%.

Inhibition of Growth

1-Methylpseudouridine and its triacetate did not inhibit a wide variety of gram-positive and gram-negative organisms. Both nucleosides showed marginal activity when tested against herpes virus (42D type 1) replication *in vitro*. There was no inhibition of L1210 cells *in vitro* (Argoudelis and Mizsak, 1976).

Biochemical Properties

The isolation of pseudouridine and 1-methylpseudouridine from the culture filtrates of many of the *Streptomyces* along with the production of potent nucleoside antibiotics (i.e., 5-azapyrimidine nucleosides and oxazinomycin) poses many interesting questions concerning the biosynthesis of C-nucleosides and their relationship to other naturally occurring nucleoside antibiotics in the culture filtrates. As reported by Uematsu and Suhadolnik (1972), Ciampi et al. (1977), and Cortese et al. (1974), the biosynthesis of pseudouridine 5′-monophosphate in tRNA occurs by an intramolecular rearrangement of UMP residues. Although this intramolecular rearrangement is clearly established in macromolecules, Uematsu and Suhadolnik (1972) showed in their studies on the biosynthesis of pseudouridine as the free nucleoside that pseudouridine in the culture filtrates of *S. ladakanus* does not proceed by an intramolecular rearrangement of [U-^{14}C]uridine. Therefore, a new mechanism for the biosynthesis of this C-nucleoside must exist. One mechanism involves an intramolecular rearrangement of UMP at the nucleotide level. A second mechanism involves the formation of the C-nucleoside that does not proceed by an intramolecular rearrangement of uridine with its 5′-phosphates.

Although 1-methylpseudouridine has not been identified in RNA, $O^{2'}$-methylpseudouridine has been identified as a component of wheat embryo cytosol rRNA (Hudson et al., 1965; Gray, 1974).

4.5 NEBULARINE

Nebularine, 9-β-D-ribofuranosyl purine (Fig. 4.15), is a naturally occurring purine riboside that is structurally similar to adenosine. It is isolated from *Agaricus* (*Clitocybe*) *nebularis batsch* and *S. yokosukanensis*. It is toxic to mice, *Mycobacterium*, and animal cells in culture and inhibits tumor growth. 7-Deazanebularine (7-β-D-ribofuranosyl pyrrolopyrimidine) has been compared with the enzymes that regulate purine ribonucleotide synthesis (Smith et al., 1974). Nebularine has been reviewed (Suhadolnik, 1970; Townsend, 1975; Bloch, 1978; Bohr, 1975).

Biochemical Properties

Even though nebularine has been studied in a number of bacterial and mammalian systems, little is known about its biochemistry. It is toxic to animals and cells in culture (Brown and Welicky, 1953; Biesele et al., 1955). The manifestation of nebularine toxicity requires entry into the cell by the nucleoside transport mechanism (Paterson et al., 1979a). Henderson (1968) and Henderson et al. (1972) have reported that nebularine and 7-deazanebularine are potent inhibitors of purine biosynthesis. Tamm et al. (1956) have shown that nebularine inhibits influenza B virus multiplication. Nebularine is enzymatically phosphorylated to the 5′-mono-, 5′-di-, and 5′-triphosphates (Lindberg et al., 1967). The biochemical effects of nucleotides of nebularine have not been determined. Smith et al. (1974) studied the effect of nebularine and 7-deazanebularine on the metabolism of purine bases in Ehrlich ascites tumor cells. The inhibition of adenosine kinase presumably occurs following the phosphorylation of nebularine or deazanebularine. PRPP accumulation in the cells was inhibited 85 and 90%, respectively, by nebularine and deazanebularine. The conversion of 7-deazanebularine to its 5′-mono-, 5′-di-, and 5′-triphosphates in mouse fibroblasts has been reported (Brdar and Reich, 1972).

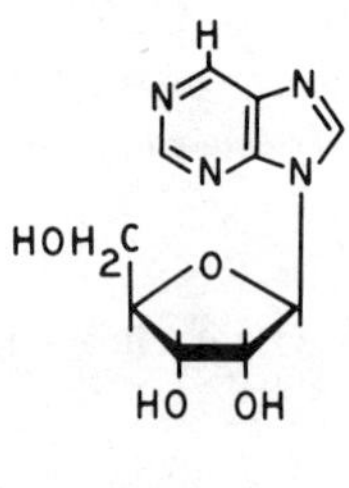

Figure 4.15 Structure of nebularine.

In a recent detailed report Bohr (1978) described the effects of nebularine on nucleic acid synthesis in ascites cells. It is readily phosphorylated and inhibits RNA and DNA synthesis in different Yoshida ascites cells. Total RNA synthesis is inhibited by 70%, 18 and 28 S ribosomal components are inhibited by 90%, and a combination of 4 and 5 S RNA are inhibited by 60% (Fig. 4.16); protein synthesis is inhibited by 60%.

Although nebularine 5′-triphosphate is not incorporated into RNA by *E. coli* RNA polymerase, it completely inhibits RNA synthesis (Bohr, 1978). The K_i is 2.4×10^{-5} *M*; $K_{m\text{ATP}}$ is 3×10^{-6} *M*. Tumor cells devoid of adenosine kinase are not inhibited by nebularine (Bennett et al., 1966).

Protection of Mice Against Nebularine Toxicity by Nitrobenzylthioinosine (MBMPR). Paterson and coworkers (Warnick et al., 1972) reported that when MBMPR was added to L5178Y cells in culture, the cytotoxic effects of nucleosides were absent or reduced. MBMPR inhibited drug entry into cultured cells. More recently, Paterson et al. (1979a) demonstrated that MBMPR protected PRMI 6410 human lymphoblastoid cells in culture and mice against lethal doses of nebularine, tubercidin, and toyocamycin when administered (i.p. or subcutaneously) in advance of nebularine. Finally, Paterson et al. (1979b) used nucleoside analogs, where the aglycon was greatly modified (i.e., showdomycin, 5-azacytidine, sangivamycin, and nebularine) to determine the specificity of the nucleoside transport mechanism. They concluded that there is a low specificity in a cultured line of human lymphoblastoid cells.

Pharmacology

The LD_{50} of nebularine is 15 mg/kg bodyweight (guinea pigs), 220 mg/kg (rats), and 200 mg/kg (mice) (Truant and D'Amoto, 1955).

4.6 5,6-DIHYDRO-5-AZATHYMIDINE

5,6-Dihydro-5-azathymidine, 1-(2-deoxy-β-D-*erythro*pentofuranosyl)-5,6-dihydro-5-methyl-*s*-triazine-2,4(1H, 3H)-dione (DHAdT) (Fig 4.17), is isolated from the culture filtrates of *S. platensis* var. *clarensis* (Bannister et al., 1976). It contains the *s*-triazine ring, as does 5-azacytidine (Fig. 3.10). Pseudouridine and 1-methylpseudouridine are found in the same culture medium. The analog is active against DNA viruses and gram-negative bacteria.

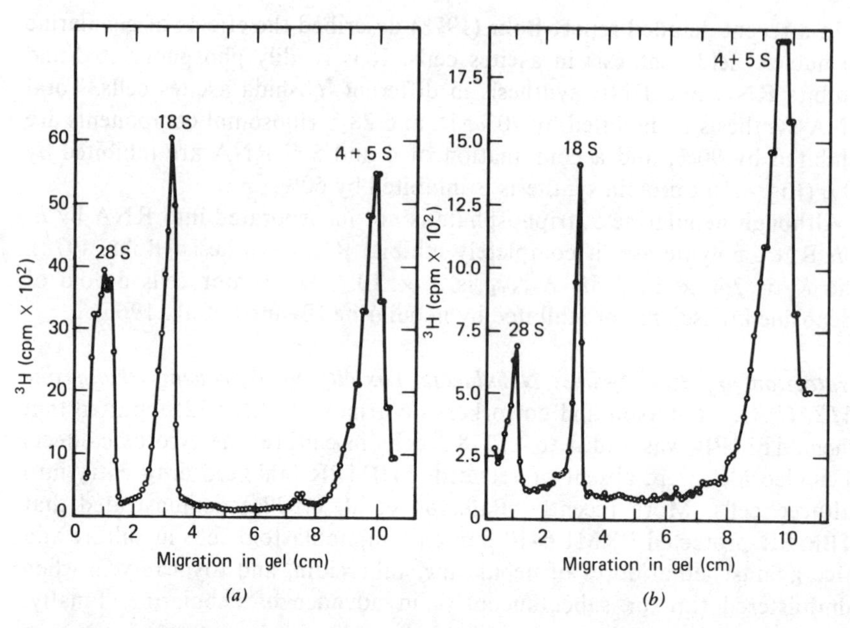

Figure 4.16 Inhibition of RNA synthesis by nebularine (purine riboside). Electropherograms showing 18, 28 and 4 plus 5 S RNA components from Yoshida ascites cells: (*a*) without nebularine; (*b*) 0.8 m*M* nebularine. Note the difference in scale in *a* and *b*. Reprinted with permission from Bohr, 1978.

Discovery, Production, and Isolation

DHAdT is isolated from the culture filtrates of *S. platensis* var. *clarensis* as described by DeBoer and Bannister (1975) and by DeBoer et al. (1976). A seed inoculum is prepared prior to the inoculation of the fermentation medium. The fermentation requires 5–12 days. The antibiotic titer of the fermentation is monitored by agar plate disc assay for *Klebsiella pneumoniae*. The nucleoside is isolated from the culture filtrate by adsorption and silica gel chromatography.

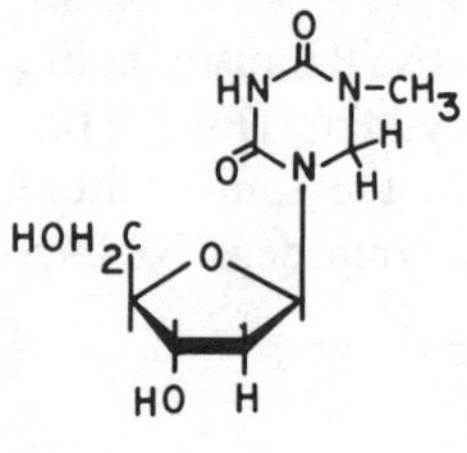

Figure 4.17 Structure of 5,6-dihydro-5-azathymidine (DHAdT).

Physical and Chemical Properties

The molecular formula for DHAdT is $C_9H_{15}N_3O_5$; mol wt 245; mp 141–142°C; $[\alpha]_D^{25}$ 5° (*c* 0.903, H_2O); it is highly soluble in water, methanol and ethanol, but insoluble in organic solvents; the ir, nmr, and mass spectra have been described (DeBoer and Bannister, 1975). The nmr and mass spectra of the antibiotic along with its di-*O*-acetyl derivative indicate that the nucleoside is a β-D-2′-deoxypentofuranoside. Degradation gave 1-methyl-5,6-dihydro-5-azauracil and 2-deoxy-D-*erythro*pentose (Bannister et al., 1976).

Inhibition of Growth

DHAdT is active against DNA viruses *in vitro* and *in vivo*; it also inhibits gram-negative bacteria, but does not inhibit pathogenic fungi, yeast, or bacteria (DeBoer and Bannister, 1975; Zurenko and Lewis, 1976); when administered subcutaneously to mice infected with *Proteus vulgaris*, *Proteus mirabilis*, *Salmonella flexneri*, or *E. coli*, it is active and well tolerated. Because it is water soluble and stable in neutral and alkaline solutions, it diffuses readily into animal tissues. Underwood and Weed (1977) reported that this analog, administered subcutaneously, is active prophylactically and therapeutically against cutaneous herpes virus infection of hairless mice. The inhibitory activity is analogous to that of ara-A. DHAdT is toxic to cats, but not to mice and rats; 80–88% of the antibiotic is detected in the blood and tissue 20 min after one subcutaneous injection; 75% of the absorbed antibiotic is excreted in the urine in 48 h and 49% in 8 h (Stern and Lewis, 1976). The antibiotic is rapidly distributed throughout all the organs of the test animal.

Biosynthesis of DHAdT

Slechta and Cialdella (1976) have reported on the biosynthesis of DHAdT. They demonstrated that 6-azauracil completely inhibited the biosynthesis of the nucleoside antibiotic. Because 6-azauracil is a competitive inhibitor of orotidylic acid decarboxylation, Slechta and Cialdella postulated that 5-azaorotic acid might be the precursor for the biosynthesis of DHAdT. The addition of 5-azaorotic acid stimulated the production of the nucleoside analog. This stimulation was completely abolished by the addition of 6-azauracil. Based on these data, Slechta and Cialdella synthesized $[^{14}C]$-5,6-dihydro-5-azaorotic acid. Addition of this *s*-triazine led to ^{14}C-labeled

DHAdT. Based on the degradation of allantoic acid by *Streptoccus allantoicius*, Slechta and Cialdella concluded that after the condensation with urea (**1**) and glyoxylic acid (**2**), glyoxylurea (**3**) would yield glyoxylbiuret (**4**) which on ring closure would give 5,6-dihydro-5-azaorotic acid (**5**), which would add PRPP to form **6**. Compound **6** loses CO_2 to form compound **7** (Fig. 4.18). [2-^{14}C]Glyoxylic acid is incorporated into DHAdT. To determine the sequence of addition of the methyl group, it was shown that 5,6-dihydro-5-azathymine is not a precursor for DHAdT. Therefore, the methylation of the aglycon appears to occur following the reduction of ribose to 2-deoxyribose.

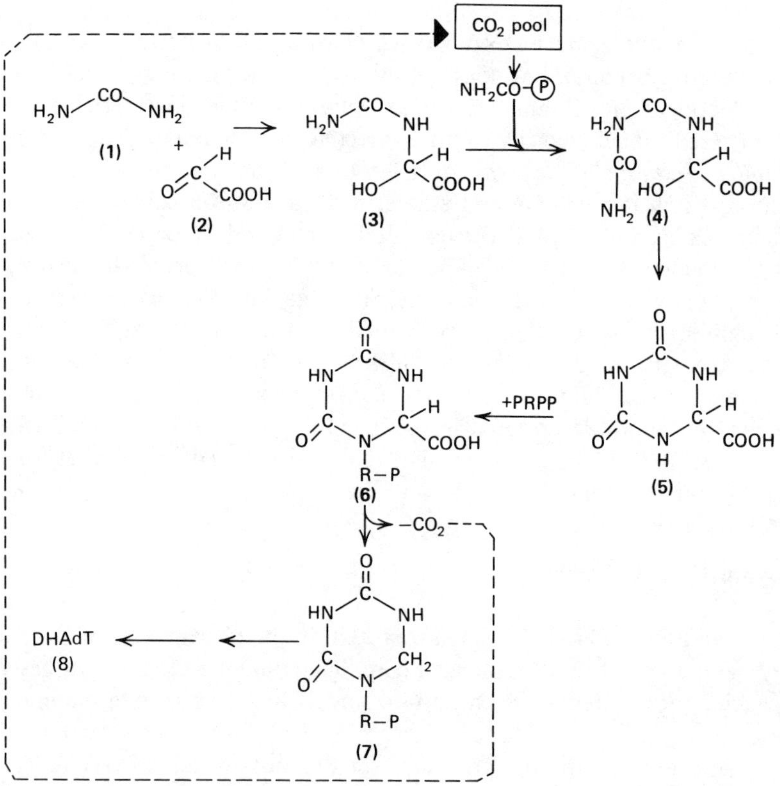

Figure 4.18 The biosynthesis of 5,6-dihydro-5-azathymidine (**8**) (DHAdT). Slechta and Cialdella, 1979b.

Biochemical Properties

Antiviral Activity. This analog inhibits the replication of herpes simplex virus type 1 (HSV-1) in primary rabbit kidney cells in culture (6–100 μg/ml). The inhibitory action of DHAdT can be completely reversed by the addition of either thymidine or deoxyuridine to the culture medium or partially reversed by addition of deoxycytidine. Addition of this inhibitor 4 h after infection (in a single cycle growth) results in a greater than 90% reduction in virus yields (Renis et al., 1976). HSV-1 was more sensitive to DHAdT than was HSV-2. Mice infected with HSV-1 were protected by 5,6-dihydro-5-azathymidine treatment (100–400 mg/kg) for 4–5 days. Antiviral activity was observed when drug therapy was initiated 48–72 h following HSV inoculation. Mice inoculated intracerebrally with HSV-1 were also protected by this analog. No toxicity was detected in mice (Renis et al., 1976).

The mode of action of DHAdT has been reported by Stroman (1976). The antibacterial activity of DHAdT in *E. coli* is reversed by the addition of thymidine. Although *E. coli* is sensitive to DHAdT, the mode of action does not involve the direct inhibition of DNA synthesis, but rather, inhibition of the phosphorylation of thymidine. This conclusion is based on the observation that thymidine and deoxyuridine can reverse the inhibition. Protection is also obtained by either increasing cAMP levels, altering 2-deoxyribose metabolism, or changing the cell membrane organization by alterations in the structure of the ribosome.

Immunosuppressive Activity. DeBoer et al. (1976) demonstrated that this analog suppresses bone marrow cells in cats and dogs, but not in rats.

Resistance to 5,6-Dihydro-5-azathymidine. Cialdella and Slechta (1976) reported on the isolation of an *E. coli* strain that was resistant to 5,6-dihydro-5-azathymidine. The resistance is not attributed to the action of thymidine phosphorylase. When the level of thymidine phosphorylase was studied in the mutant and in a susceptible strain of *E. coli*, the level of thymidine phosphorylase in the susceptible strain increased four times upon the addition of thymidine to the growth medium. In the resistant strain, the constitutive level of the enzyme was only one-third that of the susceptible strain. However, upon induction with thymidine, the level of the enzyme increased about 10 times. The analog failed to act as an inducer of thymidine phosphorylase in the resistant strain. DHAdT did not interfere with the induction of thymidine phosphorylase by thymidine. The resistant strain of *E. coli* was accompanied by altered thymidine catabolism because the resistant strain could no longer utilize thymidine for growth. The susceptible strain grew very well when thymidine was the carbon source.

SUMMARY

Ara-A is a cytostatic analog of deoxyadenosine. It possesses antibacterial activity against an *E. coli* mutant, has antiproliferation activity against animal cells in culture, and has some activity against transplantable animal tumors. Ara-A strongly suppresses viral replication and viral DNA synthesis of a broad spectrum of DNA viruses, including herpes viruses. Clinical studies show that ara-A is useful against herpes simplex keratitis, herpes simplex encephalitis, hepatic keratitis, and disseminated herpes zoster. Although ara-A is deaminated, it is transported to sensitive cells and phosphorylated to the 5′-mono-, 5′-di-, and 5′-triphosphates. The mechanism of action of ara-ATP involves the inhibition of DNA synthesis. Ara-AMP is incorporated into the internucleotide linkage of DNA and the 3′-terminal position of RNA. Ara-ATP is a competitive inhibitor of dATP of virus- or tumor cell-induced DNA polymerases. Ara-AMP is incorporated into the newly formed DNA internucleotide linkage and markedly reduces DNA of high molecular weight. This latter effect does not appear to be due to ara-AMP acting as a chain terminator. The addition of adenosine deaminase inhibitors markedly increases the toxicity of ara-A. Low concentrations of coformycin increase the ara-ATP pool; high concentrations of coformycin lower the ara-ATP pool, but increase the ara-inosine pool, probably by the ara-AMP to ara-IMP to ara-inosine pathway. Ara-AMP can slowly penetrate the cell intact. Ara-ATP is a better inhibitor of viral DNA polymerase than of host mammalian DNA polymerase-α. However, the α-anomer, α-ara-A, is intracellularly phosphorylated to α-ara-ATP and is incorporated into DNA. Unlike β-ara-ATP, α-ara-ATP is a potent inhibitor of host mammalian DNA polymerase-α, but has no effect on either DNA polymerase-β or herpes simplex virus-induced DNA polymerase. Ara-ATP is a selective inhibitor of terminal deoxynucleotidyl transferase.

The biosynthesis of ara-A has been shown to involve the direct conversion of adenosine by the enzyme adenosine-2′-epimerase.

Ara-T is a pyrimidine nucleoside analog of deoxythymidine (dT). It is a relatively selective inhibitor of herpes viral infection in cells in culture and in animals. Ara-T is phosphorylated by the viral-coded pyrimidine deoxyribonucleoside kinase (dPyK). Ara-T has low toxicity to uninfected cells and is not a substrate for uninfected host cell dTK.

The chemical synthesis of oxazinomycin has been described. Oxazinomycin inhibits bacteria, but not yeast or fungi. It has activity against tumors and has been used to promote growth in children who are underweight. The biosynthesis of oxazinomycin involves asymmetric incorporation of carbons 3, 4, and 5 of glutamate to form carbons 6, 5, and 4, respectively, of the aglycon.

1-Methylpseudouridine and its acetate show limited inhibitory properties against viruses. L1210 cells are not inhibited.

Nebularine is readily phosphorylated to its 5′-mono-, 5′-di-, and 5′-triphosphates. It inhibits bacteria, mammalian cells in culture, and influenza B virus and is toxic to animals. Nebularine 5′-triphosphate can replace ATP in the mengovirus RNA polymerase reaction. Nebularine monophosphate can only form one hydrogen bond with UMP in RNA. It inhibits RNA and DNA synthesis in Yoshida ascites cells. Protein synthesis is inhibited by 60%. Nebularine is not inhibitory in cells devoid of adenosine kinase.

5,6-Dihydro-5-azathymidine is a pyrimidine analog that is active against DNA viruses and gram-negative bacteria. Thymidine, deoxyuridine, and deoxycytidine reverse the inhibition by this analog. The action of this analog appears to involve an inhibition of the phosphorylation of thymidine and not an inhibition of DNA synthesis. Resistance to this nucleoside is due to the change in thymidine phosphorylase. The biosynthesis of the aglycon involves the condensation of urea, glyoxylic acid, and carbamoyl phosphate.

REFERENCES

Allen, L. B., and R. W. Sidwell, *Antimicrob. Agents Chemother.*, **2,** 229 (1972).

Allen, L. B., J. H. Huffman, R. L. Tolmon, G. R. Revankar, L. N. Simon, R. K. Robins, and R. W. Sidwell, 14th Interscience Conference on Antimicrobial Agents Chemotherapy, 1974.

Argoudelis, A. D., and S. A. Mizsak, *J. Antibiot.* (*Tokyo*) **29,** 818 (1976).

Aswell, J. F., and G. P. Gentry, *Ann. N.Y. Acad. Sci.*, **284,** 342 (1977).

Aswell, J. F., G. P. Allen, A. T. Jamieson, D. E. Campbell, and G. A. Gentry, *Antimicrob. Agents Chemother.*, **12,** 243 (1977).

Baker, D. C., and T. H. Haskell, *J. Med. Chem.*, **18,** 1041 (1975).

Baker, D. C., T. H. Haskell, and S. R. Putt, *J. Med. Chem.*, **21,** 1218 (1978).

Baker, D. C., T. H. Haskell, S. R. Putt, and B. J. Sloan, *J. Med. Chem.*, in press (1979).

Baltimore, D., *Nature*, **226,** 1209 (1970).

Bannister, B., L. Slechta, and A. D. Argoudelis, Abstracts, 10th International Symposium on Chemistry of Natural Products, August 1976, Abstr. C28.

Bennett, L. L., Jr., H. P. Schnebli, M. H. Vail, P. W. Allen, and J. A. Montgomery, *Mol. Pharmacol.*, **2,** 432 (1966).

Bergmann, W., and R. J. Feeney, *J. Am. Chem. Soc.*, **72,** 2809 (1950).

Bergmann, W., and R. J. Feeney, *J. Org. Chem.*, **16,** 981 (1951).

Bergmann, W., and M. F. Stempien, Jr., *J. Org. Chem.*, **22,** 1575 (1957).

Biesele, J. J., M. C. Slautterback, and M. Margolis, *Cancer*, **8,** 87 (1955).

Bloch, A., in *Encyclopedia of Chemical Technology*, Vol. 2, Wiley, New York, p. 962, 1978.

Bloch, A., M. J. Robins, and J. R. McCarthy, Jr., *J. Med. Chem.*, **10,** 908 (1967).

Bohr, V., *Biochim. Biophys. Acta*, **519,** 125 (1978).

Bohr, V., Gold Medal Thesis, University of Copenhagen, Denmark, 1975.

Brdar, B., and E. Reich, *J. Biol. Chem.*, **247,** 725 (1972).

Brink, J. J., and G. A. LePage, *Cancer Res.*, **24,** 312 (1964a).

Brink, J. J., and G. A. LePage, *Cancer Res.*, **24,** 1042 (1964b).

Brink, J. J., and G. A. LePage, *Can. J. Biochem.*, **43,** 1 (1965).

Brown, G. B., and V. S. Welicky, *J. Biol. Chem.*, **204,** 1019 (1953).

Cardeilhac, P. T., and S. S. Cohen, *Cancer Res.*, **24,** 1595 (1964).

Chassy, B. M., and R. J. Suhadolnik, *J. Biol. Chem.*, **242,** 2655 (1967).

Chaudry, I. H., and M. K. Gould, *Biochim. Biophys. Acta*, **196,** 320 (1970).

Ch'ien, L. T., A. J. Glazko, R. A. Buchanan, and C. A. Alford, *Intersci. Conf. Antimicrob. Agents Chemother.*, **11,** 94 (1971).

Ch'ien, L. T., N. J. Cannon, R. J. Whitley, A. G. Diethelm, W. E. Dismukes, C. W. Scott, R. A. Buchanan, and C. A. Alford, Jr., *J. Infect. Dis.*, **130,** 32 (1974).

Ch'ien, L. T., R. J. Whitley, L. J. Charamella, R. A. Buchanan, J. J. Cannon, W. E. Dismukes, and C. A. Alford, Jr., in *Adenine Arabinoside: An Antiviral Agent*, D. Pavan-Langston, R. A. Buchannan, and C. A. Alford, Jr., Eds., Raven Press, New York, 1975, p. 205.

Ch'ien, L. T., R. J. Whitley, C. A. Alford, Jr., G. J. Gallaso, and the Collaborative Study Group, *J. Infect. Dis.*, **133,** A184 (1976).

Chu, M. Y., and G. A. Fischer, *Biochem. Pharmacol.*, **11,** 423 (1962).

Cialdella, J., and L. Slechta, Abstracts, 16th Interscience Conference on Antimicrobiol Agents and Chemotherapy, Chicago, 1976, Abstr. 421.

Ciampi, M. S., F. Arena, and R. Cortese, *FEBS Lett.*, **77,** 75 (1977).

Cohen, S. S., *Prog. Nucleic Acid Res. Mol. Biol.*, **5,** 1 (1966).

Cohen, S. S., *Biochem. Pharmacol.*, **24,** 1929 (1975).

Cohen, S. S., *Med. Biol.* **54,** 299 (1976).

Cohen, S. S., and H. D. Barner, *J. Bacteriol.*, **21,** 588 (1956).

Cohen, S. S., and W. Plunkett, *An.. N.Y. Acad. Sci.*, **255,** 269 (1975).

Cohn, W. E., *J. Biol. Chem.*, **235,** 1488 (1961).

Connor, J. D., L. Sweetman, S. Carey, M. A. Stuckey, and R. Buchanan, *Antimicrob. Agents Chemother.*, **6,** 630 (1974).

Cortese, R., H. O. Kammen, S. J. Spengler, and B. N. Ames, *J. Biol. Chem.*, **249,** 1103 (1974).

Cory, J. G., and R. J. Suhadolnik, *Biochemistry*, **4,** 1729 (1965).

Cory, J. G., G. Weinbaum, and R. J. Suhadolnik, *Arch. Biochem. Biophys.*, **118,** 428 (1967).

Cozzarelli, N. R., *Annu. Rev. Biochem.*, **46,** 641 (1977).

De Bernardo, S., and M. Weigele, *J. Org. Chem.*, **42,** 109 (1977).

DeBoer, C., and B. Bannister, U.S. Patent 3,907,643 (1975).

DeBoer, C., B. Bannister, A. Dietz, C. Lewis, and J. E. Gray, Abstracts, Annual Meeting of the American Society for Microbiology, Atlantic City, N.J., 1976, Abstr. 029.

DeBoer, C., B. Bannister, A. Dietz, C. Lewis, and J. E. Gray, paper in preparation (1979).

DeGarilhe, M. P., and J. deRudder, *C. R. Acad. Sci.*, **259,** 2725 (1964).

DeGarilhe, M. P., and J. deRudder, *Prog. Antimicrob. Anticancer Chemother.*, **2,** 180 (1970).

DiCioccio, R. A., and B. I. Srivastava, *Eur. J. Biochem.*, **79,** 411 (1977).

Doering, A. M., J. Keller, and S. S. Cohen, *Cancer Res.*, **26,** 2444 (1966).

Drach, J. C., and C. Shipman, Jr., *Ann. N.Y. Acad. Sci.*, **284,** 396 (1977).

Drach, J. C., J. N. Sandberg, and C. Shipman, Jr., *J. Dent. Res.*, **57,** 275 (1977).

Earl, R. A., and L. B. Townsend, *J. Heterocyc. Chem.*, **14,** 699 (1977).

Falcon, M. G., and B. R. Jones, *J. Gen. Virol.*, **36,** 199 (1977).

Farmer, P. B., and R. J. Suhadolnik, *Biochemistry*, **11,** 911 (1972).

Farmer, P. B., T. Uematsu, H. P. C. Hogenkamp, and R. J. Suhadolnik, *J. Biol. Chem.*, **248,** 1844 (1973).

Fishaut, J. M., J. D. Connor, and P. W. Lampert, in *Adenine Arabinoside: An Antiviral Agent*, D. Pavan-Langston, R. A. Buchanan, and C. A. Alford, Jr., Eds., Raven Press New York, 1975, p. 159.

Furth, J. J., and S. S. Cohen, *Cancer Res.*, **27,** 1528 (1967).

Furth, J. J., and S. S. Cohen, *Cancer Res.*, **28,** 2061 (1968).

Gentry, G. A., and J. F. Aswell, *Virology*, **65,** 294 (1975).

Gentry, G. A., J. F. Aswell, G. P. Allen, and D. E. Campbell, *Proc. 18th Annu. Meet. Am. Assoc. Cancer Res.*, **18,** Abstr. 530 (1977).

Gentry, G., J. McGowan, J. Barnett, R. Nevins, and G. Allen, *Adv. Opthalmol.*, **38,** 164 (1979).

Gray, M. W., *Biochemistry*, **13,** 5433 (1974).

Haneishi, T., M. Nomura, T. Okazaki, A. Naito, I. Seki, M. Arai, T. Hata, and C. Tamura, 174th Scientific Meeting of the Japan Antibiotic Research Association, Tokyo, July 27, 1971.

Henderson, J. F., *Cancer Chemother. Rep.*, **1,** 375 (1968).

Henderson, J. F., A.R.P. Paterson, I.C. Caldwell, B. Paul, M.C. Chan, and K. F. Lau, *Cancer Chemother. Rep.*, **3,** 71 (1972).

Hershfield, M. S., *J. Biol. Chem.*, **254,** 22 (1979).

Hubert-Habart, M., and S. S. Cohen, *Biochim. Biophys. Acta*, **59,** 468 (1962).

Hudson, L., M. Gray, and B. G. Lane, *Biochemistry*, **4,** 2009 (1965).

Hyndiuk, R. A., and H. E. Kaufman, *Arch. Opthalmol.*, **78,** 600 (1967).

Hyndiuk, R. A., M. Okumoto, Damiano, R., M. Valenton, and G. Smolin, *Arch. Opthalmol.*, **94,** 1363 (1976).

Ilan, J., K. Tokuyasu, and J. Ilan, *Nature*, **228,** 1300 (1970).

Ilan, J., D. R. Pierce, and F. W. Miller, *Proc. Natl. Acad. Sci.*, **74,** 3386 (1977).

Isono, K., and R. J. Suhadolnik, *Ann. N.Y. Acad. Sci.*, **255,** 390 (1975).

Isono, K., and R. J. Suhadolnik, *J. Antibiot.* (*Tokyo*), **30,** 272 (1977).

Isono, K., and K. Uzawa, *FEBS Lett.*, **80,** 53 (1977).

Jamieson, A. T., and J. H. Subak-Sharpe, *J. Gen. Virol.*, **24,** 481 (1977).

Jamieson, A. T., G. A. Gentry, and J. H. Subak-Sharpe, *J. Gen. Virol.*, **24,** 465 (1974).

Jones, D. B., in *Adenine Arabinoside: An Antiviral Agent*, D. Pavan-Langston, R. A. Buchanan, and C. A. Alford, Jr., Eds., Raven Press, New York, 1975, p. 371.

Kaufman, H. E., A. B. Nesburn, and E. D. Maloney, *Arch., Ophthalmol.*, **67,** 583 (1962).

Kelly, R. K., and P. Loh, J. Virology, **10,** 824 (1972).

Kinkel, A. W., and R. A. Buchanan, in *Adenine Arabinoside: An Antiviral Agent*, D. Pavan-Langston, R. A. Buchanan, and C. A. Alford, Jr., Eds., Raven Press, New York, 1975, p. 197.

Kits, S., and D. R. Dubbs, *Virology*, **26,** 16 (1965).

Klein, R. J., A. E. Friedman-Kien, and E. Brady, *Antimicrob. Agents Chemother.*, **5,** 409 (1974).

Kurtz, S. M., J. E. Fitzgerald, and J. L. Schardein, *Ann. N.Y. Acad. Sci.*, **284,** 6 (1977).

Kusakabe, Y., J. Nagatau, M. Shibuya, O. Kawaguchi, C. Hirose, and S. Shirato, *J. Antibiot. (Tokyo)*, **25,** 44 (1972).

Langston, R. H. S., D. Pavan-Langston, and C. H. Dohlman, in *Adenine Arabinoside: An Antiviral Agent*, D. Pavan-Langston, R. A. Buchanan, and C. A. Alford, Jr., Eds., Raven Press, New York, 1975, p. 313.

Lee, W. W., A. Benitez, L. Goodman, and B. R. Baker, *J. Am. Chem. Soc.*, **82,** 2648 (1960).

Leonard, T. B., and S. T. Jacob, *Biochim. Biophys. Acta*, in press (1979).

LePage, G. A., *Adv. Enzyme Regul.*, **8,** 323 (1970).

LePage, G. A., Y. T. Lin, R. E. Orth, and J. A. Gottlieb, *Cancer Res.*, **32,** 2441 (1972).

LePage, G. A., A. Khaliq, and J. A. Gottlieb, *Drug Metab. Disposition*, **1,** 756 (1973).

Leung, H. B., A. M. Doering, and S. S. Cohen, *J. Bacteriol.*, **92,** 558 (1966).

Lindberg, B., H. Klenow, and K. Hansen, *J. Biol. Chem.*, **242,** 350 (1967).

Lipper, R. A., S. M. Machkovich, J. C. Brach, and W. I. Higuchi, *Mol. Pharmacol.*, **14,** 366 (1978).

Matsukage, A., T. Takahashi, C. Nakayama, and, M. Saneyoshi, *J. Biochem*, **83,** 1511 (1978).

Miller, R. L., and D. L. Adamczyk, *Biochem. Pharmacol.*, **25,** 883 (1976).

Miller, F. A., G. J. Dixon, J. Ehrlich, B. J. Sloan, and I. W. McLean, Jr., *Antimicrob. Agents Chemother.*, **1968** 136.

Miller, R. L., J. P. Iltis, and F. Rapp, *J. Virol.*, **23,** 679 (1977).

Moore, E. C., and S. S. Cohen, *J. Biol. Chem.*, **242,** 2116 (1968).

Müller, W. E. G., *Experientia*, **32,** 1572 (1976).

Müller, W. E. G., H. J. Rohde, R. Beyer, A. Maidhof, M. Lachman, H. Taschner, and R. K. Zahn, *Cancer Res.*, **35,** 2160 (1975).

Müller, W. E. G., R. K. Zahn, K. Bittlingmaier, and D. Falke, *Ann. N.Y. Acad. Sci.*, **284,** 34 (1977a).

Müller, W. E. G., A. Maidhof, R. K. Zahn, and W. M. Shannon, *Cancer Res.*, **37,** 2282 (1977b).

Müller, W. E. G., R. K. Zahn, R. Beyer, and D. Falke, *Virology*, **76,** 787 (1977c).

Müller, W. E. G., R. K. Zahn, and D. Falke, *Virology*, **84,** 320 (1978a).

Müller, W. E. G., R. K. Zahn, A. Maidhof, R. Beyer, and J. Arendes, *Biochem. Pharmacol.*, **27,** 1659 (1978b).

Müller, W. E. G., R. K. Zahn, and J. Arendes, *FEBS Lett.*, **94,** 47 (1978c).

Müller, W. E. G., R. K. Zahn, J. Arendes, A. Maidhof, and H. Umezawa, *Z. Physiol. Chem.*, **359,** 1287 (1978d).

Müller, W. E. G., in *Antiviral Mechanisms for the Control of Neoplasia*, P. Chandra, Ed., Raven Press, New York, 1979.

Nahmias, A. J., and B. Roizman, *N. Engl. J. Med.*, **289,** 667, 719, 781 (1973).

Nichols, W. E., *Cancer Res.*, **24,** 1502 (1964).

North, T. W., and S. S. Cohen, *Proc. Natl. Acad. Sci. U.S.*, **75,** 4684 (1978).

O'Day, D. M., R. H. Poirier, and J. H. Elliott, in *Adenine Arabinoside: An Antiviral Agent*, D. Pavan-Langston, R. A. Buchanan, and C. A. Alford, Jr., Eds. Raven Press, New York, 1975, p. 357.

Okami, Y., R. Utahara, S. Nakamura, and H. Umezawa, *J. Antibiot* (*Tokyo*), **A7,** 98 (1954).

Okura, A., and S. Yoshida, *J. Biochem.*, **84,** 727 (1978).

Ortiz, P. J., M. Manduka, and S. S. Cohen, *Cancer Res.*, **32,** 1512 (1972).

Parke, Davis, and Company, Belgium Patent 671,557 (1967).

Paterson, A. R. P., J. H. Paran, S. Yang, and T. P. Lynch, *Cancer Res.*, in press (1979a).

Paterson, A. R. P., S. Yang, E. Y. Lau, and C. E. Cass, *Mol. Pharmacol.*, in press (1979b).

Pavan-Langston, D., and R. A. Buchanan, *Trans. Am. Acad. Opthalmol.* **81,** OP-813 (1976).

Pavan-Langston, D., R. A. Buchanan, and C. A. Alford, Jr., Eds., *Adenine Arabinoside: An Antiviral Agent*, Raven Press, New York, 1975a.

Pavan-Langston, D., R. H. S. Langston, and P. A. Geary, in *Adenine Arabinoside: An Antiviral Agent*, D. Pavan-Langston, R. A. Buchanan, and C. A. Alford, Jr., Eds., Raven Press, New York, 1975b, p. 337.

Pavan-Langston, D., C. H. Dohlman, P. Geary, and D. Szulczewski, in *Adenine Arabinoside: An Antiviral Agent*, D. Pavan-Langston, R. A. Buchanan, and C. A. Alford, Jr., Eds., Raven Press, New York, 1975c, p. 293.

Pelling, J. C., J. C. Drach, and C. Shipman, Jr., Herpesvirus Workshop, Cambridge, England, August 1978.

Plunkett, W., and S. S. Cohen, *Cancer Res.*, **35,** 415 (1975a).

Plunkett, W., and S. S. Cohen, *Cancer Res.*, **35,** 1547 (1975b).

Plunkett, W., and S. S. Cohen, *Ann. N.Y. Acad. Sci.*, **284,** 91 (1977a).

Plunkett, W., and S. S. Cohen, *J. Cell Physiol.*, **191,** 261 (1977b).

Plunkett, W., L. Lapi, P. J. Ortiz, and S. S. Cohen, *Proc. Natl. Acad. Sci. U.S.*, **71,** 73 (1974).

Poste, G., D. F. Hawkins, and J. Thomlinson, *Obstet. Gynecol.*, **40,** 871 (1972).

Ranganathan, R. *Tetrahedron Lett.*, **1975,** 1185.

Reichman, U., K. Hirota, C. K. Chu, K. A. Watanabe, and J. J. Fox, *J. Antibiot.* (*Tokyo*), **30,** 129 (1977).

Reinke, C. M., J. C. Drach, C. Shipman, Jr., and A. Weissbach, *Onocogenesis and Herpes Viruses*, III, Lyon, France, IARC Press, p. 999 1978.

Reist, E. J., A. Benitez, L. Goodman, B. R. Baker, and W. E. Lee, *J. Org. Chem.*, **27,** 3274 (1962).

Renis, H. E., B. A. Court, and E. E. Eidson, Abstracts, 16th Interscience Conference on Antimicrobial Agents and Chemotherapy, Chicago, October 1976, Abstr. 420.

Rice, J. M., and G. D. Dudek, *Biochem. Biophys. Res. Commun.*, **35,** 383 (1969).

Rose, K. M., and S. T. Jacob, *Biochem. Biophys. Res. Commun.*, **81,** 1418 (1978).

Sasaki, K., Y. Kusakabe, and S. Esumi, *J. Antibiot.* (*Tokyo*), **25,** 151 (1972).

Scannell, J. P., A. M. Crestfield, and F. W. Allen, *Biochim. Biophys. Acta*, **32,** 406 (1959).

Schabel, F. M., Jr., *Chemotherapy*, **13,** 321 (1968).

Schmidt-Ruppen, K. H., *Chemotherapy*, **16,** 130 (1971).

Schwartz, P. M., C. Shipman, Jr., and J. C. Drach, *Antimicrob. Agents Chemother.*, **10,** 64 (1976).

Shannon, W. M., in *Adenine Arabinoside: An Antiviral Agent*, D. Pavan-Langston, R. A. Buchanan, and C. A. Alford, Jr., Eds., Raven Press, New York, 1975, p. 1.

Shannon, W. M., L. Westbrook, and F. M. Schabel, Jr., *Proc. Soc. Exp. Biol. Med.*, **145,** 542 (1974).

Shipman, C., and J. C. Drach, *Science*, **200,** 1163 (1978).

Shipman, C., S. H. Smith, R. H. Carlson, and J. C. Drach, *Antimicrob. Agents Chemother.*, **9,** 120 (1976).

Sidwell, R. W., L. B. Allen, J. H. Huffman, T. A. Khwaja, R. L. Tolman, and R. K. Robins, *Chemotherapy*, **19,** 325 (1973).

Slechta, L., and J. Cialdella, Abstracts, 10th International Congress of Biochemistry, Hamburg, Germany, No. 01-8-012 (1976).

Sloan, B. J., in *Adenine Arabinoside: An Antiviral Agent*, D. Pavan-Langston, R. A. Buchanan, and C. A. Alford, Jr., Eds., Raven Press, New York, 1975, p. 45.

Smith, C. M., F. F. Snyder, L. J. Fontenelle, and J. F. Henderson, *Biochem, Pharmacol.*, **23,** 2023 (1974).

Smith, S. H., C. Shipman, Jr., and J. C. Drach, *Cancer Res.*, **38,** 1916 (1978).

Spector, T., *J. Biol. Chem.*, **250,** 7372 (1975).

Sprinzl, M., H. Sternbach, R. von der Haar, and F. Cramer, *Eur. J. Biochem.*, **81,** 579 (1977).

Stern, K. F., and C. Lewis, Abstracts, 16th Interscience Conference on Antimicrobial Agents and Chemotherapy, Chicago, October 1976, Abstr. 420.

Stroman, D. W., Abstracts, 16th Interscience Conference on Antimicrobial Agents and Chemotherapy, Chicago, 1976, Abstr. 423.

Suhadolnik, R. J., *Nucleoside Antibiotics*, Wiley, New York, 1970.

Sweetman, L., J. D. Connor, R. Seshamani, M. A. Stuckey, S. Carey, and R. Buchanan, in *Adenine Arabinoside: An Antiviral Agent*, D. Pavan-Langston, R. A. Buchanan, and C. A. Alford, Jr., Eds., Raven Press, New York, 1975, p. 135.

Tamm, I., K. Folkers, and C. H. Shunk, *J. Bacteriol.*, **72,** 59 (1956).

Temin, H. M., and S. Mizutani, *Nature*, **226,** 1211 (1970).

Townsend, L. B., in *Handbook of Biochemistry and Molecular Biology, Nucleic Acids*, Vol. I, 3rd. ed., G. D. Fasman, Ed. CRC Press, Cleveland, 1975, p. 271.

Townsend, L. B., and R. K. Robins, *J. Heterocyc. Chem.*, **6,** 459 (1969).

Trewyn, R. W., and S. J. Kerr, *Proc. 68th Annu. Meeting Am. Assoc. Cancer Res.* **18,** Abstr, 84 (1977).

Truant, A. P., and H. E. D'Amoto, *Fed. Proc.*, **14,** 391 (1955).

Uematsu, T., and R. J. Suhadolnik, *Biochemistry*, **11,** 4669 (1972).

Underwood, G. E., and S. D. Weed, *Antimicrob. Agents Chemother.*, **11,** 765 (1977).

Vince, R., and S. Daluge, *J. Med. Chem.*, **20,** 612 (1977).

Waqar, M. A., L. A. Burgoyne, and M. R. Atkinson, *Biochem., J.*, **121,** 803 (1971).

Warnick, C. T., H. Muzik, and A. R. P. Paterson, *Cancer Res.*, **32,** 2017 (1972).

Weller, T. H., *N. Engl. J. Med.*, **285,** 203 (1971a).

Weller, T. H., *N. Engl. J. Med.*, **285,** 267 (1971b).

West German Patent 2,043,946, Kaken Chemical Co., Ltd. (1971).

Whitley, R. J., L. T. Ch'ien, R. Dolin, G. J. Galasso, C. A. Alford, Jr., and the Collaborative Study Group, *New Engl. J. Med.*, **294,** 1193 (1976).

Whitley, R. J., S. Soong, R. Dolin, G. J. Galasso, L. T. Ch'ien, and C. A. Alford, *New Eng. J. Med.*, **297,** 289 (1977).

York, J. L., and G. A. LePage, *Can. J. Biochem.*, **44,** 19 (1966).

Zurenko, G. E., and C. Lewis, Abstracts, 16th Interscience Conference on Antimicrobial Agents and Chemotherapy, Chicago, October 1976, Abstr. 420.

Chapter 5 Inhibition of Adenosine Deaminase and Immunosuppressive Activity of Nucleoside Analogs

5.1 COFORMYCIN 259

Biosynthesis of Coformycin 260

5.2 2′-DEOXYCOFORMYCIN 260

Discovery, Production, and Isolation 260
Physical and Chemical Properties, Structural Elucidation and Synthesis 260

5.3 BIOCHEMICAL PROPERTIES OF COFORMYCIN AND 2′-DEOXYCOFORMYCIN 262

Inhibitors of Adenosine Deaminase 262
Phosphorylation of Coformycin 263
Increased Pharmacological Activity of Adenosine Analogs 263
Inactivation and Reactivation of Adenosine Deaminase with Deoxycoformycin with Isolated Enzyme and Intact Human Erythrocytes and Sarcoma 180 Cells 264
Effect of dCF on Mammalian Cells in Culture, Viruses, and Tumor Bearing Animals 266
Tight Binding Inhibitors 267
Increased Toxicity of Adenosine and Deoxyadenosine in the Presence of Adenosine Deaminase Inhibitors 268
Explanation of the Biochemical Differences of Inhibitors of Adenosine Deaminase 270
Effect of dCF on the Action of Cordycepin on Nuclear RNA Synthesis in Regenerating Rat Liver 271
Inhibition of 5′-AMP Deaminase 271
Deoxyadenosine Antagonism of the Antiviral Activity of Ara-A and Ara-Hx 271
Pharmacology of CF and dCF 272

One of the most exciting developments in the enhancement of the therapeutic efficiency of the adenine nucleoside analogs is the recent discovery of the naturally occurring nucleoside inhibitors of adenosine deaminase. They are coformycin (CF) and 2′-deoxycoformycin (dCF, 2′-dCF, CoV, covidarabine, pentostatin, Co-ara-A) (Fig. 5.1). These two unique diazepin nucleosides block the deamination of adenosine, ara-A, cordycepin, formycin, and other analogs of adenosine. Additional importance of adenosine deaminase in purine regulation has been emphasized by the discovery of the inborn error of metabolism due to a deficiency of adenosine deaminase which appears to be associated with a severe combined immunodeficiency disease in which children lack T and B lymphocyte functions (Giblett et al., 1972).

5.1 COFORMYCIN

Umezawa and coworkers first reported on the isolation of coformycin (CF) (Fig. 5.1) from the culture filtrates of *Nocardia interforma* and *Streptomyces kaniharaensis* SF-557 (Niida et al., 1967; Sawa et al., 1967; Tsuruoka et al., 1967). The structure of CF has been reported by Nakamura

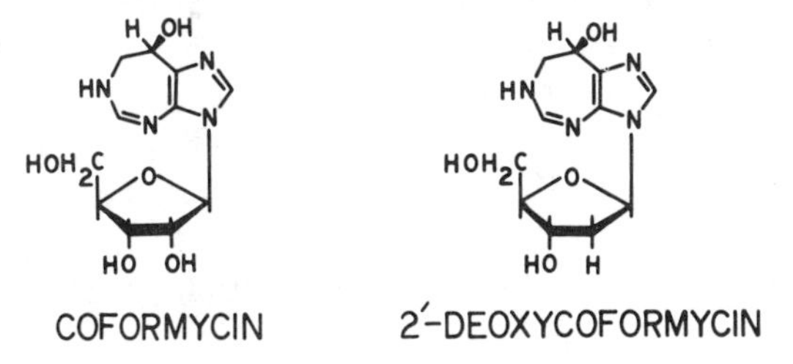

Figure 5.1 Structures of coformycin (CF) and 2′-deoxycoformycin (dCF).

et al. (1974) to be (R-3-(D-*erythro*pentofuranosyl)-3,6,7,8-tetrahydroimidazo[4,5-*d*][1,3]diazepin-8-(R)-ol. The total chemical synthesis of CF was accomplished by using the naturally occurring nucleoside nebularine (see page 244) (Ohno et al., 1974).

Coformycin has been reviewed (Suhadolnik, 1970; Townsend, 1975; Bloch, 1978).

Biosynthesis of Coformycin

Although the biosynthesis of the 1,2-diazepine ring of coformycin has not been studied, Hurley and coworkers (Hurley et al., 1975, 1976, 1979; Hurley and Gairola, 1979) and Umezawa and coworkers (Miyamoto et al., 1978) have studied the biosynthesis of the 1,4-diazepine ring of the pyrrolo(1,4)benzodiazepine nucleus of the antitumor compounds anthramycin, tomaymycin and sibiromycin (Fig. 5.2). They demonstrated that the biosynthesis of these three compounds proceeds from tryptophan by the kynurenine–anthranilic acid pathway and tyrosine by way of DOPA. These biosynthetic experiments were convincingly demonstrated by using labeled tryptophan and tyrosine. Because CF and dCF are ribo- and 2′-deoxyribonucleosides, it will be of interest to determine if there is a common precursor for the biosynthesis of the diazepine ring or if CF & dCF use the adenine ring as the carbon-nitrogen precursor in which a one carbon unit is inserted into the pyrimidine ring to form the diazepine ring.

5.2 2′-DEOXYCOFORMYCIN

Discovery, Production, and Isolation

2′-Deoxycoformycin (dCF) was isolated from the culture filtrates of *S. antibioticus* (NRRL 3238) (Woo et al., 1974; Ryder et al., 1975). dCF is extremely sensitive to pH, which requires that it be isolated between pH 7 and 9.5 (Dion et al., 1977). The isolation procedures involve concentration of the culture medium, chilling, addition of water, carbon adsorption, elution, concentration, precipitation with 80% methanol, gel filtration, and recrystallization as colorless needles from methanol and/or aqueous methanol.

Physical and Chemical Properties, Structural Elucidation and Synthesis

The molecular formula for dCF is $C_{11}H_{16}N_4O_4$; mol wt 268; $[\alpha]_D^{25}$ + 76.4° (c 1%, H_2O); pK_a' (in H_2O) 5.2; $\lambda_{max}^{pH\,7,\,H_2O}$ 282 nm (ϵ = 8000), $\lambda_{max}^{pH\,11}$ 283 nm (ϵ

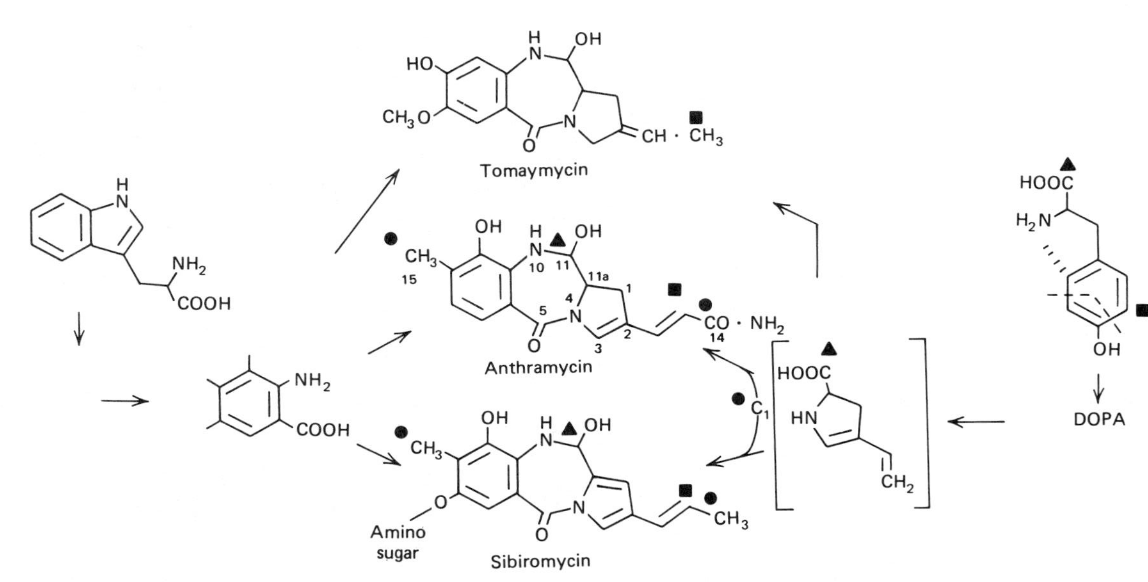

Figure 5.2 Biosynthesis of the pyrrolo(1,4)benzodiazepine antibiotics (Hurley, personal communication).

= 7970), $\lambda_{max}^{pH\,2}$ 273 nm (ϵ = 7570); maximum stability is at pH 9–11; the ir, nmr, and ^{13}C nmr spectra have been described (Dion et al., 1977). The structure of dCF has been determined to be (*R*)-3-(2-deoxy-β-D-*erythro*pentofuranosyl)-3,6,7,8-tetrahydroimidazo[4,5-*d*][1,3]diazepin-8-ol (Fig. 5.1). Baker and Putt (1979) have described the chemical synthesis of dCF.

5.3 BIOCHEMICAL PROPERTIES OF COFORMYCIN AND 2′-DEOXYCOFORMYCIN

Inhibitors of Adenosine Deaminase

Adenosine deaminase inactivates some antineoplastic agents that are analogs of adenosine. The absence of adenosine deaminase in lymphocytes and other tissues in humans is associated with the clinical syndrome severe combined immunodeficiency disease (Mills et al., 1976; Van Der Weyden and Kelly, 1977). CF and dCF are the most active adenosine deaminase inhibitors known. The primary function of the deaminase inhibitors in increasing the therapeutic efficacy of adenosine analogs is the maintenance of high concentrations of these analogs. Plunkett et al. (1978a) have shown that a single i.p. injection of EHNA or dCF into mice bearing P388 cells inhibited ara-A deamination *in vivo* and increased the recovery of ara-A over the control values within 30 min. Although dCF is a more potent inhibitor of P388 adenosine deaminase than is EHNA, there is a difference in the maximal time expression of each inhibitor. Cellular ara-ATP levels increased as did the extent and duration of inhibition of DNA synthesis capacity of P388 cells. [^{3}H]dCF appears rapidly in the plasma following i.m. or i.v. administration (Chang et al., 1975a). Parks and coworkers (Rogler-Brown et al., 1978) have demonstrated that dCF is not a substrate for purine nucleoside phosphorylase. Ara-A is protected from deamination by dCF. In the absence of dCF, nearly all of ara-A is in the form of ara-Hx. Chang and Glazko (1976) showed that [^{3}H]ara-A is rapidly deaminated by erythrocytes to ara-Hx. The ara-Hx is converted to ara-HxMP. Addition of dCF completely prevents ara-A from enzymatic deamination and results in an increased accumulation of ara-AMP, ara-ADP, and ara-ATP. The K_is for CF and dCF with purified human erythrocytic and calf intestinal adenosine deaminases are summarized in Table 5.1. Three chemically synthesized inhibitors of adenosine deaminase have been reported by Evans and Wolfenden (1970), Schaeffer and Schwender (1974), and Lipper et al. (1978). They are 1,6-dihydro-6-hydroxymethylpurine ribonucleoside

Table 5.1 Kinetic Constants of Inhibitors and Substrates of Adenosine Deaminases

	K_m ($10^{-6}M$)	K_i (M)	k_1 ($10^6\ M^{-1}$ sec^{-1})	k_2 (10^{-6} sec^{-1})
Human Erythrocytic Adenosine Deaminase				
Adenosine	25	—	—	—
Deoxyadenosine	7	—	—	—
Inosine	—	60×10^{-6}	—	—
Deoxyinosine	—	19×10^{-6}	—	—
DHMPR	—	1.3×10^{-6}	—	—
EHNA	—	1.6×10^{-9}	—	—
Coformycin	—	10×10^{-12}	2.1	24
2′-Deoxycoformycin	—	2.5×10^{-12}	2.6	6.6
Calf Intestinal Adenosine Deaminase				
Adenosine	36	—	—	—
Coformycin	—	220×10^{-12}	1.0	220
EHNA	—	6.5×10^{-9}	70.0	4500
2′-Deoxycoformycin[a]	—	6.9×10^{-9}	—	—

From Agarwal et al., 1978.
[a] Johns and Adamson, 1976.

(DHMPR), *erythro*-9-(2-hydroxy-3-nonyl)adenosine (EHNA) and 9-β-D-arabinofuranosyladenine 5′-valerate (Fig. 5.3). The K_i for DHMPR is $1.3 \times 10^{-6}\ M$, that for EHNA is $1.6 \times 10^{-9}\ M$ (Agarwal et al., 1977), and that for ara-A 5′-valerate is 11 μM (Lipper et al., 1978). The enhancement of biological activity of the naturally occurring nucleoside antibiotics by adenosine deaminase inhibitors from many laboratories is described in this chapter.

Phosphorylation of Coformycin

CF is not degraded by L5178Y cells, but is converted to its 5′-phosphates (Müller et al., 1978).

Increased Pharmacological Activity of Adenosine Analogs

The increase of ara-A as its 5′-phosphates by CF is shown in Fig. 5.4. Few or no ara-A nucleotides are detected in normal erythrocytes; however, in the presence of CF, the amount of ara-ATP approximates the amount of ATP

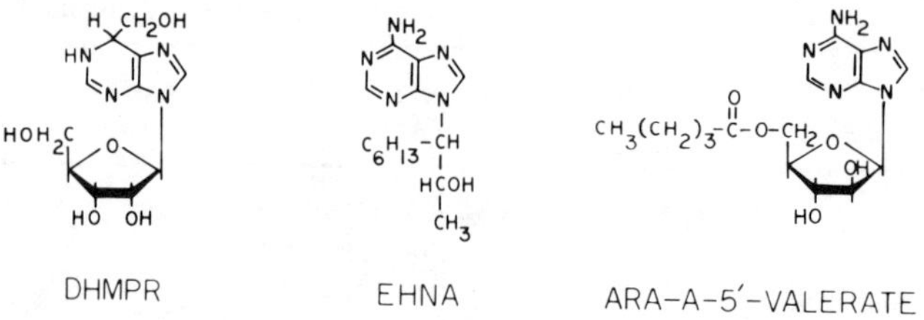

DHMPR EHNA ARA-A-5′-VALERATE

Figure 5.3 Structures of 1,6-dihydro-6-hydroxymethyl-purine ribonucleoside (DHMPR), *erythro*-9-(2-hydroxy-3-nonyl)adenosine (EHNA), and 9-β-D-arabinofuranosyladenine-5′-valerate (ara-A-5′-valerate).

after 24 h. Similar results are obtained when formycin is incubated with normal erythrocytes in the presence of CF (see page 175).

Inactivation and Reactivation of Adenosine Deaminase with Deoxycoformycin with Isolated Enzyme and Intact Human Erythrocytes and Sarcoma 180 Cells

Because there is the possibility of using dCF in combination with adenosine deaminase inhibitors, Parks and coworkers examined the factors that influence inactivation and reactivation of adenosine deaminase with the isolated enzyme and intact cells. They have demonstrated that the erythrocytic membrane plays a key role in the association of dCF with adenosine deaminase (Agarwal et al., 1977, 1978; Rogler-Brown et al., 1978). The apparent K_i values in intact erythrocytes are 300 to 500-fold lower than values obtained either with hemolysates or partially purified human erythrocytic adenosine deaminase. Therefore, the inactivation of the deaminase by dCF is several hundred times slower with intact cells than that observed with the partially purified enzyme. The erythrocytic membrane plays an important role in this inhibitory process. Since the K_i values of the purified erythrocytic enzyme and that of the hemolysates are similar, there are no factors in the hemolysates that interfere with the interaction between the deaminase and dCF (Rogler-Brown et al., 1978). With intact erythrocytes exposed to dCF and dialyzed, negligible deaminase activity was recovered after 49 hours. Enzymic activity was essentially identical in erythrocytes which were hemolyzed after 1 hour of preincubation with dCF. Therefore, the deaminase does *not* undergo permanent inactivation when inhibited by dCF for prolonged periods of time in the erythrocyte. Parks and coworkers emphasize the importance of the intact erythrocytic membrane in the inactivation of the deaminase, but also the failure of the enzyme to recover in the intact cell.

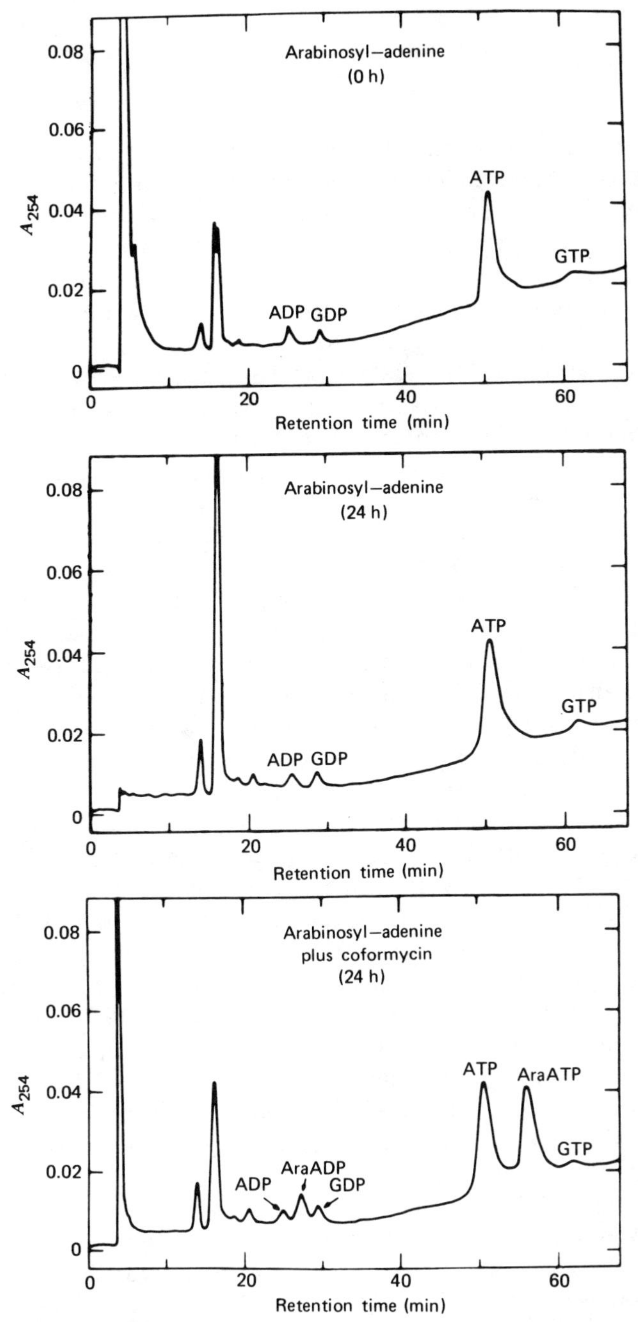

Figure 5.4 Effect of coformycin (2 μg/ml) on conversion of arabinosyladenine (ara-A) into the nucleotide pools of normal human erythrocytes. Reprinted with permission from Agarwal et al., 1978.

Effect of dCF on Mammalian Cells in Culture, Viruses, and Tumor-Bearing Animals

The addition of dCF to mammalian cell cultures infected with vaccinia virus or herpes virus hominus increases the antiviral activity of adenosine analogs ten- to fortyfold (for ara-A, see page 227; for cordycepin, see page 125; for formycin, see page 175). Mice bearing P388 ascites leukemia or solid sarcoma 180 implants survived much longer when treated with either cordycepin or ara-A plus inhibitors of adenosine deaminase (Plunkett et al., 1978b; Johns and Adamson, 1976; Koshiura and LePage, 1968).

Adamson et al. (1977) and Johns and Adamson (1976) showed that the inhibition of adenosine deaminase in a variety of tumor cells by dCF potentiated the antitumor effects of cordycepin and other analogs of adenosine. L 1210 cells have fifteenfold greater adenosine deaminase activity than rat liver and rapidly convert cordycepin to 3′-deoxyinosine (Johns and Adamson, 1976). The pharmacological significance of the observation in tissues with high adenosine deaminase activity is the ten to hundredfold enhancement by dCF of the cordycepin required for antitumor and cytotoxic activity (Johns and Adamson, 1976; Plunkett and Cohen, 1975). Adamson et al. (1977) reported that dCF potentiates the cytotoxicity of six adenosine analogs in P388 murine leukemia cell culture system. Mice with P388 ascites leukemia that were pretreated with dCF prior to treatment with xylosyladenine showed reduced dose requirements and increased survival time. More detailed reports can be found in the excellent reviews by Pavan-Langston, Buchanan, and Alford (1975) and Herrmann (1977).

EHNA in combination therapy with ara-A increases the lethality of ara-A on mice bearing the Ehrlich ascites carcinoma (Plunkett and Cohen, 1975, 1977). Schwartz et al. (1976) reported the same results with CF. Similarly, dCF potentiates the inhibitory effects of ara-A and cordycepin in cell cultures and in viral-infected animals by inhibiting the deamination of ara-A (Sloan et al., 1977; Bryson et al., 1974; Glazer et al., 1978; Savarese et al., 1978) (Table 5.2). With CF, cordycepin 5′-triphosphate, but not ara-A 5′-triphosphate, causes an accumulation of PRPP in sarcoma 180 cells (Savarese et al., 1978). These findings may be valuable in designing drug regimens with dCF and other adenosine analogs whose activation is PRPP dependent. dCF maintains high ara-A levels (Shewach and Plunkett, 1979).

The effects of CF and dCF on the metabolism of ara-A and ara-AMP *in vitro* with freshly enucleated rabbit eyes has also been studied (Chang et al., 1975b). 2′-Deoxycoformycin is a more active inhibitor than is CF. Adenosine deaminase activity is found in the corneal cells, with little enzyme activity in the aqueous humor. The diffusion of [^{3}H]dCF into the aqueous

Table 5.2 Prevention of Deamination of Ara-A to Ara-Hx in the Mouse by Simultaneous Administration of Several Subcutaneous Dose Levels of the Potent Deaminase Inhibitor Deoxycoformycin (dCF)

Drug Dosage (mg/kg)		Approximate Ara-A or Ara-Hx Concentration in Serum (μg/ml)		Plaque Reduction (%) of Vaccinia Virus — Dilution of Mouse Serum			
Ara-A	dCF			Undiluted	1:2	1:4	1:8
3000	0	Ara-Hx	40	24	27	25	23
3000	0.02	Ara-Hx	10	28	31	16	14
3000	2.0	Ara-Hx	?	35	17	0	0
		Ara-A	20				
3000	20.0	Ara-Hx	?	88	36	18	13
		Ara-A	40				
3000	40.0	Ara-A	75	100	63	33	14
Normal control serum			23	0	0	0	0

Reprinted with permission from Sloan et al., 1977.

humor increases the concentration of ara-A tenfold. Exogenously added ara-AMP could not be detected in the cornea. This indicates a rapid dephosphorylation; dCF did not protect the dephosphorylation of ara-AMP. Cass and Au-Yeung (1976) have demonstrated that the toxic effect of ara-A on L 1210 cells in culture or on subcutaneous solid tumors in mice can be greatly enhanced by the addition of CF. Ara-A alone has little toxicity (Brink and LePage, 1964a,b).

Tight Binding Inhibitors

The inhibition of adenosine deaminase by CF, dCF, and EHNA has been reported to be noncompetitive by Borondy et al. (1977). However, their calculations did not take into account the extent to which these inhibitors bind to adenosine deaminase. Using the approach of "tight-binding ligands" with the enzymes, Cha (1976) introduced a more reliable technique to study enzyme inhibition. The classical methods of kinetic analysis based on steady-state assumptions are inadequate with tight-binding inhibitors. Based on this new technique, Parks and coworkers (Cha et al., 1975; Agarwal *et al.*, 1978) speculated that CF and dCF are competitive inhibitors of adenosine deaminase.

Using the method of Cha, Agarwal and coworkers have classified DHMPR, EHNA, and CF as reversible, semi-tight-binding, and tight-binding (which is essentially irreversible binding) inhibitors (Agarwal et al., 1977, 1978). Cha et al. (1975) and Agarwal et al. (1975) calculated that CF binds 10^6–10^7 times as tightly to adenosine deaminase as does inosine. The double reciprocal plots of adenosine deaminase (human erythrocyte) in the presence of DHMPR or EHNA show a competitive-type inhibition.

The tight binding to adenosine deaminase is presumed to be due to the resemblance of the tetrahedral C-8 of dCF to the transition state compound, that formed during the deamination of adenosine (Evans and Wolfenden, 1970).

Increased Toxicity of Adenosine and Deoxyadenosine in the Presence of Adenosine Deaminase Inhibitors

Caution must be exercised when using tight-binding inhibitors of adenosine deaminase as a technique to increase the therapeutic efficacy of adenosine analogs. For example, Lapi and Cohen (1977) have demonstrated that low concentrations of 2′-deoxyadenosine are not inhibitory to L cells. However, 2′-deoxyadenosine, at these same concentrations, kills L cells that have been exposed to dCF, even though dCF is removed from the culture medium. Prolonged inhibition of adenosine deaminase by dCF and subsequent host toxicity should make EHNA, the inhibitor with a more transient activity against adenosine deaminase, the agent of choice with ara-A (Plunkett et al., 1978b). Henderson et al. (1977) also reported that dCF potentiates the toxicity of adenine in cultured mouse lymphoma cells. The cell may be affected by dCF in other ways. For example, Lowe et al. (1977) have demonstrated that the CF in L5178Y cells in culture blocks cell proliferation. The cells accumulate in the G_1- or early S_1-phase. There is a greatly reduced rate of DNA synthesis, but no effect on RNA synthesis. Levels of dCTP decrease; the addition of deoxycytidine prevents the dCTP decrease, but does not allow growth. These data suggest that the effects of dCF on cell kill are related to pool sizes of the deoxyribonucleotides. Therefore, it should be emphasized that these adenosine deaminase inhibitors are not without additional metabolic effect, so their judicious use is essential. Another explanation is that dCF affects *S*-adenosylmethionine metabolism by inhibiting *S*-adenosylhomocysteine hydrolysis (Lowe et al., 1977). Carson and Seegmiller (1976) demonstrated that the inhibition of adenosine deaminase by EHNA inhibited protein synthesis, but not DNA synthesis. Another explanation for the toxic effect of CF is the inhibition of maturation of precursor lymphocytes (Ballet et al., 1976). 2′-Deoxycoformycin does not kill mouse thymocytes or splenic cells (Burridge et al., 1977). The experimental difficulties encountered in the use of adenosine deaminase inhibitors may now be overcome because a mammalian cell line with no adenosine deaminase activity has been described (Shipman and Drach, 1978).

With L 1210 cells, Glazer et al. (1978) reported a tenfold reduction in the ID_{50} of cordycepin on RNA synthesis in the presence of dCF (Fig. 5.5). The inhibition of RNA synthesis in the presence of dCF occurred more rapidly and leveled off sooner than with cordycepin alone (Fig. 5.5*A*); with dCF, the inhibition of DNA synthesis was apparent after 2 h as opposed to 4 h with cordycepin alone (Fig. 5.5*B*). The primary effect of cordycepin is on RNA synthesis. The addition of dCF to cordycepin-treated cells causes a quantitative, but not qualitative, change in inhibition of the different species of rRNA fractions. With dCF the inhibition of rRNA and non-poly(A) HnRNA remained constant, whereas the inhibition of poly(A) HnRNA

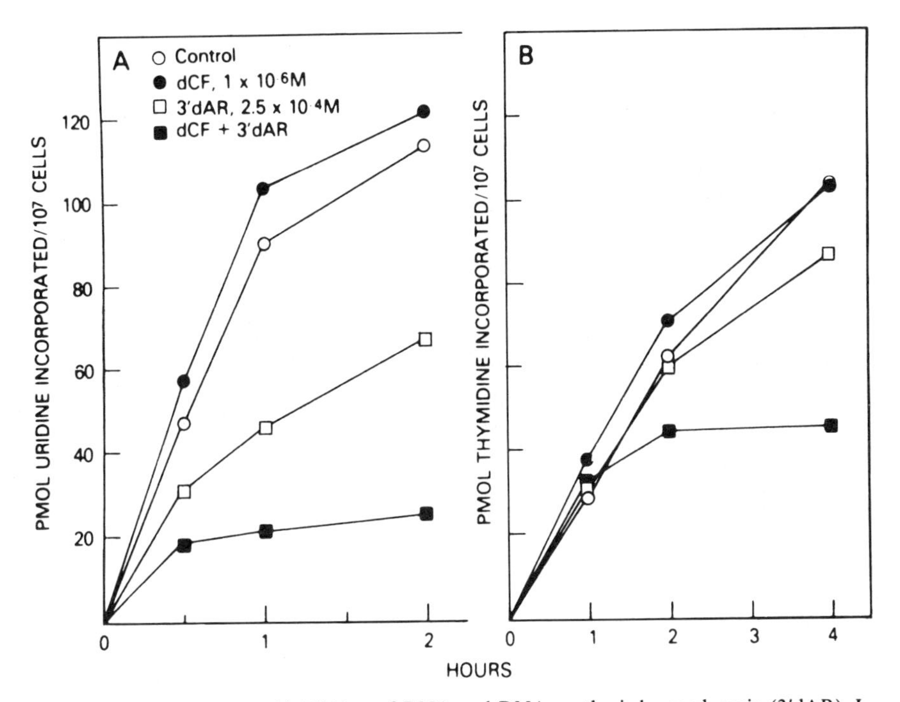

Figure 5.5 Time course of inhibition of RNA and DNA synthesis by cordycepin (3′dAR). L-1210 cells (10^7 cells/per flask) were preincubated for 30 min with (●,■) and without (○,□) 1×10^{-6} *M* dCF and then were incubated for 30 min with (□,■) and without (○,●) 2.5×10^{-4} *M* cordycepin before the addition of isotope. The abscissa shows the period of labeling after cordycepin treatment with [^{3}H]uridine (*A*) or [^{3}H]thymidine (*B*). Reprinted with permission from Glazer et al., 1978.

diminished with time. Without dCF, there was no significant inhibition by cordycepin on nonpoly(A) HnRNA at 30–60 min.

Explanation of the Biochemical Differences of Inhibitors of Adenosine Deaminase

Recent studies indicate that the inhibition of adenosine deaminase is not merely related to the inhibition of the enzyme, thereby potentiating the toxicity of the adenosine analogs. For example both EHNA and dCF inhibit adenosine deaminase. However, EHNA alone causes a 75–95% inhibition of replication of herpes simplex virus and HSV-specific DNA synthesis, while the more active inhibitor, dCF, does not (Fig. 5.6). EHNA does not inhibit uninfected HeLa cells. Furthermore, cordycepin alone (10^{-5} *M*) does not inhibit HSV replication, but in combination with EHNA (10^{-5} *M*) it inhibits HSV production by 99%; RNA synthesis is inhibited more than DNA and protein synthesis. Therefore, EHNA appears to inhibit HSV DNA replication differently when used alone and when used in combination with cordycepin. EHNA, although not as strong an inhibitor of adenosine deaminase as is dCF, may be preferred over dCF in the treatment of HSV infections in conjunction with adenosine analogs. Similar findings have been reported for (*S*)-9-(2,3-dihydroxypropyl)adenine (DeClercq et al., 1978).

Differences in the multiple forms of adenosine deaminase have also been shown to play an important role in the function of the adenosine deaminase inhibitors (Constine et al., 1979). Glazer and coworkers reported that E_S and E_I forms of adenosine deaminase could be distinguished from the E_L form on the basis of inhibition by dCF and EHNA; marked differences

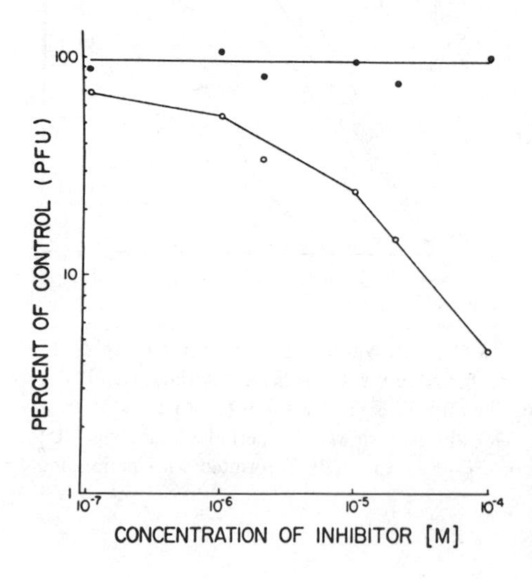

Figure 5.6 Effects of EHNA (10^{-5} *M*) (○) and dCF (10^{-5} *M*) (●) on infectious HSV. Reprinted with permission from North and Cohen, 1978.

were seen in the sensitivities of the multiple forms of the deaminase to dCF and EHNA.

In addition to the inhibition of adenosine deaminase, dCF and EHNA also inhibit the conversion of adenosine to nucleotides (Henderson et al., 1977). The entry of adenosine into the cell is blocked. dCF interactions with the nucleoside transport mechanism have been described (Rogler-Brown and Parks, 1978).

Effect of dCF on the Action of Cordycepin on Nuclear RNA Synthesis in Regenerating Rat Liver

Unlike the effect of cordycepin on tumor cell lines (see cordycepin, page 122), Glazer (1975) reported that cordycepin was equally effective in inhibiting the synthesis of nuclear rRNA, non-poly(A) HnRNA, poly(A) HnRNA, and poly(A) in regenerating rat liver. In a logical extension of his earlier findings, Glazer (1978) studied the effect of deamination on the activity of cordycepin on three species of nuclear RNA in regenerating rat liver. In contrast to neoplastic cells, regenerating rat liver has a low level adenosine deaminase which exerts a slight quantitative effect, but not a qualitative effect, on the inhibitory action of cordycepin on nuclear RNA. The complete inhibition of adenosine deaminase by dCF potentiates the activity of cordycepin by only 1.4 to 2.1 fold based on the ID_{50}. Therefore, rapidly proliferating liver may be particularly useful as a means of studying the mode of action of adenosine analogs whose pharmacological activity is greatly compromised by deamination in other tissues where deaminase activity is very high (Glazer, 1978).

Inhibition of 5′-AMP Deaminase

Whereas there is no inhibition of 5′-AMP deaminase by DHMPR, EHNA, or 2-fluoroadenosine, CF inhibits 5′-AMP deaminase by 67% (Agarwal and Parks, 1977). Preincubation of the deaminase with EHNA or DHMPR shows no inhibition, but preincubation with CF inhibits 5′-AMP deaminase 99.5% (Table 5.3). The inhibition with CF and dCF is noncompetitive. There is less deamination of AMP by purified AMP deaminase in the presence of CF (Henderson et al., 1977; Debatisse and Buttin, 1976).

Deoxyadenosine Antagonism of the Antiviral Activity of Ara-A and Ara-Hx

Until the advent of inhibitors of adenosine deaminase, it was not possible to observe the antagonistic capacity of deoxyadenosine toward ara-A and ara-

Table 5.3 Percent Inhibition of Rabbit Muscle 5′-AMP Deaminase by Various Inhibitors of Adenosine Deaminase

Inhibitors	Concentration (M)	Preincubation[a] (+)	Preincubation[a] (−)
None		0	0
Coformycin	5.5×10^{-8}	67	30
	1.3×10^{-6}	>99.5	83
Deoxycoformycin	1.3×10^{-6}	45	42
	2.8×10^{-6}	66	59
EHNA[b]	1.1×10^{-4}	0	0
DHMPR[b]	2.0×10^{-4}	0	0
2-Fluoroadenosine	1.0×10^{-3}	0	—

Reprinted with permission from Agarwal and Parks, 1977.
[a] The enzyme was preincubated with (+) or without (−) the inhibitor in the buffer mixture containing 0.5 mM ATP for 15 min at 30°C. The reaction was started by addition of AMP to (+) samples and AMP plus the inhibitor to (−) samples.
[b] See Fig. 5.3 for structure.

Hx. However, Smith et al. (1978) have demonstrated reversal of antiviral activity of ara-A and ara-Hx in HSV-infected KB cells by deoxyadenosine, but not adenosine, when CF was added to the cells.

Pharmacology of CF and dCF

Chassin et al. (1977) reported on an adenosine deaminase inhibition titration method that is sensitive to 2 nM dCF. Deoxycoformycin cleared from the plasma ($t_{1/2}$ 75–120 min) and the inhibition of adenosine deaminase in bone marrow and jejunal mucosa was 25 and 19%, respectively, which compares with only 9, 4, and 2% in spleen, liver, and kidney, respectively. McConnell (1977) has developed a sensitive photometric enzyme assay to measure dCF in tissues and body fluids of mice. The assay is based on the inhibition of calf intestinal adenosine deaminase by the dCF recovered from plasma and tissues. The $t_{1/2}$ of dCF in mouse plasma is 20 min. 2′-Deoxycoformycin is rapidly eliminated from the body through urinary excretion; approximately 87% of CF is recovered in the urine within 2 h. Another approach in the determination of CF in animal fluids and tissues is the report by Suling et al. (1977) in which they used a purine-requiring strain of *S. faecalis*. A very sensitive assay for dCF has also been described (Chassin et al., 1979).

Combination Therapy: Ara-A, Ara-C, CF, and dCF

Combination of adenosine analogs with adenosine deaminase inhibitors is effective in the therapy of viral infections. This combination also enhances the survival time of mice bearing i.p. L 1210 tumors and L 1210 in culture (Brockman et al., 1976; Kimball et al., 1976; LePage et al., 1976; Schabel et al., 1976; Cass and Au-Yeung, 1976). Because of the high adenosine deaminase activity in the L 1210 tumor, ara-A alone is ineffective (Brink and LePage, 1964a, b; LePage, 1970). However, low doses of dCF increase the efficacy of ara-A therapy (Kimball et al., 1976; LePage et al., 1976; Cass and Au-Yeung, 1976).

Kimball et al. (1977) and Lee et al. (1977) reported that a simple i.p. dose of dCF and ara-A in leukemia L 1210 cerebral implants in mice increases the toxicity of ara-A to the tumor. Brain adenosine deaminase from L 1210 cells in mice are inhibited by 90% at 5.2×10^{-10} *M* dCF. Similar inhibitions of adenosine deaminase have also been reported (Burridge et al., 1977; Woo et al., 1974; LePage et al., 1976; Cass and Au-Yeung, 1976).

The triple combination of dCF, ara-AMP, and ara-C allowed more prolonged therapy and more cures in mice with intracerebral leukemia L 1210 than with doses of ara-C alone (Lee et al., 1977). Ara-C has the advantage of decreasing the host toxicity observed with ara-A and dCF alone. Brockman et al. (1976, 1977) studied the effect of dCF and EHNA in mice bearing L 1210 tumors by treatment with either one dose of dCF plus ara-A or one dose of EHNA plus ara-A. The dCF treatment results in high intracellular levels of ara-ATP, which fell rapidly ($t_{1/2}$ = 2 h). When a single dose of dCF was given with repeated doses of ara-A, ara-ATP concentrations remained high for 9–12 h. EHNA and repeated doses of ara-A do not sustain ara-ATP levels. This indicates a rapid reversal of inhibition of adenosine deaminase *in vivo*.

To determine the biochemical basis of increased ara-A activity in the presence of adenosine deaminase inhibitors, Plunkett et al. (1977) compared the DNA synthetic capacity of mouse leukemia with that of host bone marrow and gastrointestinal mucosa. Ara-A alone inhibits DNA synthesis in mouse leukemia cells by 50% in 1 h with a return to normal after 3 h. There is no significant inhibition of DNA synthesis in the bone marrow and gastrointestinal mucosa. Ara-A plus dCF increases the inhibition in both tissues from 50 to 90% in 6 h. Recovery is delayed 18 h. The DNA synthetic capacity of bone marrow and gastrointestinal mucosa is inhibited 39–80% in 1 h. Recovery is complete in 3 h.

Immunosuppressive Activity

Another biological function for the adenosine deaminase inhibitors is their role as immunosuppressants. CF and dCF increase the success of tumor grafts, which makes them candidates for use as immunosuppressants during organ transplantation. Deoxycoformycin exhibited immunosuppressive activity in an allograft tumor test system, whereas EHNA had little immunosuppressive effect (Adamson et al., 1978). 2′-Deoxycoformycin and EHNA can be used as immunosuppressants because they are not myelosuppressive and act by way of the adenosine deaminases that are limited to lymphocytes and monocytes. Because the congenital absence of adenosine deaminase results in defects of T cell (thymus dependent lymphocyte) immune function, Lum et al. (1979a) reasoned that inhibitors of adenosine deaminase would have potential use as immunosuppressive agents. They demonstrated that *in vitro* treatment of human lymphocytes with EHNA inhibits T cells, but not B cells or macrophage–monocyte cell surface markers (Lum et al., 1978). They also showed that EHNA prolonged the survival of pancreatic islet cell allografts across a non-H-2 barrier in the mouse (Lum et al., 1979a). In combination with ara-A, dCF showed synergistic immunosuppressive effects. Equally exciting is the report on the correlation of adenosine deaminase activity in cytomegalovirus-related graft and patient loss (Lum et al., 1979b). CF and dCF mimic the rare immunodeficiency syndrome in individuals who lack adenosine deaminase (Mills et al., 1976).

SUMMARY

Naturally occurring and synthetic inhibitors of adenosine deaminase are described. They are CF, dCF, DHMPR, EHNA, and ara-A 5′-valerate. The K_i of dCF is 2.5×10^{-12} *M*. It is a tight binder and acts as a transition state intermediate. The above inhibitors of adenosine deaminase potentiate the toxicity of adenosine analogs of cells in culture, viruses and tumors in animals ten- to fortyfold by increasing the concentration of the corresponding analog 5′-triphosphates. The absence of adenosine deaminase (severe combined immunodeficiency, regenerating rat liver, or aqueous humor of rabbits) shows little or no effect on the toxicity of adenosine analogs when inhibitors of adenosine deaminase are added. Coformycin is phosphorylated by L5178Y cells.

Although these compounds inhibit adenosine deaminase, there are marked differences in their effect on viral-infected cells (i.e., EHNA inhibits HSV, but coformycin, a more potent inhibitor of adenosine deaminase, does not). EHNA and dCF have been used to distinguish the multiple forms of adenosine deaminase. Another function of these inhibitors is their role as immunosuppressants. Deoxycoformycin exhibits immunosuppressive activity, whereas EHNA has little effect.

Finally, it must be emphasized that these adenosine deaminase inhibitors have additional metabolic effects. Therefore, their judicious use is essential.

REFERENCES

Adamson, R. H., D. W. Zaharevitz, and D. G. Johns, *Pharmacology*, **15,** 84 (1977).

Adamson, R. H., M. M. Chassin, M. A. Chirigos, and D. G. Johns, *Current Chemotherapy*, American Society of Microbiology, Washington, D.C., 1978, p. 1116.

Agarwal, R. P., and R. E. Parks, Jr., *Biochem. Pharmacol.*, **26,** 663 (1977).

Agarwal, R. P., S. M. Sager, and R. E. Parks, *Biochem. Pharmacol.*, **24,** 693 (1975).

Agarwal, R. P., T. Spector, and R. E. Parks, Jr., *Biochem. Pharmacol.*, **26,** 359 (1977).

Agarwal, R. P., S. Cha, G. W. Crabtree, and R. E. Parks, Jr., in *Symposium of Chemistry and Biology of Nucleosides and Nucleotides*, (*American Chemical Society, Advances in Chemistry Series*), R. K. Robins and R. E. Harmon, Eds., Academic Press, New York, 1978, pp. 159–197.

Baker,, D. C., and S. R. Putt, *Abstracts Papers Third Biennial C. S. Marvel Symposium*, Tucson, Arizona, March 19–20 (1979), paper 11.

Ballet, J.-J., R. Insel, E. Merler, and F. S. Rosen, *J. Exp. Med.*, **143,** 1271 (1976).

Bloch, A., in *Encyclopedia of Chemical Technology*, Vol. 2, Wiley, New York, p. 962, 1978.

Borondy, P. E., T. Chang, E. Maschewske, and A. J. Glazko, *Ann. N.Y. Acad. Sci.*, **284,** 9 (1977).

Brink, J. J., and G. A. LePage, *Cancer Res.*, **24,** 312 (1964a).

Brink, J. J., and G. A. LePage, *Cancer Res.*, **24,** 1042 (1964b).

Brockman, R. W., S. C. Shaddix, L. M. Rose, and J. Carpenter, *Proc. Am. Assoc. Cancer Res.*, **17,** 52 (1976).

Brockman, R. W., L. M. Rose, F. M. Schabel, Jr., and W. R. Laster, *Proc. Am. Assoc. Cancer Res.*, **18,** Abstr. 192 (1977).

Bryson, Y., J. D. Connor, L. Sweetman, S. Carey, M. A. Stuckey, and R. A. Buchanan, *Antimicrob. Agents Chemother.*, **6,** 98 (1974).

Burridge, P. W., V. Paetkau, and J. F. Henderson, *J. Immunol.*, **119,** 675 (1977).

Carson, D. A., and J. E. Seegmiller, *J. Clin. Invest.*, **57,** 274 (1976).

Cass, C. E., and T. H. Au-Yeung, *Cancer Res.*, **36,** 1486 (1976).

Cha, S., *Biochem. Pharmacol.*, **25,** 2695 (1976).

Cha, S., R. P. Agarwal, and R. E. Parks, Jr., *Biochem. Pharmacol.*, **24,** 2187 (1975).

Chang, T., and A. J. Glazko., *Res. Commun. Chem. Pathol. Pharmacol.*, **14,** 127 (1976).

Chang, T., E. Maschewske, L. Corskey, H. Schneider, and A. J. Glazko, Abstracts, 15th Interscience Conference on Antimicrobial Agents and Chemotherapy, Washington, D.C., Sept. 24–26, 1975a, Abstr. 355.

Chang, T., E. Maschewske, and A. J. Glazko, Symposium on Antivirals with Clinical Potential, Stanford Univ., Aug. 26–29, 1975b.

Chassin, M. M., R. H. Adamson, and D. G. Johns, *Proc. Am. Assoc. Cancer Res.*, **18,** Abstr. 586 (1977).

Chassin, M. M., R. H. Adamson, D. W. Zaharevitz, and D. G. Johns, *Biochem. Pharmacol.*, in press (1979).

Constine, J., R. I. Glazer, and D. G. Johns, *Biochem. Biophys. Res. Commun.*, **85,** 198 (1979).

Debatisse, M., and G. Buttin, *J. Cell Biol.*, **70,** 348a (1976).

DeClercq, E., J. Descamps, P. De Somer, and A. Holy, *Science*, **200,** 563 (1978).

Dion, H. W., P. W. K. Woo, and A. Ryder, *Ann. N.Y. Acad. Sci.*, **284,** 21 (1977).

Evans, B., and R. Wolfenden, *J. Am. Chem. Soc.*, **92,** 4751 (1970).

Giblett, E. R., J. E. Anderson, F. Cohen, B. Pollara, and H. J. Meuwissen, *Lancet*, **ii,** 1067 (1972).

Glazer, R. I., *Biochim. Biophys. Acta*, **418,** 160 (1975).

Glazer, R. I., *Toxicol. Appl. Pharmacol.*, **46,** 191 (1978).

Glazer, R. I., T. J. Lott, and A. L. Peale, *Cancer Res.*, **38,** 2233 (1978).

Henderson, J. F., L. Brox, G. Zombor, D. Hunting, and C. A. Lomax, *Biochem. Pharmacol.*, **26,** 1967 (1977).

Herrmann, E. C., Jr., Third Conference on Antiviral Substances, *Ann. N.Y. Acad. Sci.*, **284,** (1977).

Hurley, L., and C. Gairola, *Antimicrob. Agents Chemother.*, **15,** 42 (1979).

Hurley, L., M. Zmijewski, and C.-J. Chang, *J. Am. Chem. Soc.*, **97,** 4372 (1975).

Hurley, L., C. Gairola, and N. Das, *Biochemistry*, **15,** 3760 (1976).

Hurley, L., W. L. Lasswell, R. K. Malhotra, and N. V. Das, *Biochemistry*, in press (1979).

Johns, D. G., and R. H. Adamson, *Biochem. Pharmacol.*, **25,** 1441 (1976).

Kimball, A. P., G. A. LePage, L. S. Worth, and S. H. Lee, *Proc. Am. Assoc. Cancer Res.*, **17,** 168 (1976).

Kimball, A. P., S. H. Lee, and N. Caron, *Proc. Am. Assoc. Cancer Res.*, **18,** Abstr. 499 (1977).

Koshiura, R., and G. A. LePage, *Cancer Res.*, **28,** 1014 (1968).

Lapi, L., and S. S. Cohen, *Biochem. Pharmacol.*, **26,** 71 (1977).

Lee, S. H., N. Caron, and A. P. Kimball, *Cancer Res.*, **37,** 1953 (1977).

LePage, G. A., *Adv. Enzyme Regul.*, **8,** 323 (1970).

LePage, G. A., L. S. Worth, and A. P. Kimball, *Cancer Res.*, **36,** 1481 (1976).

Lipper, R. A., S. M. Machkovech, J. C. Drach, and W. I. Higuchi, *Mol. Pharmacol.*, **14,** 366 (1978).

Lowe, J. K., B. Gowans, and L. Brox, *Cancer Res.*, **37,** 3013 (1977).

Lum, C. T., J. Schmidtke, D. E. R. Sutherland, and J. S. Najarian, *Clin. Immunol. Immunopathol.*, **10,** 258 (1978a).

Lum, C. T., D. E. R. Sutherland, W. D. Payne, P. Gorecki, and J. S. Najarian, *J. Surg. Res.*, **24,** 388 (1978b).

Lum, C. T., D. E. R. Sutherland, W. G. Yasmineh, D. S. Fryd, R. J. Howard, and J. S. Najarian, *J. Surg. Res.*, in press (1979).

McConnell, W. R., *Proc. Am. Assoc. Cancer Res.*, **18,** Abstr. 165 (1977).

Mills, G. C., F. C. Schmalstieg, K. B. Trimmer, A. S. Goldman, and R. M. Goldblum, *Proc. Natl. Acad. Sci. U.S.*, **73,** 2867 (1976).

Miyamoto, M., T. Sawa, S. Kondo, T. Takeuchi, and H. Umezawa, *J. Ferm. Technol.*, **56,** 329 (1978).

Müller, W. E. G., R. K. Zahn, J. Arendes, A. Maidhof, and H. Umezawa, *Z. Physiol. Chem.*, **359,** 1287 (1978).

Nakamura, H., G. Koyama, Y. Iitaka, M. Ohno, N. Yagisawa, S. Kondo, K. Maeda, and H. Umezawa, *J. Am. Chem. Soc.*, **96,** 4828 (1974).

Niida, T., T. Niwa, T. Tsuruoka, N. Ezaki, T. Shomura, and H. Umezawa, 153rd Scientific Meeting of Japan Antibiotics Research Association, January 1967.

North, T. W., and S. S. Cohen, *Proc. Natl. Acad. Sci. U.S.*, **75,** 4684 (1978).

Ohno, M., N. Yagisawa, S. Shibahara, S. Kondo, K. Maeda, and H. Umezawa, *J. Am. Chem. Soc.*, **96,** 4326 (1974).

Pavan-Langston, D., R. A. Buchanan, and C. A. Alford, Jr., Eds. *Adenine Arabinoside: An Antiviral Agent*, Raven Press, New York, 1975.

Plunkett, W., and S. S. Cohen, *Cancer Res.*, **35,** 1534 (1975).

Plunkett, W., and S. S. Cohen, *Ann. N.Y. Acad. Sci.*, **284,** 91 (1977).

Plunkett, W., L. Alexander, and T. L. Loo, *Proc. Am. Assoc. Cancer Res.*, **18,** Abstr. 232 (1977).

Plunkett, W., L. Alexander, S. Chubb, and T. L. Loo, *Biochem. Pharmacol.*, **28,** 201 (1979).

Plunkett, W., L. Alexander, S. Chubb, and T. L. Loo, *Proc. Am. Assoc. Cancer Res.*, **19,** Abstr. 875 (1978).

Rogler-Brown, T., R. P. Agarwal, and R. E. Parks, Jr., *Biochem. Pharmacol.*, **27,** 2289 (1978).

Rogler-Brown, T., and R. E. Parks, *Pharmacologist*, **20,** 183 (1978).

Ryder, A., H. W. Dion, P. W. K. Woo, and J. D. Howells, U.S. Patent No. 3,23,785 (1975).

Savarese, T. M., G. W. Crabtree, and R. E. Parks, Jr., *Proc. Am. Assoc. Cancer Res.*, **19,** Abstr. 483 (1978).

Sawa, T., Y. Fukagawa, I. Homma, T. Takeuchi, and H. Umezawa, *J. Antibiot.* (*Tokyo*), **20A,** 227 (1967).

Schabel, F. M., Jr., M. W. Trader, and W. R. Laster, Jr., *Proc. Am. Assoc. Cancer Res.*, **17,** 46 (1976).

Schaeffer, H. J., and C. F. Schwender, *J. Med. Chem.*, **17,** 6 (1974).

Schwartz, P. M., C. Shipman, Jr., and J. C. Drach, *Antimicrob. Agents Chemother.*, **10,** 64 (1976).

Shewach, D. S., and W. Plunkett, *Biochem. Pharmacol.*, in press (1979).

Shipman, C., Jr., and J. C. Drach, *Science*, **200,** 1163 (1978).

Sloan, B. J., J. K. Kieltz, and F. A. Miller, *Ann. N.Y. Acad. Sci.*, **284,** 60 (1977).

Smith, S. H., C. Shipman, Jr., and J. C. Drach, *Cancer Res.*, **38,** 1916 (1978).

Suhadolnik, R. J., *Nucleoside Antibiotics*, Wiley, New York, 1970.

Suling, W. J., L. S. Rice, and W. M. Shannon, *Proc. Am. Assoc. Cancer Res.*, **18,** Abstr. 167 (1977).

Townsend, L. J., in *Handbook of Biochemistry and Molecular Biology, Nucleic Acids*, Vol. F, 3rd ed., G. D. Fasman, Ed., CRC Press, Cleveland, 1975, p. 271.

Tsuruoka, T., N. Ezaki, S. Amano, C. Uchida, and T. Niida, *Meiji Shika Kenkyu Nempo*, **9,** 17 (1967).

Van Der Weyden, M. B., and W. N. Kelly, *Life Sci.*, **20,** 1645 (1977).

Woo, P. W. K., H. W. Dion, S. M. Lange, L. F. Dahl, and L. J. Durham, *J. Heterocyc. Chem.*, **11,** 641 (1974).

Chapter 6 Inhibition of Purine and Pyrimidine Interconversions

6.1 PSICOFURANINE AND DECOYININE 279

Derivatives 280
Biosynthesis 280
Biochemical Properties 280
Inhibition of Sporulation by Decoyinine 280

6.2 PYRAZOFURIN 281

Chemical Synthesis 283
Biosynthesis 283
Biochemical Properties 285
Phosphorylation to the 5′-Phosphates 285
Antitumor Activity in Humans and Animals 285
Antiviral Activity 286

SUMMARY 289

REFERENCES 289

6.1 PSICOFURANINE AND DECOYININE

Psicofuranine (angustmycin C) and decoyinine (angustmycin A), the two known adenine-ketose naturally occurring nucleoside analogs, are elaborated by *S. hygroscopicus* var. *decoyicus*. Their structures are 6-amino-9-(β-D-psicofuranosyl)purine and 9-β-D-(5,6-psicofuranoseenyl)-6-aminopurine (Fig. 6.1), respectively. They have antibacterial and antitumor activity. Psicofuranine and decoyinine are noncompetitive inhibitors of XMP aminase. Phosphorylation is not necessary for this inhibition. Guanine and guanosine reverse the inhibition. Psicofuranine is not deaminated

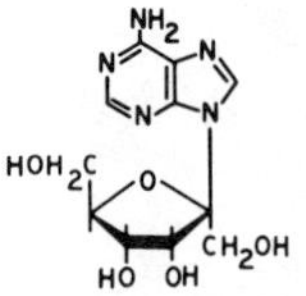

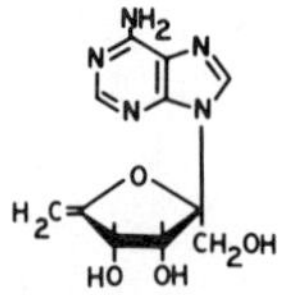

PSICOFURANINE DECOYININE

Figure 6.1 Structures of psicofuranine and decoyinine.

by adenosine deaminase; however, it is phosphorylated to the 5′-nucleotide, which then acts as a false feedback inhibitor of purine biosynthesis. Psicofuranine and decoyinine did not inhibit RNA synthesis in *Euglena* (Ebringer, 1971).

Psicofuranine and decoyinine have been reviewed (Suhadolnik, 1970; Townsend, 1975; Bloch, 1978).

Derivatives

The syntheses of psicofuranine, decoyinine, and their cytosine and isocytidine derivatives have been described (Schroeder and Hoeksema, 1959; McCarthy et al, 1968; Prisbe et al., 1976; Verheyden and Moffatt, 1966; Hrebabecky and Farkas, 1974a, b; Skaric and Matulic, 1975; Lerner, 1972a, b, 1975, 1976; Suciu and Lerner, 1975). The reason for the synthesis of these adenine nucleoside analogs was to determine their utility as antitumor agents.

Biosynthesis

The biosyntheses of psicofuranine and decoyinine have been reported by Sugimori and Suhadolnik (1965). Adenine and glucose are the carbon–nitrogen precursors for the biosynthesis of psicofuranine. The sequence in the biosynthesis of psicofuranine and decoyinine has been shown to involve the conversion of the ketohexose, psicofuranine, to the 4′,5′-unsaturated ketohexose, decoyinine (Chassy et al., 1966).

Biochemical Properties

Inhibition of Sporulation by Decoyinine. Sporulation of bacilli occurs when rapidly metabolized carbon or nitrogen compounds are replaced by more slowly metabolizable compounds. One hypothesis related to the initiation of sporulation suggests that a new synthetic balance takes place in the cell such that there is a reduced rate of RNA, protein, and cell wall synthesis and a continued rate of membrane synthesis (Freese, 1976). Therefore, Freese and coworkers (Mitani et al., 1977; Freese et al., 1978) speculated that the inhibition of the synthesis of any metabolite that maintains a proper biosynthetic balance might induce sporulation in the presence of rapidly metabolizable carbon and nitrogen sources where it would normally not occur. Indeed, they found that decoyinine and, less efficiently, hadacidin, both of which are known to block purine nucleotide biosynthesis, induced sporulation of *B. subtilis* while growing exponentially in the presence of excess ammonia, glucose, and phosphate (Fig. 6.2). Apparently, the limitation of AMP synthesis (due to the inhibition of adenylosuccinate synthetase by hadacidin) or the limitation of GMP synthesis (due to the inhibition of GMP synthesis by decoyinine) sufficient to cause sporulation of *B. subtilis* by decoyinine and hadacidin could only be reversed by guanine and adenine, respectively. The limitation of a compound required for RNA synthesis is sufficient to initiate sporulation. Sporulation with psicofuranine requires concentrations greater than decoyinine (Freese, personal communication).

6.2 PYRAZOFURIN

Pyrazofurin (previously named pyrazomycin), 3-(1-β-D-ribofuranosyl)-4-hydroxypyrazole 5-carboxamide, (Fig. 6.3) is a C-nucleoside analog isolated from the culture filtrates of *Streptòmyces candidus* (Gerzon et al., 1969; Williams and Hoehn, 1974). The chemical synthesis of pyrazofurin and 2′-deoxypyrazofurin has been described by Farkas et al. (1972) and De Bernardo and Weigele (1976). The α-form of pyrazofurin has also been reported (Gutowski et al., 1973; Wenkert et al., 1973). Pyrazofurin has been of considerable interest because of its antitumor and broad spectrum antiviral activity. Pyrazofurin suppresses the replication of vaccinnia, herpes simplex, measles, rhino, influenza viruses, and Friend and Rauscher leukemia virus in mice (Streightoff et al., 1969; Gutowski et al., 1975; Shannon, 1977; Sweeney et al., 1973). Cultured Novikoff rat hepatoma cells, HeLa cells and mouse L cells have also been tested for sensitivity to pyrazofurin (Cadman et al., 1978). Pyrazofurin is currently being tested as an antitumor agent in humans (Gutowski et al., 1975; Ohnuma and Holland, 1977). The 4-amino analog of pyrazofurin, 4-amino-3-(β-D-ribofuranosyl)pyrazole-5-carboxamide has been synthesized from the N^6-oxide of formycin or formycin B (Lewis et al., 1976). Pyrazofurin has been reviewed (Suhadolnik, 1970; Townsend, 1975; Bloch, 1978).

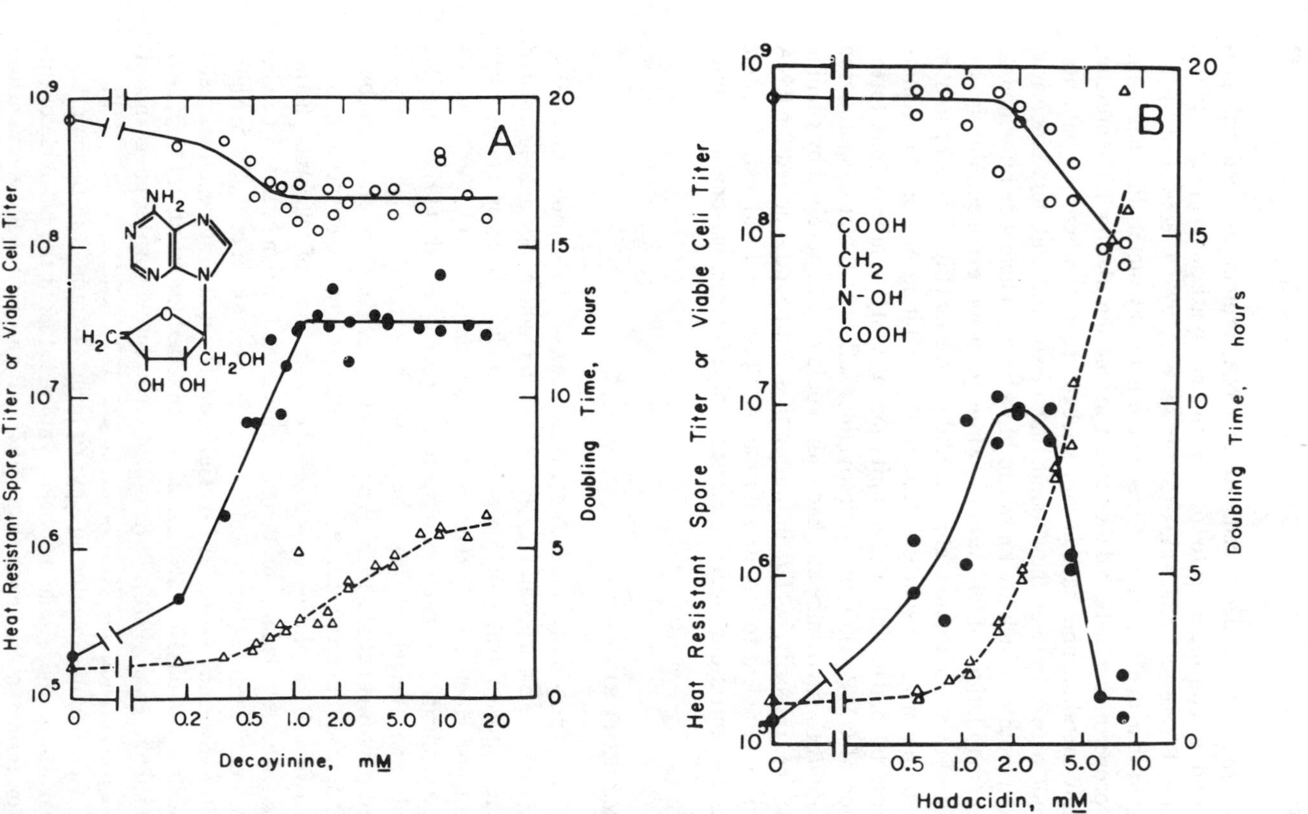

Figure 6.2 Induction of sporulation by decoyinine or hadacidin. Doubling times (Δ) were measured within the first 2 or 3 h after addition of the compound during midexponential growth (OD_{600} = 0.5). Eight hours after addition, the cells were plated to determine the total viable cell titer (○) and the heat-resistant spore titer (●). (*A*) Decoyinine; (*B*) hadacidin. Reprinted with permission from Mitani et al., 1977.

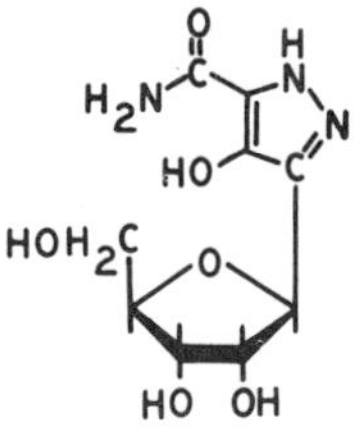

Figure 6.3 Structure of pyrazofurin.

Crain et al. (1973) reported on the mass spectrum of pyrazofurin. The most abundant ion is at m/e 156, which corresponds to B + 30 (M-103). This fragmentation has been suggested to be a definite feature in the mass spectrum of C-nucleosides (Townsend and Robins, 1969; Rice and Dudek, 1969).

Chemical Synthesis

The chemical synthesis of 1′-β- and 1′-α-pyrazofurin (**1** and **2**) (Fig. 6.4) described by De Bernardo and Weigele (1976) starts with the α- and β-anomers of 2,3-*O*-isopropylidene-D-ribofuranose (**3**), which are converted to the di-*p*-nitrobenzoates, **4** and **5**. The mixture of **4** and **5** is converted to the crystalline ribosyl bromide, **6**, which is alkylated with diethyl acetonedicarboxylate to form **7** (43% yield) (Fig. 6.4). The sodium salt of **7** is diazotized by *p*-toluenesulfonyl azide to form the cyclic derivative, **11**. Treatment of **11** with sodium ethoxide in ethanol produces **12**. The amide, **13**, is formed by heating **12** in methanolic ammonia for a short period. Upon prolonged heating **13** is isomerized to **14**. Removal of the isopropylidene group from **13** and **14** gives α- (**2**) and β-pyrazofurin (**1**), respectively.

Biosynthesis

The enrichment of C-4 of pyrazofurin with ^{13}C from [2-^{13}C]acetate was demonstrated following the incorporation by *S. hygroscopicus* (Fig. 6.5). Similarly, C-3 of pyrazofurin was enriched with ^{13}C following the incorporation of [1-^{13}C]acetate into pyrazofurin. These findings strongly support the notion that a two- or three-carbon fragment from acetate and/or glutamate serves as the precursor in the biosynthesis of the carbon skeleton of the aglycon of the naturally occurring C-nucleoside analogs, showdomycin, formycin, oxazinomycin, and pyrazofurin (see Fig. 1.34, pg. 56).

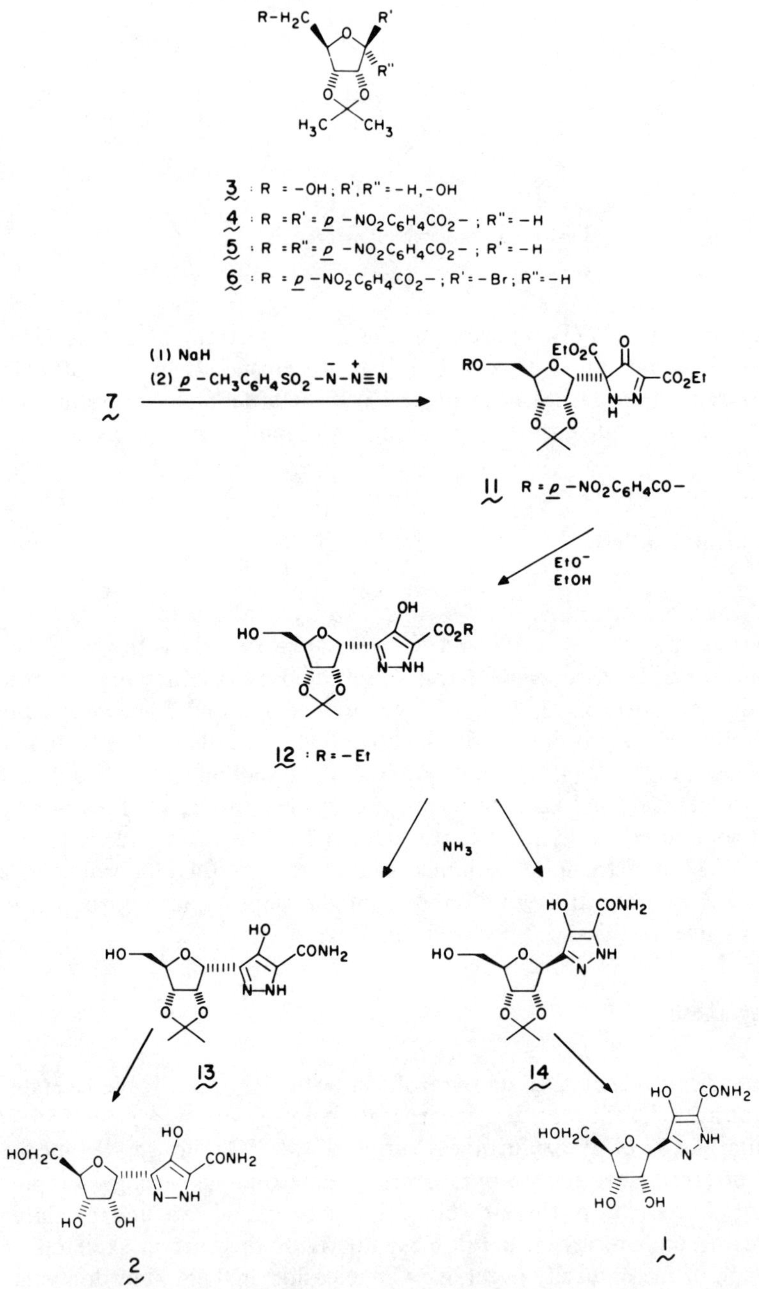

3 : R = −OH; R′,R″ = −H, −OH
4 : R = R′ = p−$NO_2C_6H_4CO_2$−; R″ = −H
5 : R = R″ = p−$NO_2C_6H_4CO_2$−; R′ = −H
6 : R = p−$NO_2C_6H_4CO_2$−; R′ = −Br; R″ = −H

(1) NaH
(2) p−$CH_3C_6H_4SO_2$−$\overset{-}{N}$−$\overset{+}{N}$≡N

11 R = p−$NO_2C_6H_4CO$−

EtO^-
EtOH

12 : R = −Et

NH_3

13

14

2

1

Figure 6.4 Chemical synthesis of α-(**2**) and β-pyrazofurin (**1**). Reprinted (in modified form) with permission from DeBernardo and Weigele, *J. Org. Chem.*, **41**, 287 (1977). Copyright by the American Chemical Society.

Biochemical Properties

Phosphorylation to the 5′-Phosphates. Administration of pyrazofurin (i.v.) in humans increases urinary orotidine and orotic acid, which supports the hypothesis that pyrazofurin is converted to its 5′-monophosphate by a cellular kinase(s). The monophosphate then acts as an inhibitor of orotidylic acid decarboxylase (Jakubowski et al., 1977; Ohnuma et al., 1977; Cadman et al., 1978). Pyrazofurin 5′-di- and 5′-triphosphates react with AMP and myokinase to give ADP and ATP. Although pyrazofurin 5′-diphosphate is a substrate for pyruvate kinase, the V_{max} is only one two-hundredth of that observed for UDP.

Antitumor Activity in Humans and Animals. Pyrazofurin has been studied clinically with 25 patients with inoperable carcinoma and lymphoma. Pyrazofurin was well tolerated by most patients at doses of 100 mg/m^2 following i.v. administration. Infusion of pyrazofurin to leukemic patients resulted in severe, but reversible, toxicity. In several patients, mucositis, leucopemia, and anemia were observed. The toxicity of pyrazofurin is primarily on the oral mucosa and not the bone marrow or intestinal epithelium (Cadman et al., 1978). The toxic reactions of pyrazofurin were more pronounced in those patients who had previously received radiotherapy. Of four patients with breast carcinoma, two responded (Ohnuma and Holland, 1977). Of 17 given pyrazofurin (i.v., 200 mg/m^2), only one patient each with mycosis fungoides, adenocarcinoma of the lung, and erythroleukemia showed improvement (Cadman et al., 1978). Sweeney et al. (1973) reported that pyrazofurin completely inhibited the growth of Walker carcinosarcoma 256. Mammary carcinoma 755, Gardner lymphosarcoma, and X5563 plasma cell myeloma were inhibited more than 50% by pyrazofurin. The murine leukemias did not respond to the nucleoside analog; pyrazofurin markedly decreases the incorporation of deoxyuridine into DNA of Novikoff hepatoma cells in culture (Fig. 6.6). The reversal of inhibition of 10 μM deoxyuridine suggests that the inhibition by pyrazofurin is mostly the result of the inhibition of DNA rather than RNA due to a rapid depletion of the TTP and dCTP soluble pool. These results support the view that pyrazofurin inhibits *de novo* synthesis of pyrimidine nucleosides. Pyrazofurin (1 mM) does not affect the conversion of UTP to CTP, nor does it affect the nucleoside transport system (Plagemann and Behrens, 1976). Because some cell lines are not as sensitive to pyrazofurin, Cadman et al. (1978) studied its effect with other drugs. L5178Y cells in culture are not sensitive to pyrazofurin or 5-fluorouracil. Depletion of the pool of pyri-

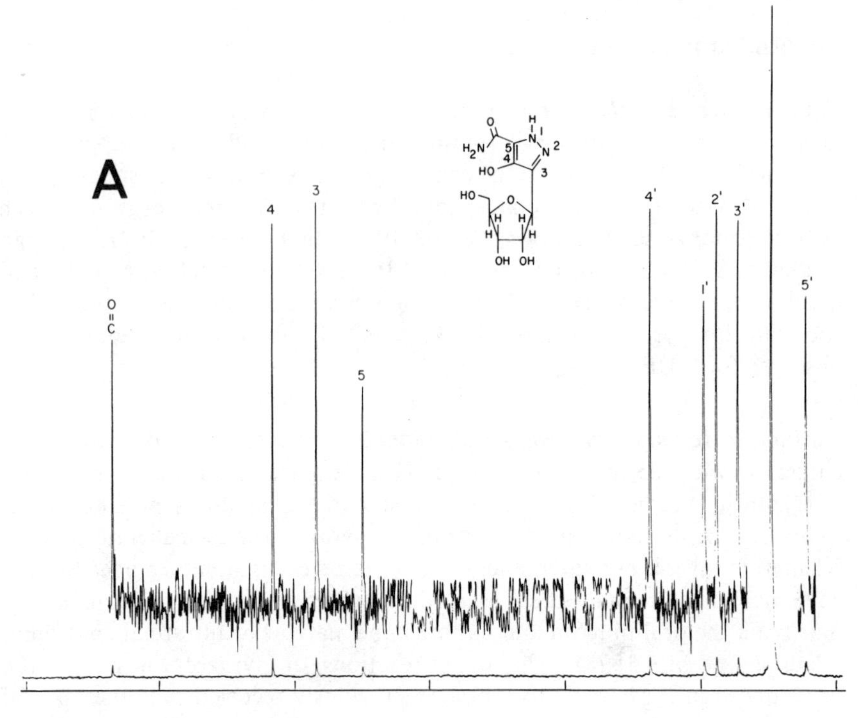

Figure 6.5 Enrichment of C-4 of pyrazofurin by ^{13}C following the incorporation of [2- (Suhadolnik, unpublished results).

midine nucleotides by pyrazofurin might be expected to enhance the inhibitory effect on other pyrimidine analogs. Indeed, exposure to pyrazofurin prior to ara-C increased the rate of cell kill tenfold.

Antiviral Activity. Descamps and DeClercq (1978) evaluated the antiviral potential of pyrazofurin in various cell cultures inoculated with viruses representative of the major virus families (Pox-, Herpeto-, Rhabdo-, Paramyxo-, Toga-, and Picornaviridae). In these experiments, pyrazofurin was found to inhibit the cytopathic effects of vaccinia, herpes simplex, vesicular stomatitis, Newcastle disease, measles, Sindbis, polio, and coxsackie viruses, although an earlier report (Gutowski et al., 1975) indicated that pyrazofurin had little or no effect against polio or coxsackie viruses. This C-nucleoside was compared with ribavirin (virazole), a structural analog of pyrazofurin. Virazole has been shown to be a broad spectrum antiviral agent (Sidwell et al., 1972) and is currently being investigated in human clinical studies. Whereas pyrazofurin blocks pyrimidine nucleotide

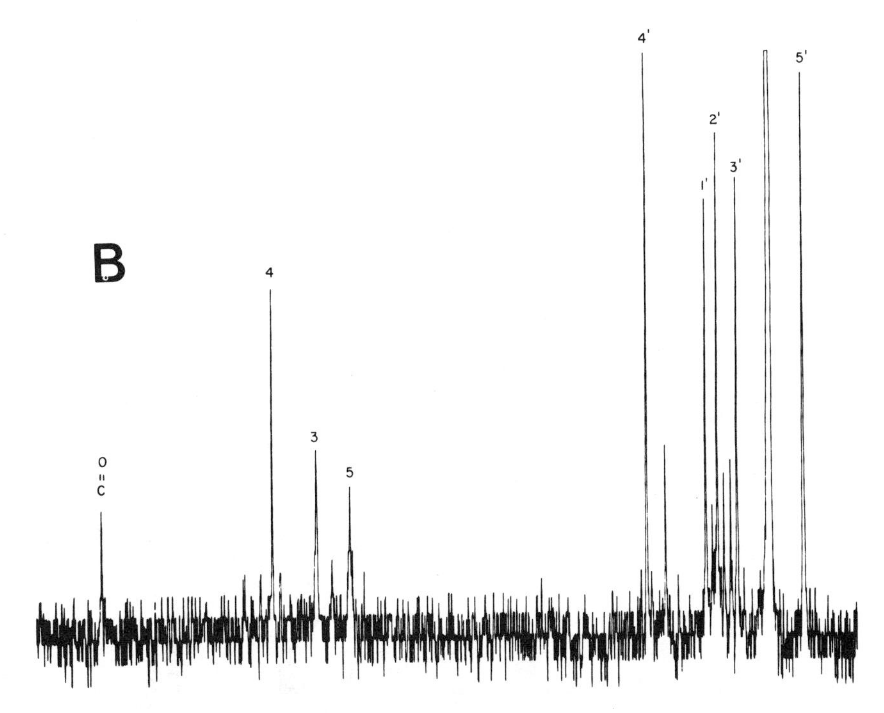

^{13}C]acetate by *S. hygroscopicus*. (*A*) Natural abundance; (*B*) [2-^{13}C]acetate incorporation.

synthesis by inhibiting orotidylic acid decarboxylase, virazole inhibits IMP dehydrogenase (Streeter et al., 1973). Uridine completely reverses the inhibitory properties of pyrazofurin for virus multiplication. Herpes simplex virus types 1 and 2 are equally sensitive to the action of pyrazofurin and virazole. The isopropylidene derivative of pyrazofurin is less effective. Descamps and DeClercq (1978) reported that the antiviral potential of pyrazofurin (with some viruses) is about a thousand fold greater than that noted for virazole. Virazole inhibits RNA and DNA synthesis of the host cells at concentrations that correspond closely to those required for the inhibition of viral replication (DeClercq et al., 1975). Pyrazofurin was not inhibitory to RNA, DNA, or protein synthesis up to concentrations 10^3–10^4 times as high as those needed to inhibit viral replication. Finally, Shannon (1977) reported that pyrazofurin (at 0.01 μg/ml) caused a 50% inhibition of (Gross) murine leukemia replication in Swiss mouse embryo cells. This is about 300 times lower than the concentration at which virazole effected a 50% inhibition of murine leukemia virus replication.

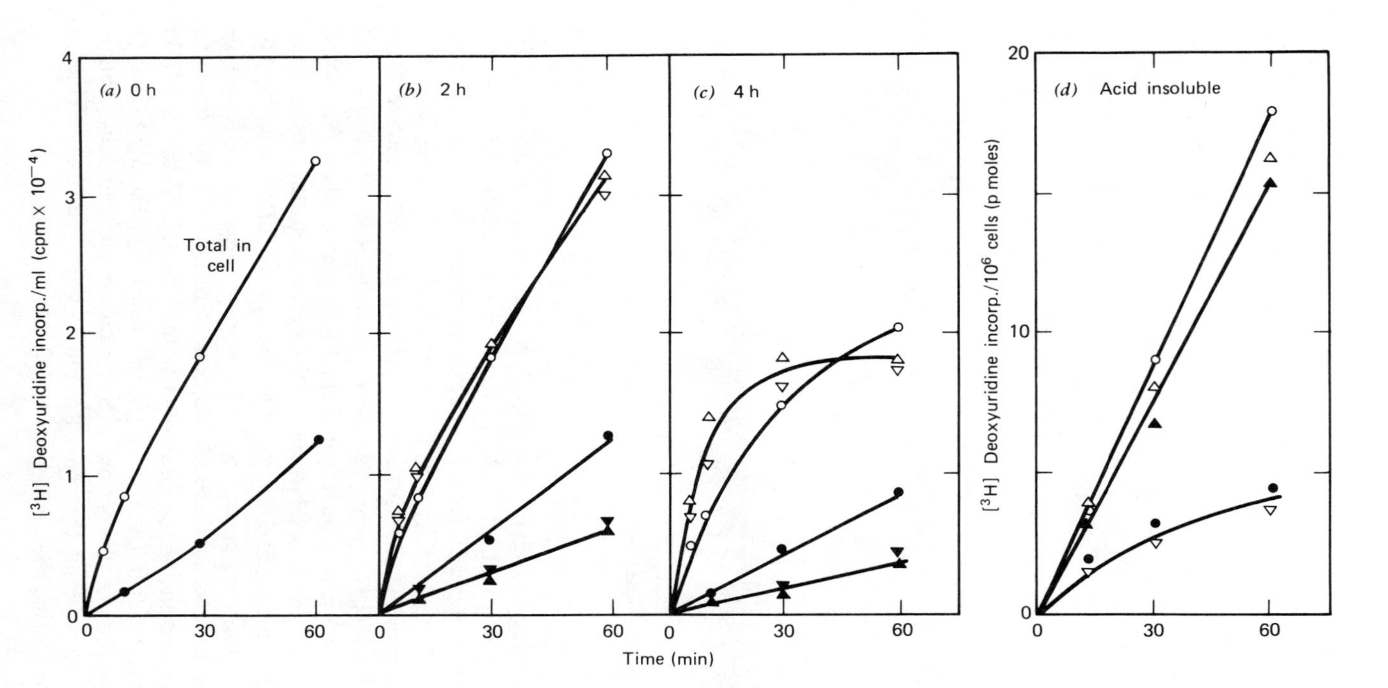

Figure 6.6 Effect of pyrazofurin on the incorporation of deoxyuridine into DNA by N1S1-67 cells (*a-c*) and the reversal by nucleosides (*d*). Samples were analyzed for radioactivity in total cell material (○, △, ▽) and radioactivity in DNA (●, ▲, ▼). [*a-c*: concentration of pyrazofurin: 0 (○, ●), 1 (△, ▲), and 10 μM (▽, ▼)]. The experiment in *d* was conducted as in *c* with 10 μM pyrazofurin and 10 μM [^{3}H]deoxyuridine and contained: no other additions (●), 10 μM deoxycytidine (△), 10 μM uridine (▲), or 10 μM hypoxanthine (▽); control (○). Reprinted with permission from Plagemann and Behrens, 1976.

Because of the excellent *in vitro* antiviral activity of pyrazofurin, this analog was tested for its antiviral properties in newborn mice inoculated subcutaneously with Coxsackie B-4 virus. Pyrazofurin (i.p.) failed to reduce the mortality rate of the inoculated newborn mice. With mice inoculated intravenously with vaccinnia virus, pyrazofurin caused a marked reduction in the number of tail lesions (Descamps and DeClercq, 1978). These results confirm earlier results reported by Gutowski et al. (1975). Pyrazofurin could not be administered at doses greater than 1 mg/kg/day because of its toxicity. Because the toxicity in mice may be associated with structural features that are not necessarily related to the antiviral potency of the nucleoside, it should be possible to alter the structure of pyrazofurin in such a way as to decrease its toxic properties and yet retain antiviral activity. The carbocyclic analog of pyrazofurin does not show antiviral, antibacterial, or antifungal activity (Just and Kim, 1976).

SUMMARY

Psicofuranine and decoyinine are adenineketohexose analogs that inhibit XMP aminase at the level of the nucleoside. Guanine and guanosine reverse their inhibition. Sporulation of *B. subtilis* occurs rapidly when decoyinine is added to exponentially growing cells.

Pyrazofurin is a C-nucleoside analog that is converted to its 5′-phosphate and inhibits orotidylic acid decarboxylase. It inhibits leukemia and breast carcinoma in mammalian cells in culture. Pyrazofurin is a potent inhibitor of viruses. However, at 1 mg/kg/day, it is toxic to mice.

REFERENCES

Bloch, A., *Encyclopedia of Chemical Technology*, Vol. 2, Wiley, New York, p. 962, 1978.

Cadman, E. C., D. E. Dix, and R. E. Handschumacher, *Cancer Res.*, **38,** 682 (1978).

Chassy, B. M., T. Sugimori, and R. J. Suhadolnik, *Biochim. Biophys. Acta*, **130,** 12 (1966).

Crain, P. F., J. A., McCloskey, A. F. Lewis, K. H. Schram, and L. B. Townsend, *J. Heterocyc. Chem.*, **10,** 843 (1973).

De Bernardo, S., and M. Weigele, *J. Org. Chem.*, **41,** 287 (1976).

De Clercq, E., M. Luczak, J. C. Reepmeyer, K. L. Kirk, and L. A. Cohen, *Life Sci.*, **17,** 187 (1975).

Descamps, J., and E. De Clercq, Siegenthaler, W., and R. Luthy, Eds., in *Current Chemotherapy*, American Society for Microbiology, Washington, D.C., 1978, p. 354.

Ebringer, L., *Experientia*, **27,** 586 (1971).

Farkaš, J., A. Flegelova, and F. Šorm, *Tetrahedron Lett.*, **1972,** 2279.

Freese, E., in *Spore Research 1976;* A. N. Barker, G. W. Gould, and J. Wolf, Eds., Academic Press, London, 1976, p. 1.

Freese, E., J. Heinze, T. Mitani, and E. S. Freese, in *Spores VII*, G. Chambliss, and J. C. Vary, Eds., American Society for Microbiology, Washington, D.C., 1978, p. 277.

Gerzon, K., R. H. Williams, M. Hoehn, M. Gorman, and D. C. DeLong, Abstracts, 2nd International Congress of Heterocyclic Chemistry, Montpelier, France, 1969, Abstr. C-30.

Gutowski, G. E., M. O. Chaney, N. D. Jones, R. L. Hamill, F. A. Davis, and R. D. Miller, *Biochem. Biophys. Res. Commun.*, **51,** 312 (1973).

Gutowski, G. E., M. J. Sweeney, D. C. DeLong, R. L. Hamill, K. Gerzon, and R. W. Dyke, *Ann. N.Y. Acad. Sci.*, **225,** 544 (1975).

Hrebabecky, H., and J. Farkas, *Collec. Czech. Chem. Commun.*, **39,** 2115 (1974a).

Hrebabecky, H., and J. Farkas, *Collec. Czech. Chem. Commun.*, **39,** 1098 (1974b).

Jakubowski, A., C. Lehman, J. Moyer, and R. E. Handschumacher, *Proc. Am. Assoc. Cancer Res.*, **18,** Abstr. 865 (1977).

Just, G., and S. Kim, *Tetrahedron Lett.*, **1976,** 1063.

Lerner, L. M., *J. Org. Chem.*, **37,** 4386 (1972a).

Lerner, L. M., *J. Org. Chem.*, **37,** 473 477 (1972b).

Lerner, L. M., *Carbohydr. Res.*, **44,** 13 (1975).

Lerner, L. M., *J. Org. Chem.*, **41,** 306 (1976).

Lewis, A. F., R. A. Long, L. W. Roti-Roti, and L. B. Townsend, *J. Heterocyc. Chem.*, **13,** 1359 (1976).

McCarthy, Jr., J. R., R. K. Robins, and M. J. Robins, *J. Am. Chem. Soc.*, **90,** 4993 (1968).

Mitani, T., J. E. Heinze, and E. Freese, *Biochim. Biophys. Res. Commun.*, **77,** 1118 (1977).

Ohnuma, T., and J. F. Holland, *Cancer Treat. Rep.*, **61,** 389 (1977).

Ohnuma, T., J. Roboz, M. L. Shapiro, and J. F. Holland, *Cancer Res.*, **37,** 2043 (1977).

Plagemann, P. G. W., and Behrens, M., *Cancer Res.*, **36,** 3807 (1976).

Prisbe, E. J., J. Smejkal, J. P. H. Verheyden, and J. G. Moffatt, *J. Org. Chem.*, **41,** 1836 (1976).

Rice, J. M., and G. O. Dudek, *Biochem. Biophys. Res. Commun.*, **35,** 383 (1969).

Schroeder, W., and H. Hoeksema, *J. Am. Chem. Soc.*, **81,** 1767 (1959).

Shannon, W. M., *Ann N.Y. Acad. Sci.*, **284,** 472 (1977).

Sidwell, R. W., J. H. Huffman, G. P. Khare, L. B. Allen, J. T. Witkowski, and R. K. Robins, *Science*, **177,** 705 (1972).

Skaric, V., and J. Matulic, *Croat. Chem. Acta*, **47,** 159 (1975).

Streeter, D. G., J. T. Witkowski, G. O. Khard, R. W. Sidwell, R. J. Bauer, R. K. Robins, and L. N. Simon, *Proc. Natl. Acad. Sci. U.S.*, **70,** 1174 (1973).

Streightoff, F., J. D. Nelson, J. C. Cline, K. Gerzon, R. H. Williams, and D. C. DeLong, 9th Interscience Conference on Antimicrobial Agents and Chemotherapy, Washington, D.C., 1969, p. 18.

Suciu, N., and L. M. Lerner, *Carbohydr. Res.*, **44,** 112 (1975).

Sugimori, T., and R. J. Suhadolnik, *J. Am. Chem. Soc.*, **87,** 1136, (1965).

Suhadolnik, R. J., *Nucleoside Antibiotics*, Wiley, New York, 1970.

Sweeney, M. J., F. A. Davis, G. E. Gutowski, R. L. Hamill, D. H. Hoffman, and G. A. Poore, *Cancer Res.*, **33,** 2619 (1973).

Townsend, L. B., "Nucleoside Antibiotics: Physicochemical Constants, Spectral, Chemotherapeutic and Biological Properties," in *Handbook of Biochemistry and Molecular Biology, Nucleic Acids*, Vol. I, 3rd ed., G. D. Fasman, Ed., CRC Press, Cleveland, 1975, p. 271.

Townsend, L. B., and R. K. Robins, *J. Heterocyc. Chem.*, **6,** 459 (1969).

Verheyden, J. P. H., and J. G. Moffatt, *J. Am. Chem. Soc.*, **88,** 5684 (1966).

Wenkert, E., E. W. Hagaman, and G. E. Gutowski, *Biochem. Biophys. Res. Commun.*, **51,** 318 (1973).

Williams, R. H., and M. M. Hoehn, U.S. Patent, 3,802,999 (1974).

Chapter 7 Hyperesthetic and Hyperemic Nucleosides

7.1 CLITIDINE

It has long been known that ingestion of the toadstool, *Clitocybe acromelalga*, results in an increased sensitivity of the skin and an excess of blood in body parts (hyperesthesia, hyperemia). As part of the study on physiologically active substances produced by mushrooms, Konno et al. (1977) isolated and characterized a new pyridine nucleoside, clitidine [1,4-dihydro-4-imino-1-(β-D-ribofuranosyl)pyridine-3-carboxylic acid] (Fig. 7.1).

Discovery, Production, and Isolation

The discovery of clitidine as the physiologically active substance of *C. acromelalga* was reported by Konno et al. (1977). Clitidine was obtained in a 0.036% yield from the frozen fruit bodies by extraction with water, precipitation with acetone, dialysis, and chromatography on Sephadex G-10. Final purification was obtained by crystallization from water.

Physical and Chemical Properties

The molecular formula for clitidine is $C_{11}H_{14}O_6N_2 \cdot H_2O$; mol wt 288; mp 189–191°C; $[\alpha]_D^{25}$ −50.6° (*c* 10.6, H_2O); ultraviolet spectral properties: $\lambda_{max}^{H_2O}$

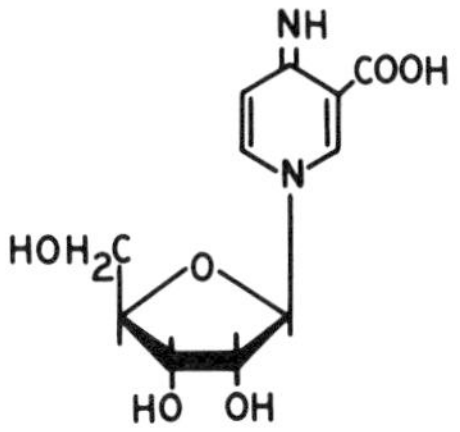

Figure 7.1 Structure of clitidine (Konno et al., 1977).

271 nm (ϵ = 4.09); $\lambda_{max}^{pH\,2}$ 267 nm (ϵ = 4.16); $\lambda_{max}^{pH\,12}$ 271 nm (ϵ = 4.18); ir absorption bands at 3300–2200, 1660, 1585, 1065, 1030 cm^{-1} (Nujol) (Konno et al., 1977). The nmr spectrum (DMSO) showed a pentose moiety [δ 3.70 (2H, bs), 4.08 (3H, bs), 5.59 (1H, J = 5 Hz)] and a 3,4-disubstituted pyridine ring [δ 6.90 (1H, d, J = 7 Hz), 8.24 (1H, dd, J = 7, 1.5), 8.7 (H, d, J = 1.5)]. The δ values of aromatic protons suggest that the substituents at C-3 and C-4 are electron attracting and electron donating, respectively. The mass spectrum of clitidine shows m/e of 130.0423 (100%, **A**), 121.0346 (13%, **B**), 93.0476 (13%, **C**) (Fig. 7.2).

Chemical Synthesis

The chemical synthesis of clitidine was accomplished by the condensation of methyl 4-aminonicotinate with 3,5-di-*O*-benzoyl-D-ribofuranosyl chloride. The product was deblocked with triethylamine–water–methanol; yield 89%. The ribosylation reaction is known to give predominantly β-anomers (Jorman and Ross, 1969).

Biochemical Properties

Clitidine can be synthesized enzymatically. Tono-Oka et al. (1977) reported that the incubation of NAD^+ with methyl 4-aminonicotinate with pig brain

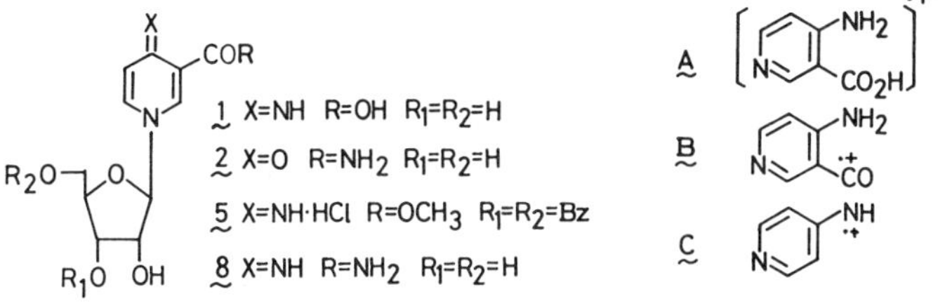

Figure 7.2 Mass spectral fragmentation of clitidine. Reprinted with permission from Konno et al., 1977.

NADase yielded the 4-amino NAD^+ analog. Enzymatic hydrolysis produced clitidine. Although NAD^+ and clitidine are the first naturally occurring pyridine nucleotide and nucleoside, respectively, a number of other pyridine nucleosides have been synthesized and their biological activities have been reported (Gregor et al., 1976; Trommer et al., 1976; Walter and Kaplan, 1963).

REFERENCES

Gregor, I., U. Sequin, and C. Tamm, *Helv. Chim. Acta*, **58,** 712 (1976).

Jorman, M., and W. C. J. Ross, *J. Chem. Soc.* (*C*), **1969,** 199.

Konno, K., K. Hayano, H. Shirahama, H. Saito, and T. Matsumoto, *Tetrahedron Lett.*, **1977,** 481.

Tono-Oka, S., A. Sasaki, H. Shirahama, T. Matsumoto, and S. Kakimoto, *Chem. Lett.*, **1977,** 1449.

Trommer, W. E., H. Blume, and H. Kapmeyer, *Liebigs Ann. Chem.*, **1976,** 848.

Walter, P., and N. O. Kaplan, *J. Biol. Chem.*, **238,** 2823 (1963).

Chapter 8 Inhibition of Cyclic-AMP Phosphodiesterase

8.1 OCTOSYL ACIDS

While studying the biosynthesis of the polyoxins (Fig 1.16, page 30), Isono et al. (1975) isolated three unusual 5-substituted uracil nucleosides from *Streptomyces cacaoi* var. *asoensis*, namely, octosyl acids A, B, and C (Fig. 8.1).

The octosyl acids are naturally occurring trans-fused anhydrooctouronic acid nucleosides. The fused sugar skeleton of the octosyl acids is the same as that of the ezomycins (page 24).

Discovery, Production, and Isolation

Octosyl acids A, B, and C are isolated from the culture filtrates of *S. cacaoi* var. *asoensis* by ion exchange and partition chromatography (Isono et al., 1975); yields from 3 liters: A, 300 mg; B, 8 mg; C, 130 mg.

Physical and Chemical Properties

Octosyl acid A (a tribasic acid): molecular formula $C_{13}H_{14}N_2O_{10} \cdot H_2O$; mol wt 358 (without H_2O); mp 290–295°C (decomp.); $[\alpha]_D^{20}$ +13.3° (*c* 0.425, 1 *N*

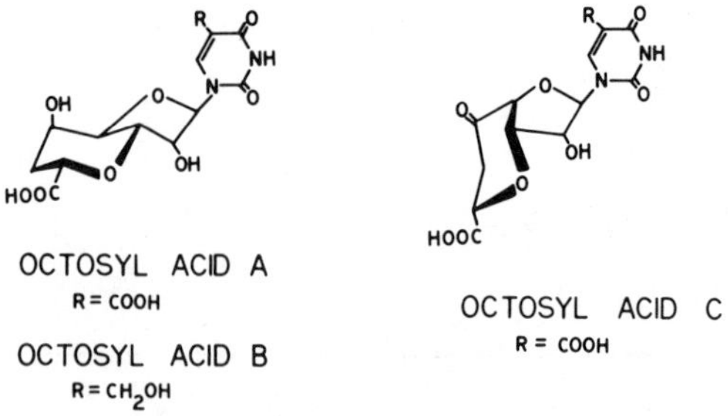

OCTOSYL ACID A
R = COOH

OCTOSYL ACID B
R = CH_2OH

OCTOSYL ACID C
R = COOH

Figure 8.1 Structures of octosyl acids.

NaOH); ultraviolet spectral properties: $\lambda_{max}^{H_2O \text{ and } 0.1\,N\,HCl}$ 220 nm (ϵ = 9900) and 276 nm (ϵ = 10,700); $\lambda_{max}^{0.1\,N\,NaOH}$ 272 nm (ϵ = 7000); pKa' = 3.0, 4.3, and 9.4.

Octosyl acid B: molecular formula $C_{13}H_{16}N_2O_9$; mol wt 344; mp 200°C (decomp.); ultraviolet spectral properties: $\lambda_{max}^{H_2O \text{ and } 0.1\,N\,HCl}$ 265 nm (ϵ = 7700); $\lambda_{max}^{0.1\,N\,NaOH}$ 265 nm (ϵ = 5500); the nmr and mass spectra indicate the presence of the 5-hydroxymethyluracil chromophore with the same sugar moiety as octosyl acid A. Oxidation of octosyl acid B gave octosyl acid A.

Octosyl acid C (a tribasic acid): molecular formula $C_{13}H_{12}N_2O_{10} \cdot H_2O$; mol wt 356 (without water); mp 192–198°C; ultraviolet spectral properties: $\lambda_{max}^{H_2O \text{ and } 0.1\,N\,HCl}$ 220 nm (ϵ = 9200) and 275 nm (ϵ = 9200); $\lambda_{max}^{0.1\,N\,NaOH}$ 272 nm (ϵ = 6400); pKa'–3.1, 4.5, and 9.9; the nmr and mass spectrum plus the decarboxylation to the uracil nucleoside proved the structure of octosyl acid C to be as shown in Fig. 8.1 (Isono et al., 1975).

Structural Elucidation

The three octosyl acids are isolated from the culture filtrates of *Streptomyces cacaoi* var. *asoensis* and are purified using Dowex 50 W, Amberlite IR-4B, DEAE cellulose, and Avicel cellulose (Isono et al., 1975).

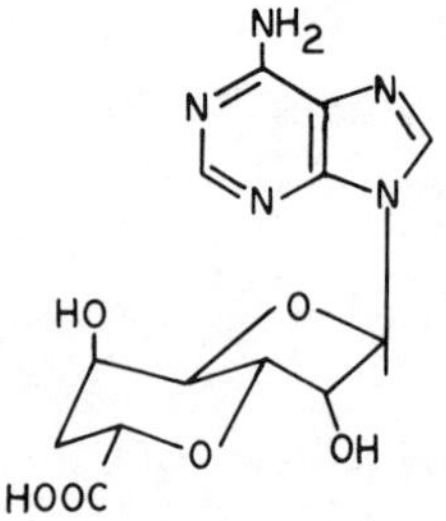

Figure 8.2 Compound resulting from the chemical transglycosylation of octosyl acid A. Reprinted with permission from Azuma et al., 1976.

The structure of octosyl acid A is 1-β-(3,7-anhydro-6-deoxy-D-*glycero*-D-*allo*-octofuranosyluronic acid)uracil-5-carboxylic acid; octosyl acid B is 1-β-(3,7-anhydro-6-deoxy-D-*glycero*-D-*allo*-octofuranosyluronic acid)-5-hydroxymethyluracil; octosyl acid C is 1-β-(3,7-anhydro-6-deoxy-D-*glycero*-L-*lyxo*-octofuranos-5-urosyluronic acid)uracil-5-carboxylic acid (Isono et al., 1975).

Chemical Synthesis

Although the chemical synthesis of the octosyl acids has not been reported, a very efficient chemical transglycosylation of octosyl acid A to the adenine derivative has been reported by Isono and coworkers (Azuma et al., 1976). This improved method may be a versatile procedure for the interconversion of purine and pyrimidine nucleosides.

Biosynthesis of the Octosyl Acids

The biosynthesis of the octosyl acids is combined with the biosynthesis of the polyoxins (pages 30–37) because the polyoxins and the octosyl acids are produced by the same *Streptomyces*.

Biochemical Properties

Octosyl acid A may be regarded as an analog of cCMP. Bloch (1975) was the first to report on the stimulatory effect of a cyclic pyrimidine nucleotide. He showed that cCMP stimulated growth of L1210 cells in culture. The octosyl acids (Fig 8.1) might be considered as analogs of cCMP. In addition, replacement of the pyrimidine ring of octosyl acid A with adenine might produce an adenine analog of cAMP. Azuma and Isono (1976) have reported on the chemical transglycosylation of octosyl acid A to the adenine derivative, 9-β-(3,7-anhydro-6-deoxy-2,5-*O*-diacetyl-D-*glycero*-D-*allo*-octofuranosyluronic acid) adenine (Fig. 8.2). Replacement of the uracil-5-carboxylic acid of octosyl acid A with adenine has produced a compound that is a competitive inhibitor of cAMP for cAMP phosphodiesterase (Isono, personal communication).

REFERENCES

Azuma, T., K. Isono, P. F. Crain, and J. A. McCloskey, *Tetrahedron Lett.*, **1976,** 1687.

Bloch, A., *Biochem. Biophys. Res. Commun.*, **64,** 210 (1975).

Isono, K., P. F. Crain, and J. A. McCloskey, *J. Am. Chem. Soc.*, **97,** 943 (1975).

Chapter 9 Induction of Hypocholesterolemia

9.1 ERITADENINE

Kaneda and coworkers (Tokuda et al., 1964) were the first to describe the marked reduction of plasma cholesterol by a hypolipidemic substance in the edible mushroom "Shiitake" *Lentinus edodes* Donko or Sing. Tokuda and Kaneda (1966a, b) and Shibukawa et al. (1967) subsequently reported on the first partial purification of the substance in the dried mushroom that is responsible for the reduction of plasma cholesterol. The biologically active substance was shown to be eritadenine, 4-(6-amino-9*H*-purine-9-yl)-2(R), 3(R)-dihydroxybutyric acid (Fig. 9.1), an adenine N_9-dihydroxybutyric acid

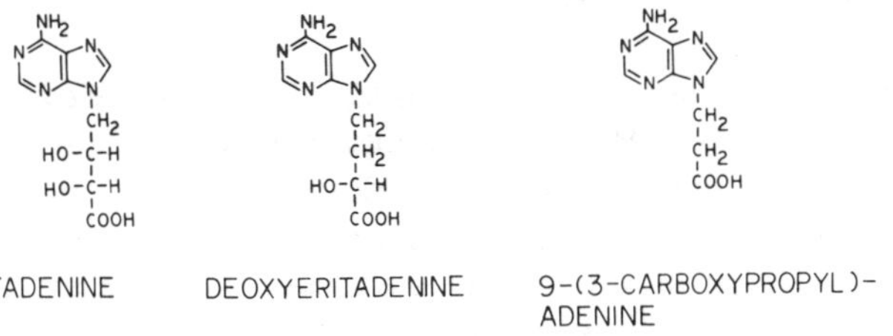

ERITADENINE DEOXYERITADENINE 9-(3-CARBOXYPROPYL)-ADENINE

Figure 9.1 Structures of eritadenine, deoxyeritadenine, and 9-(3-carboxypropyl)adenine.

derivative of adenosine. Although three provisional names were given for eritadenine, "Shii-ta-ke" by Tokuda et al. (1974), "lentysine" by Kamiya et al. (1969), and "lentinacin" by Chibata et al. (1969), the name "eritadenine" is in current use. This edible species of mushroom has long been a delicacy in Japan because of its delicious flavor and sweet fragrance (Nakajima et al., 1961; Morita and Kobayashi, 1967). Deoxyeritadenine and 9-(3-carboxypropyl)adenine have also been isolated from *L. edodes* (Fig. 9.1). It has been the elegant, detailed studies of Kaneda and his coworkers on the isolation, species variation, structural elucidation, physical and chemical properties, and biological properties of eritadenine and deoxyeritadenine that have supplied the current knowledge of this cholesterol-reducing substance isolated from mushrooms.

Discovery, Production, and Isolation

Following the earlier reports of Kaneda and coworkers (Kaneda and Tokuda, 1966a, b; Tokuda and Kaneda, 1966; Shibukawa et al., 1967, 1968), Chibata et al. (1969) crystallized 483 mg of eritadenine following the extraction of 1 kg of the caps of the dried mushroom, *Lentinus edodes*. The extract was applied to a cation exchange column, eluted with ammonium hydroxide, evporated to dryness, dissolved in water, passed through an anion exchange column, and eluted with 1 *N* acetic acid. Final purification was achieved by an amino acid analyzer. Eritadenine crystallizes from water. Tokita et al. (1971a, b) and Saito et al. (1975) used a slightly modified procedure to isolate and crystallize eritadenine and deoxyeritadenine. Tokuda and Kaneda (1966b) and Saito et al. (1975) screened other species of mushrooms to determine their effectiveness in lowering the plasma cholesterol levels; *Lentinus edodes* Donko was more effective than *Auricularia polythricha* (Jews-ear); *Flammulia velutipes* was less effective.

Physical and Chemical Properties

The molecular formula of eritadenine is $C_9H_{11}O_4N_5$; mol wt 253; mp 279°C; $[\alpha]_D^{20}$ + 50°, 0.1 *N* NaOH; + 42° (sodium salt, water); ultraviolet spectral properties: $\lambda_{max}^{0.5\,N\,HCl}$ 260 nm (ϵ = 14,000); $\lambda_{max}^{H_2O}$ 261.0 nm (ϵ = 14,300); $\lambda_{max}^{0.1\,N\,NaOH}$ 261 nm (ϵ = 14,300); ir spectrum reveals hydroxyl groups at 3500–2200 cm^{-1}; carboxylic acid at 1698 cm^{-1} (Tokuda et al., 1976); the nmr spectrum has the following signals: (100 MHz in D_2O) singlets at δ (ppm) 8.04 and 8.03 (two noncoupled protons, 2H, 8H) and multiplet at 4.40–4.10 (four protons) (Tokita et al., 1971a; Rokujo et al., 1970; Chibata et al., 1969). The mass spectra of ^{14}C-eritadenine and eritadenine have been described by Tokuda et al. (1976) and Tokita et al. (1971a).

Hydrolysis of eritadenine (6 *N* HCl, 100°C, 72 h) resulted in glycine and 4-amino-2,3-dihydroxybutyric acid. Chibata et al. (1969) also described the chemical synthesis of eritadenine. The synthetic and naturally occurring derivative showed high cholesterolemic activity. Simultaneously, Kamiya et al. (1969) reported on the structure and synthesis of eritadenine. The ultraviolet, nmr, and ir spectra were essentially the same as described by Chibata et al. (1969). Kamiya et al. (1969) provided evidence for the *erythro* configuration by reduction of the methyl ester of eritadenine with sodium borohydride in isopropanol.

A third isolation of eritadenine from *L. edodes* was described by Kamiya et al. (1972a). They crystallized eritadenine from acetic acid; mp 270°C. The physical and chemical properties were the same as described above.

Isolation of Deoxyeritadenine and 9-(3-Carboxypropyl)adenine

While isolating eritadenine, Kamiya et al. (1972b), Saito et al. (1970), and Tokita et al. (1971a, b) observed two other substances. They reported on the isolation of deoxyeritadenine and 9-(3-carboxypropyl)adenine (Fig. 9.1). Their physical and chemical properties are described. The most useful technique in establishing the structure of deoxyeritadenine was the nmr spectrum, especially with respect to the hydroxy group on the side chain. Four protons appeared at τ 7.29 (2H, broad quartet) and τ 5.72 (2H, triplet) indicating a N—CH_2—CH_2 group. A broad triplet at τ 6.03 for one proton is assigned to the C-1-proton. These data helped establish the structure of deoxyeritadenine. The absolute configuration of the hydroxy group is the R-configuration. The presence of the ethylene group of the propionic acid derivative was demonstrated by the presence of the ethylene group from the nmr spectrum. The structure of deoxyeritadenine and 9-(3-carboxypropyl)adenine was confirmed by chemical synthesis (Saito et al.,

1970). Neither deoxyeritadenine nor 9-(3-carboxypropyl)adenine lowered plasma cholesterol in rats (Tokita et al., 1971a, b; Saito et al., 1970).

Chemical Syntheses of Labeled and Unlabeled Eritadenine, Deoxyeritadenine, and 9-(3-Carboxypropyl)adenine

Several unique methods have been used for the synthesis of these three adenine derivatives. The first synthesis followed the procedure of Chibata et al. (1969) in which they condensed 4-amino-6-chloro-5-nitropyrimidine with 4-amino-2,3-dihydroxyerythrobutyric acid. More convenient methods for the synthesis of eritadenine have been described by Kamiya et al. (1972b), Saito et al. (1970), Kamiya et al. (1970), and Okumura et al. (1971). In one synthesis, adenine is treated with sodium hydride to form sodium adenide. The sodium salt of adenine is then condensed with either 2(*R*), 3(*R*)-*O*-protected dihydroxybutyrolactone (Okumura et al., 1971) or the 2,3-*O*-isopropylidine-D-erythronolactone (II) (Kamiya et al., 1972a). The final products are eritadenine (I) and its N^3-isomer (VI) (Fig. 9.2). A two-step synthesis of [8-^{14}C]eritadenine has been described by Tokuda et al. (1976).

Analogs of Eritadenine

Because eritadenine shows excellent hypolipidemic activity in rats and man, a logical extension of the studies with eritadenine would be the synthesis of analogs in which the aglycon or the dihydroxybutyric acid moiety was altered (Okumura et al., 1971, 1974; Tensho et al., 1974; Kanno and Kawazu, 1974). Kawazu et al. (1973) and Takamura et al. (1973) reported on the synthesis of homoeritadenine compounds, α-D-alkyleritadenine, the 5-deoxy-D- and 5-deoxy-L-arabinosyl, ribosyl, glucosyl, and fructosyl derivatives, noreritadenine, and the D- and L-*threo*- and D- and L-*erythro*-eritadenine. The N^3-derivative of eritadenine showed retention of limited activity while the N^7 isomer was inactive (Kawazu et al., 1973; Takamura et al., 1973).

Biosynthesis of Eritadenine

Because eritadenine is the first adenine with a C-4 acid at carbon 9, its biosynthesis is of special interest. When [8-^{14}C]adenine was infused into the cap of growing shiitake on trunks of *Quercus acutissima*, ^{14}C was incorporated into eritadenine and deoxyeritadenine. Itoh et al. (1973) suggest that eritadenine was synthesized from adenine via deoxyeritadenine.

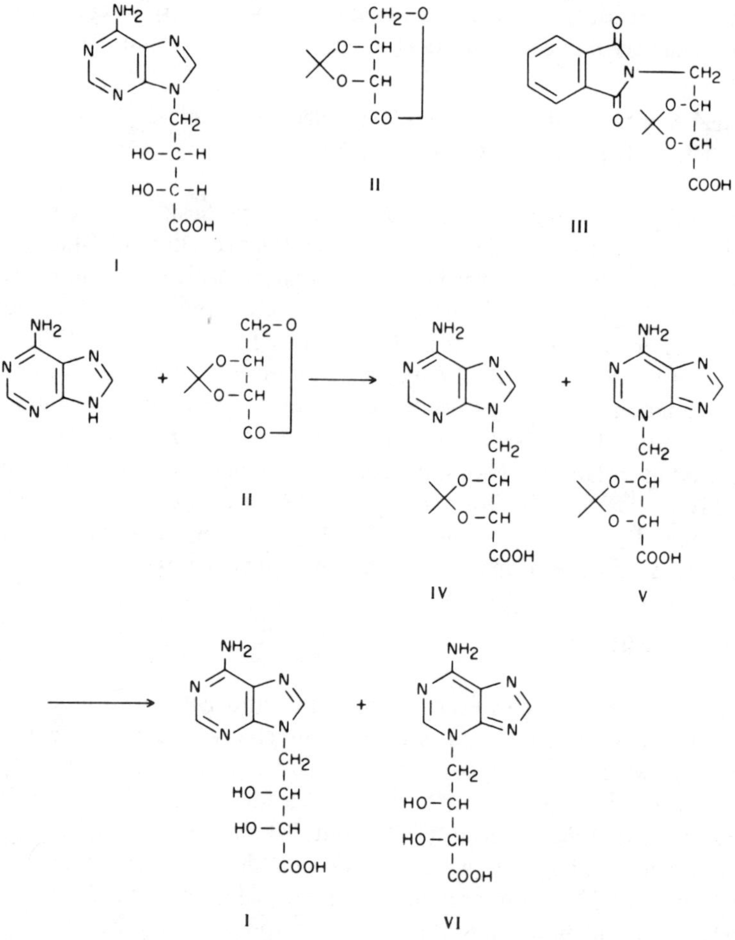

Figure 9.2 Synthesis of eritadenine (I) (Kamiya et al., 1972b).

Biochemical Properties

Effect of Eritadenine on Plasma Lipids in Animals and Man. Of the three substances isolated from dried Shiitake mushrooms, only eritadenine lowers plasma cholesterol levels in rats (Tokita et al., 1971a, b).

One of the more serious problems in maintaining proper blood supply to peripheral tissues is concerned with the pathological deposition of cholesterol-containing plaques in the intima of the aorta. This deposition of cholesterol is the biochemical criteria associated with atherosclerosis and coronary heart disease. Therefore, there have been many attempts to find drugs that diminish cholesterol formation without the accumulation of intermediates of cholesterol synthesis in either the liver or intima of the large blood vessels. One such compound that has been highly successful is clofibrate. Therefore, the first report by Kaneda et al. (1964) that a diet containing 5% of the ground dried mushroom, *L. edodes*, markedly reduced blood plasma cholesterol levels in rats supplied additional impetus for the belief that the mushroom is an "elixir of life" (Kaneda and Tokuda, 1966). Rats fed on diets of other mushrooms did not show a reduction of plasma cholesterol. The first extractions of the dried mushroom to determine the nature of the substance in the dried mushrooms were done with ether, water, and ethanol; the water-soluble fraction and the 30% ethanol-soluble fraction contained substance(s) effective in lowering plasma cholesterol (Tokuda and Kaneda, 1966a; Shibukawa et al., 1967, 1968). When a diet fed to rats contained 5% of ground, dried mushroom (*L. edodes*), neutral sterols in the feces increased, whereas bile acids did not increase; the total amount of cholesterol in the whole body was less than that of the control (Tokuda et al., 1971). Their data indicated that a substance in the mushroom (eritadenine) increased the excretion of administered cholesterol (Table 9.1). The subsequent isolation and crystallization of eritadenine by Chibata et al. (1969), Tokita et al. (1971a), and Matsuo and Hashimoto

Table 9.1 Gas Chromatographic Analysis of Sterols and Bile Acids in Rat Feces from Diets with and without Mushrooms (Containing Eritadenine)

	Intake (mg/day/rat)		Excreted (mg/day/rat)		
Group	Cholesterol	Ergosterol (in Mushroom)	Sterols (Ex/Intake (%))	Bile Acids	Total
1% cholesterol diet					
Rat I	169	—	16.3 (9.6)	15.8	32.1
Rat II	180	—	18.0 (10.0)	19.3	37.3
Average	175	—	17.1 (9.8) (100)[a]	17.6	34.7
5% mushroom and 1% cholesterol diet					
Rat III	196	2.6	30.9 (15.7)	19.8	50.7
Rat IV	202	2.6	26.2 (13.0)	21.9	49.1
Average	199	2.6	28.5 (14.4) (147)[b]	20.9	49.9

Reprinted with permission from Tokuda et al., 1971.

[a] Control (%).

[b] 5% mushroom diets (% increase).

(1971) showed that eritadenine lowers all lipid components of plasma lipoproteins in animals and man. The toxicity of eritadenine in rats is very low, while in rabbits eritadenine had no hypocholesterolemic activity (Saitoh, 1974). In addition to the hypolipidemic effect of eritadenine, Arima et al. (1970) reported that adenosine has the same properties when administered i.p. to rabbits.

Although eritadenine as a naturally occurring metabolite is a hypocholesterolemic agent, there are other fungal metabolites produced by *P. brevicompactum* and *P. citrinum* (compactin, ML-236A, and ML-236B) that have the same activity. However, in contrast to the action of eritadenine, these latter analogs are potent inhibitors of 3-hydroxy-3-methylglutaryl-CoA reductase and inhibitors of hepatic cholesterol synthesis following oral administration to rats or in liver slices (Endo et al., 1976, 1977; Brown et al., 1976).

Effect of Eritadenine on Cholesterol Transport by Plasma Lipoproteins. Tokuda et al. (1972) reported that eritadenine does not effect cholesterol synthesis *de novo* from [1-^{14}C]acetate; however, the administration of ^{14}C-cholesterol orally to rats resulted in a more rapid disappearance of the [^{14}C]cholesterol from the plasma of rats when compared to control animals. Their findings suggest that eritadenine accelerates the transport of cholesterol in the plasma. Tokuda et al. (1973) in carefully designed experiments demonstrated that eritadenine effected the composition of lipoproteins which carry cholesterol in the plasma. The protein and lipid contents of the plasma lipoprotein of rats fed Shiitake mushrooms for 30 days is shown in Table 9.2. These data indicate that the cholesterol in all lipoproteins is reduced. Eritadenine reduced the amount of cholesterol in the very low density lipoprotein (VLDL) fraction and high density lipoprotein fraction. Moreover, the transfer of [^{14}C]cholesterol from chylomicron plus VLDL to low density lipoprotein and the increased excretion of the [^{14}C]cholesterol in the feces was increased by eritadenine. Lecithin:cholesterol acyl transferase activity in the plasma was increased. After feeding eritadenine, the transport of cholesterol from chylomicron plus VLDL to β-lipoprotein is faster than in the control (Table 9.2).

In the publication by Okumura et al. (1974), 124 derivatives of eritadenine were synthesized and their hypocholesterolemic activities were evaluated. Of the 124 derivatives of eritadenine tested, the most active derivatives were the carboxylic acid esters with short-chain monohydroxy alcohols. They were as much as 50 times more active in lowering serum cholesterol of rats at a dose of 0.0001% in the diet when compared with eritadenine. The biological data suggest that carboxyl function and one

Table 9.2 Protein and Lipid Contents in Plasma Lipoproteins of Rats on Diet with and without Mushrooms[a]

	$d < 1.019$ (includ. chylomicron, VLDL)					$d = 1.019 \sim 1.063$ (β-lipoprotein) LDL				
		Cholesterol (mg/dl)					Cholesterol (mg/dl)			
Group	Protein (mg/dl)	Total	Free	Ester	Phospholipid (mg/dl)	Protein (mg/dl)	Total	Free	Ester	Phospholipid (mg/dl)
Cholesterol-free	361	14.7	6.6	8.1	36.7	198	12.3	—	—	19.2
+ mushroom	253	5.3	2.9	2.4	22.7	198	7.4	—	—	19.4
1% cholesterol	369	39.0	12.6	16.4	33.4	247	10.1	2.6	7.6	21.5
+ mushroom	339	23.8	8.8	15.0	29.8	283	6.9	1.9	5.0	20.5

	$d = 1.063 \sim 1.21$ (α-lipoprotein) HDL					$d > 1.21$				
		Cholesterol (mg/dl)					Cholesterol (mg/dl)			
Group	Protein (mg/dl)	Total	Free	Ester	Phospholipid (mg/dl)	Protein (mg/dl)	Total	Free	Ester	Phospholipid (mg/dl)
Cholesterol-free	333	32.9	6.8	26.0	52.9	624	12.5	4.5	8.1	38.7
+ mushroom	217	20.5	4.3	16.2	42.7	648	10.4	1.9	8.5	38.5
1% cholesterol	272	15.9	4.3	11.6	37.9	621	9.3	1.9	7.5	37.4
+ mushroom	126	16.7	4.5	12.2	33.3	622	10.3	1.9	8.4	40.1

Reprinted with permission from Tokuda et al., 1973.

[a] Each value is the mean of six rats after administration of [4-^{14}C]cholesterol.

hydroxyl group along with the intact adenine ring is necessary for biological activity. Tensho et al. (1974) also reported that the purine ring was essential for biological activity as well as the amino group at position 6. They too reported that the presence of the butyric acid with a hydroxyl at the α or β position induced strong hypocholesterolemic activity.

The effect of eritadenine on lipids in hepatic bile of choledochostomy surgical patients has been described by Saitoh (1974). In these patients, eritadenine decreased bile cholesterol content, but had no effect on serum cholesterol levels. Total bile acid concentration was increased by eritadenine. When the enterohepatic circulation was reestablished, there was no change in either the sera or bile lipid levels or in the bile lipid ratio. Eritadenine markedly increased biliary deoxycholic acid.

Because of the hypocholesterolemic effect of eritadenine, Takashima et al. (1974) studied the effect of eritadenine on the metabolism of ^{14}C-labeled cholesterol in the rat. The data they presented indicate that the mechanism of the hypocholesterolemic action of eritadenine does not involve an inhibition of hepatic cholesterol biosynthesis from [1-^{14}C]-acetate or [2-^{14}C]mevalonate. However, eritadenine has a marked effect in the shifting of the equilibrium of cholesterol between plasma and tissues. It was observed that the total quantity of plasma cholesterol was considerably lower than the total amount of cholesterol in the liver. Takashima et al. (1974) suggested that a possible mechanism for the shift in the equilibrium in plasma cholesterol toward tissues in the presence of eritadenine may be due to the relation of its chemical structure to the adenine nucleotide and the interference of metabolic processes influenced by cAMP-dependent protein kinase.

Although the increased activity of these esters is, at least partially, due to the increased absorbability from the intestinal tract (Okumura et al., 1974), it is possible that the esters of eritadenine act *per se* at the receptor site. To test the possibility that eritadenine and/or its esters affect the cAMP-dependent protein kinase system, Iwai (1974) reported on a detailed study of the isoamyl ester of eritadenine on cAMP-dependent biological systems from rat liver and adipose tissue. The isoamyl ester of eritadenine showed a twofold effect on stimulated lipolysis in rat epididymal fat cells. The esters inhibited theophylline-induced glycerol release, but potentiated the lipolytic action of epinephrine. The isoamyl ester inhibited cAMP-dependent protein kinase in fat pad infranatant and liver supernatant fractions. In addition, the isoamyl ester of eritadenine also inhibited cAMP phosphodiesterase of fat cells. Kinetic studies with partially purified liver enzyme showed that eritadenine isoamyl ester inhibits protein kinase by acting as a competitive inhibitor of ATP (Fig. 9.3). The accumulation of intracellular cAMP in fat

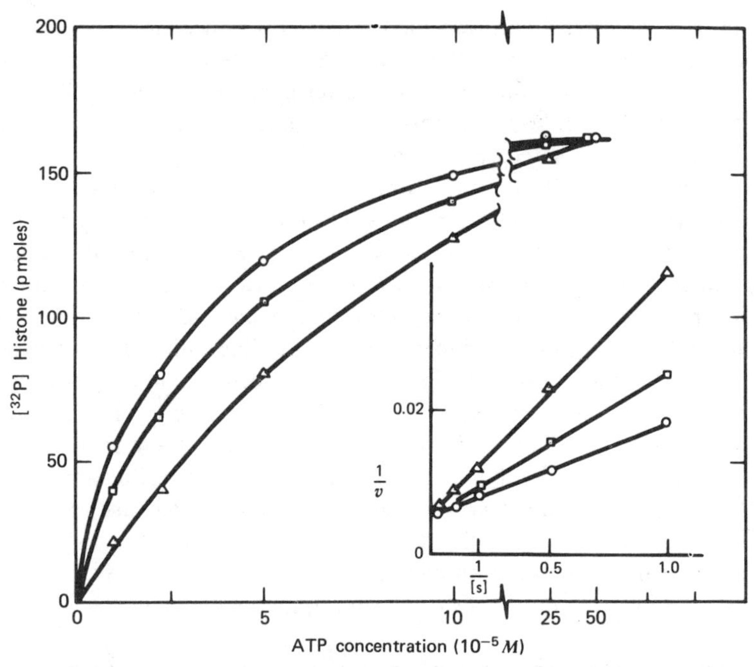

Figure 9.3 Effect of ATP concentration on cAMP-dependent protein kinase in the presence of eritadenine isoamyl ester. After incubation for 5 min at 30°C, ^{32}P transfer from [γ-^{32}P]ATP to histone was estimated. Control (○); with 2.5×10^{-4} M ester (□); with 10^{-3} M ester (△). $\frac{1}{V}$, V = velocity; $\frac{1}{S}$, S = Substrate concentration. Reprinted with permission from Iwai, 1974.

cells treated with the isoamyl ester of eritadenine plus epinephrine was believed to be due to the ester inhibiting cAMP phosphodiesterase.

Hypolipidemic Activity of Eritadenine Administered Orally or I.V. The first studies on the effects of the oral administration of eritadenine on cholesterol metabolism in rats were the reports by Kaneda and coworkers (Kaneda et al., 1974; Tokuda and Kaneda, 1966a, b; Shibukawa et al., 1967, 1968).

Takashima et al. (1973) subsequently reported that eritadenine exerts its powerful hypolipidemic activity without causing fatty livers in the rat. Only 10% of eritadenine is absorbed from the intestinal tract of the rat. However, when the dose is high enough, a hypocholesterolemic effect can be observed within several hours following oral administration. The effect continues when eritadenine is removed from the diet. Eritadenine (4 mg/kg/day) maintains the hypocholesterolemic state for 1 year. Examination of the plasma free cholesterol and esterified cholesterol showed that the free cholesterol is lowered more than the esterified cholesterol. These data suggest that the lipoproteins rich in free cholesterol are more susceptible to the action of eritadenine. Administration of eritadenine i.v. does not elicit the hypocholesterolemic effect observed in oral administration of eritadenine. Eritadenine administered i.v. is rapidly cleared from the circulation and secreted through the kidney. The intestinal wall of the liver may be the site of action of eritadenine. The suppression of triton-induced hyperlipidemia by eritadenine is much less than that reported for orotic acid or 4-aminopyrazolopyrimidine. The effect of exogenously supplied adenine and adenosine on the orotic acid and 4-aminopyrazolopyrimidine induced fatty livers as well as on the action of eritadenine showed that orotic acid and 4-aminopyrazolopyrimidine induced fatty livers are competitively blocked by the simultaneous addition of adenine, but not adenosine (Windmueller, 1964; Puddu et al., 1969). The hypocholesterolemic action of eritadenine may be affected when a tenfold excess of adenine is added. Because exogenously supplied adenine is better utilized by animals than is exogenously supplied adenosine, eritadenine may antagonize one of the adenine metabolites that is important in the homeostasis of plasma lipid metabolism.

Metabolism of [8-^{14}C]Eritadenine in Rats. Tokuda et al. (1976) reported on the metabolism of [8-^{14}C]eritadenine following oral administration to rats. Some of the eritadenine was found in the liver; however, most was found in the feces. When the liver was fractionated, ^{14}C was found in the supernatant, microsomes, and mitochondria. There was little difference in protein and RNA content in the control and eritadenine-treated animals.

SUMMARY

Eritadenine and deoxyeritadenine, as well as 9-(3-carboxypropyl)adenine acid have been isolated from the edible mushroom "Shiitake" *Lentinus edodes* Sing. Only eritadenine has hypocholesterolemic activity in that eritadenine increases the excretion of cholesterol into the feces. Eritadenine has no effect on the endogenous synthesis of cholesterol from [1-^{14}C] acetate. Eritadenine maintains the hypocholesterolemic state in rats for 1 year. The isoamyl ester of eritadenine inhibits protein kinase by acting as a competitive inhibitor of ATP which indicates that these esters act at the receptor site. Eritadenine decreases the amount of cholesterol in all lipoprotein fractions (i.e., chylomicron, VLDL, and HDL). Lecithin:cholesterol acyl transferase activity is increased by eritadenine. Most of the orally administered [8-^{14}C]-eritadenine is excreted into the feces; some carbon-14 was found in various fractions of the liver.

REFERENCES

Arima, T., S. Yamazaki, M. Mambu, and T. Fukumoto, *Jap. J. Geriatr.*, **7,** 46 (1970).

Brown, A. G., T. C. Smale, T. J. King, R. Hasenkamp, and R. H. Thomson, *J. Chem. Soc., Perkin I,* 1165 (1976).

Chibata, I., K. Okumura, S. Taneyama, and K. Kotera, *Experientia,* **25,** 1237 (1969).

Endo, A., M. Kuroda, and K. Tanzawa, *FEBS Lett.*, **72,** 323 (1976).

Endo, A., Y. Tsujita, M. Kuroda, and K. Tanzawa, *Eur. J. Biochem.*, **77,** 31 (1977).

Itoh, H., T. Morimoto, K. Kawashima, and I. Chibata, *Experientia,* **29,** 271 (1973).

Iwai, H., *J. Biochem.*, **76,** 419 (1974).

Kamiya, T., Y. Saito, M. Hashimoto, and H. Seki, *Tetrahedron Lett.*, **1969,** 4729.

Kamiya, T., Y. Saito, M. Hashimoto, and H. Seki, *Chem. Ind.* (*London*), **1970,** 652.

Kamiya, T., Y. Saito, M. Hashimoto, and H. Seki, *J. Heterocycl. Chem.*, **9,** 359 (1972a).

Kamiya, T., Y. Saito, M., Hashimoto, and H. Seki, *Tetrahedron,* **28,** 899 (1972b).

Kaneda, T., K. Arai, and S. Tokuda, *J. Jap. Soc. Food Nutr.*, **16,** 466 (1964).

Kaneda, T., and S. Tokuda, *J. Nutr.*, **90,** 371 (1966).

Kanno, T., and M. Kawazu, *Chem. Pharm. Bull.*, **22,** 2836 (1974).

Kawazu, M., T. Kanno, S. Yamamura, T. Mizoguchi, and S. Saito, *J. Org. Chem.*, **38,** 2887 (1973).

Matsuo, M., and K. Hashimoto, *Hippon Yakurigaku Zasshi,* **67,** 11p (1971).

Morita, K., and S. Kobayashi, *Chem. Pharm. Bull.*, **15,** 988 (1967).

Nakajima, N., K. Kchikawa, M. Kamada, and E. Fujita, *J. Agr. Chem. Soc. Jap.*, **35,** 797 (1961).

Okumura, K., T. Oine, Y. Yamada, M. Tonie, T. Adachi, T. Hagura, M. Kawazu, T. Mizoguchi, and I. Inoue, *J. Org. Chem.*, **36,** 1573 (1971).

Okumura, L., K. Matsumoto, M. Fukamizu, H. Yasao, Y. Taguchi, Y. Sugihara, I. Inoue, M. Seto, Y. Sato, N. Takamura, T. Kanno, M. Kawazu, T. Mizoguchi, S. Saito, K. Takashima, and S. Takeyama, *J. Med. Chem.*, **17,** 846 (1974).

Puddu, P., V. Ottani, P. Zanetti, and M. Marchetti, *Proc. Soc. Exp. Biol. Med.*, **130,** 493 (1969).

Rokujo, R., H. Kikuchi, A. Tensho, Y. Tseukitani, T. Takenawa, K. Yashida, and T. Kamiya, *Life Sci.*, **9,** 379 (1970).

Saito, Y., M. Hashimoto, H. Seki, and T. Kamiya, *Tetrahedron Lett.*, **1970,** 4863.

Saito, M., T. Yasumoto, and T. Kaneda, *J. Jap. Soc. Food Nutr.*, **28,** 503 (1975).

Saitoh, T., *Mushroom Sci.*, **9,** 469 (1974).

Shibukawa, N., S. Tokuda, and T. Kaneda, *J. Jap. Soc. Food Nutr.*, **19,** 373 (1967).

Shibukawa, N., S. Tokuda, and T. Kaneda, *J. Jap. Soc. Food Nutr.*, **20,** 101 (1968).

Takamura, N., N. Taga, T. Kanno, and M. Karazu, *J. Org. Chem.*, **38,** 2891 (1973).

Takashima, K., K. Izumi, H. Iwai, and S. Takeyama, *Atherosclerosis*, **17,** 491 (1973).

Takashima, K., C. Sato, Y. Sasaki, T. Morita, and S. Takeyama, *Biochem. Pharmacol.*, **23,** 433 (1974).

Tensho, A., I. Shimizu, T. Takenawa, H. Kikuchi, T. Rokujo, and T. Kamiya, *Yakugaku Zasshi*, **94,** 708 (1974).

Tokita, F., N. Shibukawa, T. Yasumoto, and T. Kaneda, *Mushroom Sci.*, **8,** 783 (1971a).

Tokita, F., N. Shibukawa, T. Yasumoto, and T. Kaneda, *J. Jap. Soc. Food Nutr.*, **24,** 92 (1971b).

Tokuda, S., and T. Kaneda, *J. Jap. Soc. Food Nutr.*, **17,** 297 (1966a).

Tokuda, S., and T. Kaneda, *J. Jap. Soc. Food Nutr.*, **19,** 222 (1966b).

Tokuda, S., A. Tagiri, E. Kano, and T. Kaneda, *J. Jap. Soc. Food Nutr.*, **24,** 477 (1971).

Tokuda, S., E. Kano, and T. Kaneda, *J. Jap. Soc. Food Nutr.*, **25,** 609 (1972).

Tokuda, S., Y. Sugawara, and T. Kaneda, *J. Jap. Soc. Food Nutr.*, **26,** 113 (1973).

Tokuda, S., A. Tagiri, E. Kano, Y. Sugawara, S. Suzuki, H. Sato, and T. Kaneda, *Mushroom Sci.*, **9** (Part I), 445 (1974).

Tokuda, S., S. Suzuki, and T. Kaneda, *J. Jap. Soc. Food Nutr.*, **29,** 95 (1976).

Windmueller, H. G., *J. Biol. Chem.*, **239,** 530 (1964).

Chapter 10 Naturally Occurring Nucleosides With Limited Biological Activity

10.1 HERBICIDINS A AND B

Two adenine nucleoside antibiotics, herbicidin A and herbicidin B, have been isolated from the culture filtrates of *Streptomyces saganonensis* (Haneishi et al., 1976; Arai et al., 1976). Herbicidins A and B exhibit potent herbicidal activity against dicotyledonous plants, thereby showing promise as weed killers. Herbicidins A and B are not toxic to mice.

Discovery, Production, and Isolation

The medium, isolation, and purification of herbicidins A and B have been described; yield of herbicidin A: 29 g of 80% purity per 350 liters of medium; yield of herbicidin B: 30.5 g of 80% purity per 350 liters of medium (Arai et al., 1976; Haneishi et al., 1976).

Physical and Chemical Properties

The physical and chemical properties of herbicidins A and B are given in Table 10.1. Adenine is isolated from herbicidins A and B following acid hydrolysis. These two new antibiotics are closely related in structure (Haneishi et al., 1976).

Biochemical Properties

The outstanding properties of the herbicidins are their selective activity as herbicides and protective activity against bacterial leaf blight and the inhibition of germination of plant seeds. In biosynthetic studies, herbicidin A is converted to herbicidin B by mycelial mats of *S. saganonensis* (Haneishi et al., 1976).

10.2 5′-*O*-GLYCOSYL-*RIBO*-NUCLEOSIDES

Munakata (1971) and Hayashi et al. (1973) have isolated a group of mono- and disaccharide nucleosides from *Brevibacterium ammoniagenes* and

Table 10.1 Physicochemical Properties of Herbicidins

	Herbicidin A	Herbicidin B
Nature	Basic, white powder	Basic, white crystal
Solubility	Sol. in H_2O, MeOH, EtOH, $(CH_3)_2CO$ and EtOAc.	Sol. in H_2O, MeOH, EtOH and $(CH_3)_2CO$
M.P.	133°C (dec.)	155°C (dec.)
$[\alpha]_D^{20}$	+57.7° (*c* 1, MeOH)	+63° (*c* 1, MeOH)
Elemental analysis	Found	Found
	C, 49.59; H, 5.32; N, 12.56%	C, 46.68; H, 4.99; N, 14.08%
	Calcd. for	Calcd. for
	$C_{23}N_{29}O_{11}N_5 \cdot \frac{1}{2} H_2O$	$C_{18}H_{23}O_9N_5 \cdot \frac{1}{2} H_2O$
M.W.	551 (calcd.)[a]	453 (calcd.)
UV max (ϵ)	258 nm (11,500) in MeOH	258 nm[a] (9200) in MeOH
IR (Nujol)	1750, 1725 cm^{-1}	1740~1750 cm^{-1}

Reprinted with permission from Haneishi et al., 1976.

[a] Calculated from the analyses of mass spectra of each methanolysis product.

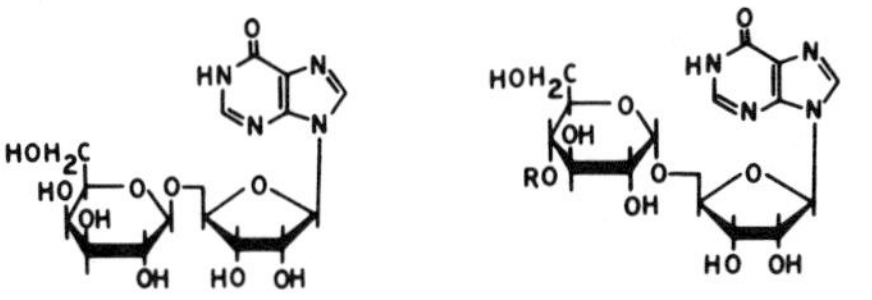

1

2 R=H
3 R=α-D-GLUCOSYL

O-GLYCOSYL RIBONUCLEOSIDES

Figure 10.1 Structures of the O-glycosyl ribonucleosides (Kitahara et al., 1976).

Bacillus sp. No. 102. These bacterial nucleosides have the structural elements of 5′-*O*-hexosylated nucleosides of the type shown in Fig. 10.1 with either hypoxanthine or guanine as the purine base and either galactose, glucose, or maltose with the α- or β-configuration of the intersaccharide bond.

Chemical Synthesis

Lichtenthaler and coworkers (Kitahara et al., 1976; Lichtenthaler et al., 1978a) have described the synthesis of the 5′-*O*-glycosyl-*ribo*-inosine (**2;** Fig. 10.1) by the *O*-glycosylation of **4** with **5** to yield **8** (Fig. 10.2). Compound **8** is converted to the crystalline α-hexaacetates (**11**). De-*O*-acetylation of **11** gave 5′-*O*-(α-D-glycosyl)inosine (**2;** Fig. 10.1) which is identical with the product from *Bacillus* sp. No. 102. The syntheses of the *O*-6 nucleoside (**6**), the N-1 and the N-7 glycosylated nucleosides have also been described (Lichtenthaler et al., 1978b; Riess and Lichtenthaler, 1978).

10.3 RAPHANATIN AND 6-BENZYLAMINO-7-β-D-GLUCOPYRANOSYLPURINE

Raphanatin and 6-benzylamino-7-β-D-glucopyranosylpurine are active cytokinins and metabolites of zeatin, 6-(4-hydroxy-3-methylbut-*trans*-2-enylamino)purine, and 6-benzylaminopurine, respectively (Deleuze et al., 1972; Parker et al., 1972). The metabolite of 6-benzylaminopurine was first assigned the structure 6-benzylamino-7-glucofuranosylpurine (Deleuze et al., 1972). However, Duke et al. (1975) have shown by chemical syntheses that raphanatin and its metabolite are 6,4-(hydroxy-3-methylbut-*trans*-2-enylamino)-7-β-D-glucopyranosylpurine and 6-benzylamino-7-β-D-glucopyranosylpurine, respectively (Fig. 10.3).

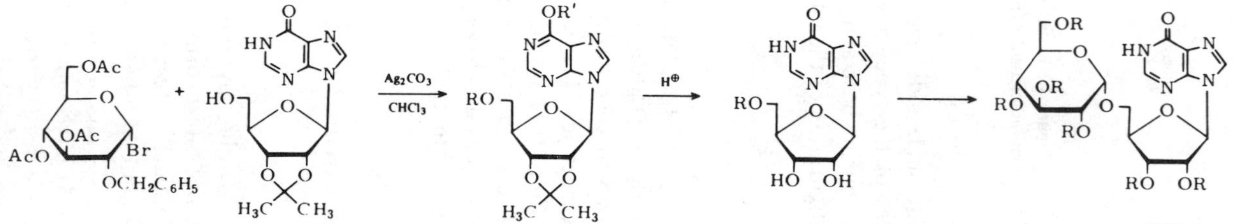

Ag_2CO_3, $CHCl_3$; $H^{\oplus}$

(4) + (5) → (6), R = H; R′ = β-TBG
(7), R = α/β-TBG; R′ = β-TBG
(8), R = α/β-TBG; R′ = H
→ (9), R = α-TBG
(10), R = **β-TBG**
→ (11), R = Ac

Figure 10.2 Chemical synthesis of the O-glycosyl ribonucleosides. TBG = 2-benzyl-3,4,6-triacetylglucopyranosyl. After Lichtenthaler et al., 1978a). Reprinted with permission from Verlag Chemie International Inc.

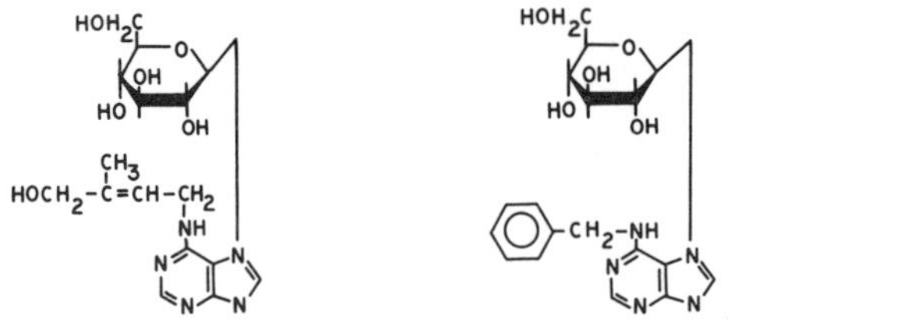

RAPHANATIN

6-BENZYLAMINO-7-β-D-GLUCO-PYRANOSYLPURINE

Figure 10.3 Structures of raphanatin and 6-benzyl-7-β-D-glucopyranosylpurine.

REFERENCES

Arai, M., T. Haneishi, N. Kitahara, R. Enokita, K. Kawakubo, and Y. Kondo, *J. Antibiot.*, (*Tokyo*), **29,** 863 (1976).

Deleuze, G. G., J. D. McChesney, and J. E. Fox, *Biochem. Biophys. Res. Commun.*, **48,** 1426 (1972).

Duke, C. C., A. J. Liepa, J. K. MacLeod, D. S. Letham, and C. W. Parker, *J. Chem. Soc. Chem. Commun.*, 964 (1975).

Haneishi, T., A. Terahara, H. Kayamori, J. Yabe, and M. Arai, *J. Antibiot.* (*Tokyo*), **29,** 870 (1976).

Hayashi, T., K. Shirahata, and I. Matsubara, Abstracts, Japan Chemical Society Meeting, Tokyo, 1973, Abstr. L44.

Kitahara, K., B. Kraska, Y. Sanemitsu, and F. W. Lichtenthaler, *Nucleic Acids Res.*, Spec. Publ. No. 1, S521 (1976).

Lichtenthaler, F. W., Y. Sanemitsu, and T. Nohara, *Angew. Chem. Int. Ed. Engl.*, **17,** 772 (1978a).

Lichtenthaler, F. W., K. Kitahara, and W. Riess, *Nucleic Acids Res.*, Spec. Publ. No. 4, s115 (1978b).

Munakata, T., *Jap. Kokai* 72-23599 (1971).

Parker, C. W., D. S. Letham, D. E. Cowley, and J. K. MacLeod, *Biochem. Biophys. Res. Commun.*, **49,** 460 (1972).

Riess, W., and F. W. Lichtenthaler, *Nucleic Acids Res.*, Spec. Publ. No. 4, s191 (1978).

Author Index

Numbers in roman numerals indicate the page on which an author's work is cited. Numbers in *italics* show the page on which the complete reference is listed.

Acs, G., 158, *198*
Adamson, R. H., 266, 274, *275*
Agarwal, K. C., 174, *198*
Agarwal, R. P., 162, 174, 175, 176, 177, 180, *198*, 263, 264, 265, 267, 268, 271, 272, *275*
Agranoff, B. W., 100, *108*
Aharonowitz, Y., 97, *108*
Aizawa, S., 169, *198*
Alford, B., 178, *198*
Allen, L. B., 226, 231, *251*
Altman, S., 177, *198*
Angus, T. A., 184, *198*
Arai, M., 84, *108*, 311, *315*
Argoudelis, A. D., 46, 53, *63*, 104, 105, 106, 107, *108*, 241, 242, 243, *251*
Arima, T., 304, *309*
Asano, S., 173, *198*
Aswell, J. F., 232, 234, 235, 236, *251*
August, J. T., 119, *198*
Azuma, T., 32, 37, *63*, 296, 297, *297*
Azzam, M. E., 98, 99, *108*

Bablanian, R., 128, *198*
Baddiley, J., 15
Baker, D. C., 219, *251*, 262, *275*
Baksht, E., 177, *198*
Balis, M. E., 61, *63*
Ballet, J. J., 175, *198*, 268, *275*
Baltimore, D., 229, *251*
Bannister, B., 245, 247, *251*
Barbacid, M., 88, *108*
Barondes, S. H., 100, *108*
Bartnicki-Garcia, S., 12, 37, *63*
Bassleer, R., 161, *198*
Bateman, J. R., 143, *198*
Battaner, E., 89, *108*
Baxter, C., 161, *198*
Bayley, H., 100, *108*
Beach, L. R., 124, *198*
Beebee, T. J. C., 184, 188, 189, 190, 191, 192, 193, *198*
Beisler, J. A., 142, *198*
Bellet, R. E., 144, *198*
Beneviste, R. E., 107, *108*
Benitez, T., 35, 36, 37, *63*
Benne, R., 97, *108*
Bennett, L. L., Jr., 147, *198*, 245, *251*
Benz, G., 184, 188, *198*
Bergmann, W., 217, 233, *252*
Bergy, M. E., 135, *198*
Berlin, Y. A., 82, *108*
Bermek, E., 52, *63*
Berry, D. R., 21, 22, *63*
Bettinger, G. E., 14, 15, *63*
Bhalla, R. B., 124, *198*
Bhuyan, B. K., 141, 162, *199*
Biesele, J. J., 244, *252*
Bisel, H. F., 158, *199*
Blecher, M., 165, *199*
Bloch, A., 73, *108*, 158, 161, 162, 170, *199*, 219, 225, 244, *252*, 260, *275*, 280, 281, *289*, 297, *297*
Boeck, L. D., 19, 20, *63*
Bohr, V., 244, 245, 246, *252*

Bollum, F. J., 124, *199*
Bond, R. P. M., 183, 184, 185, 186, 187, 188, 192, *199*
Borchardt, R. T., 147, *199*
Borondy, P. E., 267, *275*
Both, G. W., 122, 148, *199*
Bracha, R., 15, *63*
Brandner, G., 34, *63*
Brawerman, G., 120, *199*
Brdar, B., 166, 167, 168, *199*, 244, *252*
Brevet, J., 191, *199*
Brimacombe, J. S., 16, *63*
Brink, A. J., 37, *63*, 218, 225, *252*, 267, 273, *275*
Brockman, R. W., 273, *275*
Brodniewicz-Proba, T., 124, *199*
Brown, A. G., 304, *309*
Brown, G. B., 244, *252*
Bruzel, A., 133, 134, *199*
Bryson, Y., 266, *275*
Burchenal, J. H., 143, *199*
Burgerjon, A., 187, *199*
Burgess, G. H., 165, *199*
Burridge, P. W., 176, *199*, 268, 273, *275*

Cabib, E., 36, *63*
Caboche, M., 122, *199*
Cadman, E. C., 281, 285, *289*
Cairns, J. A., 169, *199*
Camiener, G. W., 136, *199*
Cantoni, G. L., 162, *199*
Çantwell, G. E., 184, *199*
Čapek, A., 53, *63*
Cardeilhac, P. T., 224, *252*
Carrasco, L., 88, 89, *109*
Carson, D. A., 175, *199*, 268, *275*
Casini, G., 49, *63*
Caskey, C. T., 89, *109*
Cass, C. E., 267, 273, *275*
Çelma, M. L., 86, *109*
Černá, J., 73, 86, 87, *109*
Cha, S., 266, 268, *275*
Chabner, B. A., 136, 142, 144, *199*
Chamberlain, M. J., 194, *199*
Chambon, P., 131, *199*
Chang, L. M. S., 124, 163, *199*
Chang, T., 262, 266, *276*
Chassin, M. M., 272, 273, *276*
Chassy, B. M., 162, *206*, 225, *252*, 280, *289*
Chaudry, I. H., 226, *252*
Chen, W. W., 7, *63*
Cheng, C. S., 159, *200*
Chenon, M.-T., 52, *63*, 170, *200*
Cherbuliez, E., 61, *64*
Cheung, C. P., 98, *109*
Chibata, I., 299, 300, 301, 303, *309*
Ch'ien, L. T., 230, 231, 232, *252*
Chiu, T. M. K., 154, 157, *200*
Chu, M. Y., 235, *252*
Chung, H. L., 170, 174, *200*
Cialdella, J., 249, *252*
Çiampi, M. S., 243, *252*
Čihák, A., 136, 137, 139, 141, 142, 143, 189, *200*
Clark, J. M., 73, *109*
Coats, J. H., 104, 107, *109*
Cohen, A., 182, *200*
Cohen, S. S., 217, 218, 219, 221, 226, 233, *252*
Cohn, W. E., 241, *252*
Cole, F. X., 170, *200*
Coleman, M. S., 124, 136, *200*
Coles, E., 143, *200*
Connor, J. D., 225, *252*
Connor, R. M., 184, *200*
Constine, J., 270, *276*
Cooperman, B. S., 99, 100, *109*
Cortese, R., 243, *252*
Cory, J. G., 118, 125, *200*, 225, *252*
Coutsogeorgopoulos, C., 87, *109*
Coward, J. K., 148, 163, 164, 165, *200*, *201*
Cozzarelli, N. R., 219, 224, *252*
Crabtree, G. W. 172, 173, *201*
Crain, P. F., 283, *289*
Cramer, F., 132
Crystal, R. G., 97, 98, *109*
Cunningham, K. G., 118, *201*
Cunningham, T. J., 144, *201*

Dabeva, M. D., 122, *201*
Dähn, U., 43, 44, 45, *64*
Daluge, S., 101, 102, *109*, 149, *201*
Damodaran, N. P., 31, *64*
Darlix, J. L., 166, *201*
Darnall, K. R., 52, *64*
Darnell, J. E., 119, 120, 121, 124, 126, 127, 128, *201*
Das, B. C., 78, *109*
Daves, G. D., 170, *201*
Davies, R. J. H., 183, *201*
deBarjac, H., 184, 185, 188, *201*
Debatisse, M., 271, *276*

DeBernardo, S., 237, 238, 239, *252*, 281, 283, 284, *289*
DeBoer, C., 241, 242, 246, 247, 249, *253*
DeClercq, E., 165, *201*, 270, *276*, 287, *289*
deGarilhe, M. P., 218, 235, *253*
Deleuze, G. G., 313, *315*
Delseny, M., 128, *201*
Descamps, J., 286, 287, 289, *289*
Deutcher, J. D., 46, *64*
deVries, O. M. H., 37, *64*
DiCioccio, R. A., 124, *201*, 221, *253*
Diez, J., 120, *201*
Dion, H. W., 260, 262, *276*
Doering, A. M., 217, 218, *253*
Doskočil, J., 57, 58, 61, *64*, 136, 137, 142, *201*
Drach, J. C., 224, 227, *253*
Dreyer, C., 133, *201*
Drysdale, J. W., 127, *201*
Dubnoff, J. S., 97, 98, *109*
Duda, E., 9, *64*
Duke, C. C., 313, *315*
Duksin, D., 11, 16, *64*
Dunn, J. J., 194, *201*

Earl, R. A., 243, *253*
Ebringer, L., 280, *289*
Eckardt, K., 5, 7, 12, *64*
Edens, B., 97, *109*
Edmonds, M., 119, 120, *202*
Eguchi, J., 35, *64*
Elbein, A. B., 7
Elstner, E. F., 52, 55, *64*, 160, *202*
Endo, A., 12, 31, 36, *64*, 304, *309*
Ennifar, S., 78, *109*
Ensminger, P. W., 91, *109*
Epp, J. K., 91, 92, *109*
Ericson, M. C., 16, *64*
Evans, B., 262, 268, *276*
Evans, J. E., 162, *202*
Evans, J. R., 80, 81, *109*

Falcon, M. G., 231, *253*
Farkaš, J., 184, 185, 186, 192, *202*, 281, *289*
Farmer, P. B., 221, *253*
Fedorinchik, N. A., 195, *202*
Ferscht, A. R., 133, *202*
Fishaut, J. M., 232, *253*
Flexner, L. B., 100, *109*
Florini, J. R., 102, 103, *109*
Forsee, W. T., 16, *64*
Fouquet, H., 120, 127, *202*
Fox, J. J., 84, *109*, 154, 175, *202*
Franze-deFernandez, M. T., 167
Fraser, T. H., 132, 146, *202*
Fredericksen, S., 123, *202*
Freese, E., 281, *290*
Fresno, M., 98, *109*
Fučík, V., 139, *202*
Fujisawa, J., 12, *64*
Fuller, R. W., 19, 22, *64*
Funayama, S., 33, 34, 35, *64*
Furth, J. J., 218, 224, *253*
Furuichi, Y., 122, *202*

Gallo, R. C., 180, *202*
Garoff, H., 11, *64*
Gentry, G. A., 232, 234, 235, 236, *253*
Georgiev, G. P., 119, *202*
Gerzon, K., 281, *290*
Giblett, E. R., 259, *276*
Gibson, R., 11, *64*
Giziewicz, J., 172, 173, *202*
Glazer, R. I., 122, 124, 126, 127, 129, *202*, 266, 269, 270, 271, *276*
Goldberg, J. H., 97, 98, *109*
Gooday, G. W., 36, *64*
Goodman, L., 217
Gordee, R. S., 19, 20, *64*
Gotoh, S., 166, *202*
Gottesman, M. E., 119, *202*
Grage, T. B., 158, *202*
Grahame-Smith, D. G., 184, 193, *202*
Gray, M. W., 243, *253*
Gregor, I., 294, *294*
Gregory, J. D., 57, *64*
Guarino, A. J., 103, *109*
Gumport, R. I., 133, 134, *202*
Gutowski, G. E., 281, 286, *290*

Haar, F. von der, 132, 133, 178, *213*
Hadjivassiliou, A., 119, *202*
Hadler, H. I., 52, 60, *64*
Hagimoto, H., 148, *203*
Hamill, R. L., 19, 20, *65*, 77, 78, 79, 89, 90, 93, *109*
Hamilton, R. H., 50, *65*
Hammett, J. R., 127, *203*
Hampton, A., 129, *203*
Hancock, I. C., 15, *65*
Haneishi, T., 84, *109*, 236, 237, *253*, 311, 312, *315*
Hanka, L. J., 135, 145, *203*

Hannay, C. L., 184, *203*
Harada, S., 41, 42, *65*, 79, 80, *109*, *110*
Hardesty, C. T., 169, *203*
Harris, B., 127, 128, *203*
Harris, R. J., 97, *110*
Hart, G. W., 17, 19, *65*
Hashimoto, T., 43, *65*
Hasilik, A., 13, *65*
Haskell, T. H., 81, *110*
Hatanaka, S., 34, *65*
Hayaishi, O., 181, *203*
Hayashi, M., 150, 151, 153, *203*
Hayashi, T., 312, *315*
Heby, O., 141, *203*
Hecht, S. M., 132, 180, *203*
Heimpel, A. M., 184, *203*
Heine, U., 166, *203*
Heip, J., 48, *65*
Heisetz, A., 7
Henderson, J. F., 162, *203*, 244, *253*, 268, 271, *276*
Henry, K. B., 100, *110*
Henshaw, E., 119, *203*
Herrmann, E. C., Jr., 266, *276*
Hershfield, M. S., 176, *203*, 227, *253*
Hickman, S., 16, *65*
Hilbert, G. E., 154, 157, *203*
Hirasawa, K., 160, *203*
Hirsch, J., 162, *203*
Hishizawa, T., 86, *110*
Hobbs, J. B., 94, *110*, 145, *203*
Hochstadt-Ozer, J., 94, *110*
Honke, T., 46, 47, *65*
Hopwood, J. J., 19, *65*
Hori, M., 32, 36, 37, *65*, 169, *203*
Horowitz, J. P., 181, *203*
Horská, K., 185, 186, 188, 191, *203*
Hovi, T., 175, *203*
Hoyer, B. H., 119, *203*
Hrebabecky, H., 280, *290*
Hrodek, O., 143, *203*
Huang, M., 61, *64*
Hubert-Habart, M., 217, 218, 224, *253*
Hudson, L., 243, *253*
Hurley, L., 260, 261, *276*
Hyndiuk, R. A., 231, *253*

Iapalucci-Espinoza, S., 166, 167, *203*
Ikehara, M., 94, 95, *110*, 149, 162, 182, *203*, *204*
Ikeuchi, T., 84, *110*
Ilan, J., 227, *253*
Ingoglia, N. A., 125, *204*
Ishizaki, H., 35, 37, *65*
Isono, K., 23, 30, 31, 32, 33, 34, 35, 36, 37, 43, 47, *65*, *66*, 238, 240, 241, *253*, 295, 296, 297, *297*
Israili, Z. H., 143, *204*
Ito, H., 187, *204*
Ito, T., 4, 5, 6, *66*
Itoh, H., 301, *309*
Iwai, H., 306, 307, *309*
Iwasa, T., 41, 42, 43, *66*, 79, 80, *110*
Iwata, H., 151, *204*

Jaffe, J. J., 165, *204*
Jain, K., 182, *204*
Jakubowski, A., 285, *290*
Jamieson, A. T., 235, *254*
Jayaraman, J., 97, 98, *110*
Jaynes, E. N., Jr., 99, 101, *110*
Jelinek, W., 120, *204*
Jenkins, I. D., 102, *110*
Johns, D. G., 125, 182, *204*, 266, *276*
Johnson, A. E., 99, *110*
Johnson, D. E., 184, 191, 192, 193, *204*
Jones, D. B., 231, *254*
Jorman, M., 293, *294*
Jurovčík, M., 136, 143, *204*
Just, G., 53, *66*, 289, *290*

Kaehler, M., 162, *204*
Kaempfer, R., 98, *110*
Kalman, T. I., 52, 53, 59, *66*
Kalousek, F., 136, *204*
Kaluza, G., 9, *66*
Kalvoda, L., 53, *66*, 170, 171, 186, *204*
Kamiya, T., 299, 300, 301, 302, *309*
Kaneda, T., 298, 299, 303, 307, *309*
Kann, H. E., Jr., 124, *204*
Kanno, T., 301, *309*
Kappen, L. S., 98, *110*
Karnofsky, D. A., 195, *204*
Karon, M., 139, 143, *204*
Kasai, H., 159, 161, *204*
Kates, J., 119, 120, *204*
Katoh, Y., 12, 13, *66*
Katz, F. N., 11, *66*
Kaufman, H. E., 230, *254*
Kawazu, M., 301, *309*
Keefer, R. D., 145, *204*
Keiley, M. L., 7, *66*

Keller, F. A., 36, *66*
Keller, R. K., 16, *66*
Kelly, R. K., 230, *254*
Kerr, A., 47, 48, 49, *66*
Kim, Y. T., 185, 187, 192, *205*
Kimball, A. P., 272, *276*
Kinkel, A. W., 232, *254*
Kirst, H. A., 90, *110*
Kishi, T., 80, *110*
Kisselev, L. L., 131, *205*
Kitahara, K., 313, *315*
Kits, S., 235, *254*
Klassen, W., 195, *205*
Klein, E., 165, *204*
Klein, R. J., 231, *254*
Klenk, H.-D., 10, *66*
Klenow, H., 118, *205*
Klier, A. F., 191, 192, *205*
Kmetec, E., 195, *205*
Kneifel, H., 38, 39, 40, *66*
Kohler, R. E., 98, *110*
Kohls, R. E., 184, *205*
Komatsu, Y., 52, 57, 58, *66*
Konishi, M., 81, 82, 83, 84, *110*
Konno, K., 292, 293, *294*
Koshiura, R., 266, *276*
Koyama, G., 183, *205*
Krieg, A., 187, *205*
Kuchino, Y., 159, 161, *205*
Kumar, S. A., 162, 172, *205*
Kunimoto, T., 172, *205*
Kuo, S.-C., 5, 9, 13, 16, *66*
Kupper, H. A., 194, *205*
Kuroyanagi, T., 152, *205*
Kurtz, S. M., 232, *254*
Kusaka, T., 147, 148, 151, *205*
Kusakabe, Y., 237, 240, *254*
Kuzuhara, H., 31, *66*

Lambert, P. A., 5, 13, *66*
Langston, R. H. S., 230, *254*
Lapi, L., 268, *276*
Larsen, P. O., 34, *66*
Lavers, G. C., 127, *205*
Lawrence, F., 162, *205*
Leavitt, R., 10, *66*
Lee, S. H., 273, *276*
Lee, S. Y., 120, *205*
Lee, T. T., 136, 137, *205*
Lee, W. W., 217, 231, *254*
Legraverend, M., 129, *205*
Lehle, L., 9, 13, *66*
Leinwand, L., 129, 130, *205*
Lennon, M. B., 119, 165, *205*
Leonard, N. J., 96, *110*
Leonard, T. B., 224, *254*
LePage, G. A., 218, 232, *254*, 273, *276*
Lerner, L. M., 280, *290*
Leslie, L., 57, *67*
Leung, H. B., 217, 218, *254*
Leung, K.-K., 57, 58, *67*
Levey, I. L., 127, *206*
Levi, J. A., 143, *206*
Levitan, I. B., 137, 141, *206*
Lewin, P. K., 195, *206*
Lewis, A. F., 169, 172, 182, *206*, 281, *289*
Li, L. H., 136, 139, 141, 143, *206*
Lichtenthaler, F. W., 73, 84, 85, 86, 87, 103, *110*, 313, 314, *315*
Lieff, B. D., 100, *110*
Lim, L., 119, 120, *206*
Lim, M., 53, *67*
Lindahl, U., 17, *67*
Lindberg, B., 244, *254*
Linder-Horowitz, M., 127, *206*
Linn, T. G., 191, *206*
Lipper, R. A., 221, 225, *254*, 262, 263, *276*
Loftfield, R. B., 131, *206*
Lomen, P. L., 144, *206*
Lon, U., 177, *206*
Long, R. A., 172, 182, *206*
Lopez-Romero, E., 36, *67*
Lowe, J. K., 268, *276*
Lowy, B. A., 61, *67*
Lucas, J. J., 7, *67*
Lum, C. T., 274, *276*

Maale, G., 123, *206*
McArthur, H. A. I., 15, *67*
McCardell, B. A., 49, 50, *67*
McCarthy, J. R., Jr., 280, *290*
McConnell, E., 184, *206*
McConnell, W. R., 272, *277*
McCredie, K. B., 143, *206*
McGuire, W., 139, 140, 143, 144, *206*
Mackedonski, V. V., 189, 190, 191, 192, 193, *206*
McNamara, D. J., 141, *206*
Maelicke, A., 173, 177, *206*
Mahy, B. W. J., 128, *206*
Majima, R., 177, *206*
Makabe, O., 174, *206*

Mantsch, H. H., 61, *67*
Martinez, A. P., 49, *67*
Marumoto, R., 149, *207*
Matsukage, A., 236, *254*
Matsuo, M., 303, *309*
Matsuura, S., 57, *67*
Mendecki, J., 119, 120, 124, *207*
Menzel, H. M., 73, 86, *110*
Michaels, A. F., Jr., 195, *207*
Michelot, R. J., 163, *207*
Miko, M., 161, *207*
Miller, F. A., 229, *254*
Miller, R. L., 221, 235, *254*
Mills, G. C., 262, 274, *277*
Milne, G. H., 172, *207*
Mintz, B., 127, *207*
Mitani, T., 281, 282, *290*
Miyamoto, M., 82, *110*, 260, *277*
Mizrahi, A., 12, *67*
Mizuno, K., 149, 150, 151, 152, 153, *207*
Mizuno, M., 43, *67*
Modak, M. J., 177, *207*
Moertel, C. G., 144, *207*
Moffatt, J. G., 31
Momparler, R. L., 137, 145, *207*
Monesi, V., 127, *207*
Monro, R. E., 86, 97, 98, *110*, *111*
Moore, E. C., 218, *254*
Morell, S. A., 57, *67*
Morita, K., 299, *309*
Morton, G. O., 102, *111*
Moss, B., 128, *207*
Moss, D. R., 100, *111*
Much, R. P., 165, *207*
Müller, W. E. G., 122, 123, 124, 131, 181, *207*, 219, 220, 221, 222, 223, 224, 225, 226, *254*, 263, *277*
Munakata, T., 312, *315*
Muthukrishnan, S., 122, 148, *207*
Myslovataya, M. L., 188, *207*

Nagarajan, R., 19, 20, 21, *67*, 78, *111*
Nahmias, A. J., 231, *255*
Nair, C. N., 195, *207*
Nair, V., 96, 97, *111*, 128, 129, *207*
Nakagawa, Y., 52, 53, *67*
Nakajima, N., 299, *309*
Nakamura, H., 259, *277*
Nakamura, K., 11, *67*
Nakamura, S., 5, *67*, 73, *111*
Nakanishi, T., 93, 94, 95, 96, *111*, 145, *207*
Nakazato, H., 120, *207*
Nathans, D., 96, *111*
Neil, G. L., 144, 145, *207*, *208*
Nester, E. W., 50
Nevins, J. R., 119, 121, 128, *208*
Nichols, W. E., 218, 224, *255*
Nicholson, A. W., 99, *111*
Niessing. J., 120, 123, *208*
Niida, T., 259, *277*
Nishimura, H., 52, 57, *67*
Nishimura, S., 159, 161
Nitschke, R., 141, *208*
North, T. W., 225, 227, 228, 229, *255*, 270, *277*

O'Brien, S. J., 128, *208*
O'Brien, T. G., 141, *208*
Ochi, K., 172, *208*
O'Day, D. M., 231, *255*
Ofengand, J., 132, *208*
Ohno, M., 260, *277*
Ohnuma, T., 281, 285, *290*
Ohta, N., 36, *67*
Okada, N., 159, 161, *208*
Okami, Y., 237, *255*
Okumura, K., 301, 304, 306, *309*
Okura, A., 221, *255*
Olden, K., 11, *67*
Omura, G. A., 144, *208*
Ortiz, P. J., 219, *255*
Ozaki, M., 52, 53, *67*

Pačes, V., 137, *208*
Paggi, P., 100, *111*
Paik, W. K., 162, *208*
Pais, M., 186, *208*
Palm, P. E., 143, *208*
Palmiter, R. D., 7, *67*
Panicali, D. L., 129, *208*
Parke, Davis, & Co., 217, 218, *255*
Parker, C. W., 313, *315*
Parkman, R., 175, *208*
Parks, R. E., 161, 162, 172, 177, 180, *208*, 262, 264
Paterson, A. R. P., 244, 245, *255*
Pavan-Langston, D., 219, 230, *255*, 266, *277*
Pelling, J. C., 222, *255*
Penman, S., 119, 120, 127, *208*
Périès, J., 167, *208*
Perlman, D., 53, *67*
Perry, R. P., 122, 167, *208*
Person, A., 128, *208*

Pestka, S., 87, 97, *111*
Pettit, G. R., 61, *67*
Philipson, L., 120, 127, 128, *208*
Phillips, S. G., 166, *208*
Píškala, A., 135, *208*
Pitha, J., 137, *208*
Plagemann, P. G. W., 285, 288, *290*
Plaut, G. W. E., 62, *67*
Plunkett, W., 125, *208*, 218, 219, 224, 225, 226, 228, *255*, 262, 266, 268, 273, *277*
Poste, G., 231, *255*
Presant, C. A., 143, 145, *208*
Prisbe, E. J., 280, *290*
Prokop'ev, V. N., 187, *209*
Prusiner, P., 174, 209
Prystaš, M., 184, 185, 186, 187, *209*
Puddu, P., 308, *309*
Pugh, C. S. G., 21, 23, *67*, 147, 148, *209*

Quagliana, J. M., 144, *209*

Ralph, R. K., 49, *67*
Ranganathan, R., 219, *255*
Rao, M. S., 122, *209*
Raška, K., 137, *209*
Reichenbach, N. L., 135, *209*
Reichman, M., 137, *209*
Reichman, U., 243, *255*
Reinke, C. M., 222, 228, *255*
Reist, E. J., 217, *255*
Renis, H. E., 249, *255*
Rice, J. M., 242, *255*, 283, *290*
Rich, M. A., 118, *209*
Richardson, L. S., 128, *209*
Richmond, M. E., 17, *67*
Riess, W., 313, *315*
Riquelme, P. T., 130, *209*
Rizzo, A. J., 120, 125, 126, *209*
Roberts, F. M., 15, *67*
Roberts, W. P., 47, 48, 49, *67*
Robins, R. K., 152
Rodén, L., 17, *67*
Rogler-Brown, T., 262, 264, 271, *277*
Rokujo, R., 300, *309*
Rose, K. M., 123, 124, *209*, 224, *256*
Rosenberg, G., 185, *209*
Ross, A. F., 162, *209*
Rothman, E. J., 11, *67*
Rottman, F., 118, *209*
Rowinski, J., 127, *209*
Roy-Burman, S., 52, 57, *68*
Ryder, A., 260, *277*

Sabol, S., 98, *111*
Saiki, J. H., 143, 144, *209*
Saito, Y., 299, 300, 301, *309*
Saitoh, T., 304, 306, *309*
Sakagami, Y., 84, *111*
Sakaguchi, K., 150, 152, 153, *209*, *210*
Sakata, K., 24, 25, 26, 27, 28, 29, *68*
Salvatore, F., 162, *210*
Salzman, N. P., 119, *210*
Sasaki, K., 237, 238, 240, *256*
Savarese, T. M., 266, *277*
Sawa, T., 169, *210*, 259, *277*
Sawicki, S. G., 120, 121, *210*
Scannell, J. P., 241, *256*
Schabel, F. M., Jr., 229, 231, *256*, 273, *277*
Schaeffer, H. J., 262, *277*
Schell, J., 47, 51, *68*
Schmidt-Ruppen, K. H., 231, *256*
Scholar, E. M., 180, *210*
Scholtissek, C., 10, *68*
Schramm, K. H., 169, *210*
Schramm, V. L., 174, *210*
Schroeder, W., 280, *290*
Schwartz, P. M., 224, *256*, 266, *277*
Schwarz, R. J., 16, *68*
Schwarz, R. T., 10, *68*
Schweizer, M. P., 49, *68*
Şeal, S. N., 98, *111*
Šebesta, K., 183, 184, 185, 186, 187, 188, 191, 192, *210*
Sehgel, P. B., 133, *210*
Sekeris, C. E., 192, *210*
Senft, A. W., 165, *210*
Seto, H., 73, 74, 75, 76, 77, *111*, 154, 155, 156, 157, 158, *210*
Shannon, W. M., 229, *256*, 281, 287, *290*
Shapiro, S. K., 148, 162, *210*
Sharma, C. B., 184, 192, 193, *210*
Sharpless, N. E., 57, *68*
Shatkin, A. J., 148, *211*
Shchepetil'nikova, V. A., 195, *211*
Shealy, Y. F., 147, *211*
Sheen, M. R,. 182, 183, *211*
Shewach, D. S., 266, *277*
Shibukawa, N., 298, 299, 303, 307, *309*
Shibuya, K., 32, 35, *68*
Shieh, T. R., 188, *211*
Shigeura, H. T., 118, *211*
Shimke, R. T., 7

Shimotohno, K., 122, *211*
Shipman, C., Jr., 224, 227, *256*, 269, *277*
Shirato, S., 53, *68*
Shoji, J., 43, *68*
Shomura, T., 42, *68*
Shutt, R., 137, *209*
Sibatani, A., 119, *211*
Sidwell, R. W., 231, *256*, 286, *290*
Siev, M., 119, 124, 125, 127, *211*
Sikorski, N. M., 86, *111*
Silbert, J. E., 17, *68*
Skalko, R. G., 127, *211*
Skaric, V., 280, *290*
Skoog, F., 180, *211*
Slechta, L., 247, 248, *256*
Sloan, B. J., 226, 231, *256*, 266, 267, *277*
Smetana, K., 184, 194, *211*
Smith, C. G., 158, 159, 161, 162, 169, *211*
Smith, C. M., 242, 244, *256*
Smith, J. D., 177, *211*
Smith, S. H., 229, *256*, 272, *277*
Smuckler, E. A., 189, 193, *211*
Smulson, M., 159, *211*
Snyder, F. F., 175, *211*
Söll, D., 131, *211*
Somerville, H. J., 192, *211*
Šorm, F., 143, *209*
Spector, T., 221, *256*
Spiegel, S., 128, *211*
Sprinzl, M., 132, 146, 178, *211*, 224, *256*
Srivastava, B. I. S., 124, *211*
Stegman, R. J., 161, *211*
Stern, K. F., 247, *256*
Streeter, D. G., 152, *211*, 287, *290*
Streightoff, F., 281, *290*
Stringer, E. A., 98, *111*
Stroman, D. W., 249, *256*
Struck, D. K., 7, 8, 9, 16, *68*
Subramanian, A. R., 98, 99, *111*
Suciu, N., 280, *290*
Sugimori, T., 280, *290*
Suhadolnik, R. J., 4, 30, 46, 52, 61, *68*, 73 84, 89, 96, 101, *111*, 118, 130, 131, 135, 136, 137, 146, 147, 159 162, 165, 166, 169, 170, 179, 181, *211*, *212*, 219, 241, 244, *256*, 260, 277, 280, 281, *290*
Suling, W. J., 272, *278*
Sulkowski, E., 12
Sundaralingam, M., 97, *111*
Sussman, M., 182, *212*
Swart, C., 167, 168, *212*
Sweeney, M. J., 281, 285, *291*
Sweetman, L., 221, *256*
Symons, R. H., 102, *111*

Tabor, C. W., 162, *212*
Tabor, H., 162, *212*
Takamura, N., 301, *309*
Takaoka, K., 23, *68*
Takashima, K., 306, 307, *310*
Takatsuki, A., 4, 5, 6, 7, 8, 9, 10, 11, 12, 13, 14, 17, 18, 19, 62, *68*, *69*
Tamm, I., 244, *256*
Tamura, G., 13, 14, *69*
Tan, C., 143, 144, *212*
Tasca, R. J., 127, *212*
Tavitian, A., 166, *212*
Taylor, J. M., 195, *212*
Temin, H. M., 229, *256*
Tensho, A., 301, 306, *310*
Thompson, H. A., 100, 102, *111*
Thompson, T. J., 47, 48, 49, *69*
Thomson, J. L., 127, *212*
Thrum, H., 12, *69*
Tilghman, S. M., 127, *212*
Titani, Y., 52, 59, *69*
Tkacz, J. S., 5, 7, 8, 12, 13, *69*
Tobey, R. A., 141, *212*
Todde, P. S., 192, *212*
Tokita, F., 300, 301, 302, 303, *310*
Tokuda, S., 298, 299, 300, 301, 303, 304, 305, 307, 308, *310*
Tomita, K., 82, *112*
Tonew, E., 12, *69*
Tono-Oka, S., 293, *294*
Torrence, P. F., 165, 182, *212*
Townsend, L. B., 46, 52, *69*, 73, 84, 89, 112, 143, 169, 170, 172, *212*, 219, 242, 244, *256*, 260, *278*, 280, 281, 283, *291*
Trewyn, R. W., 143, *212*
Troetel, W. M., 225, *256*
Trommer, W. E., 294, *294*
Truant, A. P., 245, *256*
Truman, J. T., 125, 146, *212*
Trummlitz, G., 53, *69*
Tsukuda, Y., 52, *69*
Tsuruoka, T., 42, *69*, 259, *278*
Tsutsumi, K., 178, *212*
Turco, S. J., 19, *69*
Turner, J. R., 20, 22, *69*

Uchida, K., 242, *256*

Uematsu, T., 77, 78, 79, *112*
Uesugi, S., 174, *212*
Umezawa, H., 107, *112*, 260
Underwood, G. E., 246, *256*
Uramoto, M., 30, *69*
Uretsky, S. C., 169, *212*

Vadlamudi, S., 143, *212*
Valk, P. van der, 37, *69*
Van der Weyden, M. B., 262, *278*
Vaňková, J., 188, *212*
Van Larebeke, N., 50, *69*
Van Oortmerssen, E. A. C., 128, *213*
Vardanis, A., 35, *69*
Vazquez, D., 89, *112*
Venkov, P. V., 166, *213*
Verheyden, J. P. H., 280, *291*
Veselý, J., 135, 136, 137, 142, 145, *213*
Vince, R., 101, 102, *112*, 195, *213*, 226, *256*
Virtanen, A. I., 34, *69*
Vogler, W. R., 143, 144, *213*
Von Hoff, D. D., 136, 143, 144, *213*
Voytek, P., 142, *213*
Vuilhorgne, M., 78, *112*

Waechter, C. J., 5, 7, *70*
Wainfan, E., 161, *213*
Walter, P., 294, *294*
Walter, R. D., 129, 162, *213*
Wagar, M. A., 222, *257*
Ward, D. C., 162, 174, *213*
Ward, J. B., 13, 14, *70*
Warnick, C. T., 245 , *257*
Watanabe, K. A., 154, 157, *213*
Watanabe, S., 57, *70*, 84, *112*
Watson, B., 50, 51, *70*
Webb, T. E., 125
Weinberg, R. A., 122, *213*
Weisblum, B., 107, *112*
Weiss, A. J., 144, *213*
Weiss, J. W., 136, 137, 138, 166, *213*
Weiss, S. R., 129, *213*
Weissbach, H., 97, 98, *112*
Weller, T. H., 231, *257*
Wenkert, E., 281, *291*
West German Patent, 237, 240, *257*
Weston, A., 15, *70*
Whitley, R. J., 219, 230, *257*
Williams, R. H., 280, *291*
Wilson, S. G. F., 195, *214*
Windmueller, H. G., 308, *310*
Winicov, I., 133, *214*
Wittman, H. G., 99, *112*
Wolberg, G., 175, *213*
Wolf, S. F., 122, *213*
Wolfenberger, D. A., 187, *214*
Woo, P. W. K., 260, 273, *278*
Wood, W. A., 35, *70*
Wu, A. M., 128, *214*
Wu, J., 98, *112*
Wulff, V. J., 100, *112*
Wyke, A. W., 15, *70*

Yamaguchi, I., 89, *112*
Yonehara, H., 77, *112*
York, J. L., 218, 231, *257*
Yoshikawa, M., 53, *70*
Yoshioka, H., 150, *215*

Zadražil, S., 139, 141, 143, *214*
Zähner, H., 43
Zaenen, I., 50, *70*
Zahringer, J., 127, *214*
Zain, B. S., 137, *213*
Zamir, A., 100, *112*
Žemlička, J., 174, 179, *214*
Zurenko, G. E., 247, *257*

Subject Index

Page numbers in *italics* indicate illustrations. Page numbers followed by the letter "t" indicate tabular information.

A9145, A, C, E, and H, *see* Sinefungin
Ablastimycin, nikkomycin and, 43
N-Acetoxy-*N*-acetyl-2-aminofluorene, 60, *60*
N-Acetylglucosamine, nikkomycin and, 45
N-Acetylglucosaminyl-lipids, tunicamycin and, 5
N-Acetylglucosaminyl pyrophosphate, tunicamycin and, 8
Actinomycin D, 127
Adenosine deaminase, anhydroformycins and, 174
  ara-A and, 225-226
  ara-A 5'-valerate inhibition of, 221, 225-226, 263, *264*
  coformycin inhibition of, 225-226, 259-275
  cordycepin and, 266
  deoxycoformycin inhibition of, 175-176, 225-226, 259-275
  DHMPR inhibition of, 225-226, 267-268
  EHNA inhibition of, 225-226, 267-268
  formycin and, 175
  immunodeficiency disease and, 262
  inactivation and reactivation, with deoxycoformycin, 264
  N^6-methyladenosine and, 225
  N-methylformycins and, 173
  multiple forms of, 270-271
  tubercidin and, 162, 163
Adenosine deaminase deficiency, 174-177
  coformycin and ATP increase in, 174, 175
  formycin 5'-triphosphate accumulation in, 175, *176*
  terminal deoxynucleotidyl transferase and, 176-177
Adenosine deaminase inhibitors, *see* Adenosine deaminase
S-Adenosylhomocysteine hydrolase, ara-A and, 227
  cordycepin and, 227
S-Adenosylmethionine (SAM), role in cellular reactions, *164*
  sinefungin, A9145A, A9145C and, 19, 22
  transmethylation, 162-165
S-Adenosylmethionine decarboxylase, 141
Adenylate cyclase, thuringiensin and, 184, 193
ADP-ribosylation, cordycepin, formycin and tubercidin and, 129-131, *131*, 165, 179-180
Agrocin 84, 47-51
  Agrobacterium, inhibited by, 47, 49
  biochemical properties of, 49-51
  chemical properties of, 48
  discovery, production, isolation, 47-48
  DNA synthesis inhibited by, 49-50, *50*
  growth inhibition by, 49
  physical properties of, 48
  plasmid Ti, requirement of, 50-51
  protein synthesis inhibited by, 49-50, *50*
  RNA synthesis inhibited by, 49-50, *50*
  structure of, 48-49, *48*
  tumor induction inhibited by, 49
Amicetin (amicetin A), 86-89
  structure of, *74*
Amicetin A, *see* Amicetin

Amicetin B, *see* Norplicacetin
Amicetin C, *see* Bamicetin
α-Aminitin, RNA polymerases effected by, 194-294, *194*
Aminoacylaminohexosylcytosine analogs, 73-89
  biochemical properties of, 86-89
  structures of, *74*, *77*, *79*, *81*, *82*
Aminoacyl-tRNA, analogs of, 86, *97*, 103
3′-Aminoadenosine, 146
  biochemical properties, 146
  biosynthesis of, 146
  ezomycins similar to, 23
  incorporation into 3′-terminus of tRNA, 146
  isolation of, 146
  structure of, *146*
3′-Aminoaristeromycin, antiviral activity of, 149
2′-Aminoguanosine (2′-amino-2′-deoxyguanosine), 92-96, 145
  antibacterial activity of, guanosine reversal of, 94
  biochemical properties of, 94-96
  chemical properties of, 93-94
  chemical synthesis of, 94, *95*
  discovery, production, isolation, 93
  DNA synthesis inhibited by, 95, *96*
  E. coli inhibited by, 94
  growth inhibition of, 94
  molecular formula of, 93
  phosphorylation of, 95, 145
  physical properties of, 93-94
  protein synthesis inhibited by, 95, *96*
  RNA synthesis inhibited by, 95, *96*
  spectra, ^{13}C nmr, infrared, nmr, 93-94
  structure of, *93*
4-Aminohexose pyrimidine nucleosides, *see* Aminoacylaminohexosylcytosine analogs
Amipurimycin, 40-43
  biochemical properties of, 42-43
  chemical properties of, 41
  discovery, production, isolation, 41
  fungi inhibited by, 42
  growth inhibition of, 42
  miharamycins A, B similar to, 42
  molecular formula of, 41
  physical properties of, 41
  rice blast inhibited by, 42
  spectra, ^{13}Cnmr, nmr ultraviolet, 41, *42*
  structure of, 41-42, *41*
  toxicity of, 42-43
5′-AMP deaminase, coformycin inhibition of, 271, 272t
  deoxycoformycin inhibition of, 271-272t
AMP kinase, analogs and, 129
AMP nucleosidase, formycin 5′-phosphate inhibition of, 174
Angustmycin A, *see* Decoyinine
Angustmycin C, *see* Psicofuranine
Anhydroformycins, 170, 172, 174, 178, *179*
Anthelmintic agents, hikizimycin, 78-79
  tubercidin, 165
Anthelmycin, *see* Hikizimycin
Antibiotic A201A, 89-93
  biochemical properties of, 91-93
  chemical and physical properties of, 90
  dipeptide synthesis inhibited by, 92
  discovery, production, isolation, 89-90
  DNA synthesis, no inhibition by, 91, *91*
  fungi inhibited by, 91
  growth inhibition of, 91
  molecular formula of, 90
  protein synthesis inhibited by, 91, *91*
  RNA synthesis, no inhibition, 91, *91*
  spectra, infrared, nmr, 90
  structure of, *90*
  toxicity of, 91
Antibiotic A201B, antibacterial activity of, 93
  antifungal activity of, 93
  chemical and physical properties of, 93
Antibiotic 24010, nikkomycin similar to, 43
Ara-A, (β-ara-A), 217-233
  adenosine deaminase and, 225-226
  adenosine deaminase inhibitors and, 227, 266
  antitumor effect and deoxycoformycin, 273-274
  biochemical properties of, 217-219, *220*, 221-233
  biosynthesis of, 221
  cancer chemotherapy with, 231-232
  carboxyclic analog of, 226-227, *226*
    antiviral activity of, 226-227
  cell cycle and, 224
  chain terminator, DNA, 222
  chemical synthesis of, 217, 219
  chromosome damage by, 218, 224
  coformycin, increased potency, 226
  cytostatic analog of 2′-deoxyadenosine, 221
  deamination of, 218, 221, 225-227, 232

deoxycoformycin inhibition of, 259, 266, 267t
EHNA inhibition of, 259
deoxyadenosine, antagonism of, 271-272
reversal of, 271-272
derivatives of, 219, 221
DNA, viral inhibition by, 222-224, *223*, 227
DNA synthesis inhibited by, 218, 227
DNA viruses in culture, inhibition by, 227-228, *228*, *229*
with coformycin, 227-228, *228*
with EHNA, 227-228, *229*
E. coli inhibited by, 217
reversal of, 217
effects of, *220*, 221
EHNA, increase of antitumor effect, 266, 268
viral inhibition of, 227-228, *229*
herpes simplex encephalitis in humans, inhibited by, 218, 219
herpes simplex keratitis in humans inhibited by, 218, 230-231
inhibition as nucleoside, 227
isolation of, 217
LD_{50} of, 232
leukemia, human, inhibited by, 219, 221
mammalian cells and, 217
metabolism in humans, 232
pharmacology of, 232
phosphorylation of, 218, 227
protein synthesis effected by, 227
RNA viruses, effected by, 224, 229-230
structure of, *217*
synthesis of 5'deutero and tritium labeled, 219
teratogenic effects of, 232-233
toxicity of, 230, 232, 263-264, *265*
tumor-bearing animals effected by, 218
tRNA and, 225
viral infections in animals, inhibited by, 231
viral infections in humans, inhibited by, 230-231
viruses inhibited by, 218, 224, 227-228, *228*, *229*
with coformycin, 227-228, *228*
with EHNA, 227-228, *229*
α-Ara-A, cytostatic activity of, 223
DNA synthesis inhibited by, 223-224
phosphorylation of, in L5178Y cells, 224
protein synthesis not inhibited by, 223
RNA synthesis not inhibited by, 223
Ara-AMP, antiviral effects of, 226, 231
cell penetration by, 226
chemical synthesis of, 219
conversion of ara-ATP, 226
3',5'-cyclic ara-AMP, 219
herpes simplex encephalitis in humans, use in, 219
incorporation into viral DNA, 222, *223*
L-cells and, 226
LD_{50} of, 232
in mammalian cells, 219
metabolism of, 232
synthesis of 5'-deutero and tritium labeled, 219
toxicity in cornea, 231
toxicology of, 232-233
Ara-ATP (β-ara-ATP), ascites cells effected by, 218
dATP inhibition by, 218, 221
DNA polymerases α, β, and γ effected by, 218, 221-224, *222*
intracellular pool of, with coformycin, 226
polyadenylation inhibited by, 224
poly(ADP-ribose) polymerase effected by, 224
poly(A) polymerase effected by, 224
RNA, incorporation of, 224
RNA polymerases, effected by, 224
terminal deoxynucleotidyl transferase effected by, 225
tRNA, incorporation into 3'-terminus, 224
ribonucleotide reductase inhibited by, 218
α-Ara-ATP, DNA, incorporation of, 223-224
DNA polymerase, viral, no inhibition by, 223-224
DNA polymerase-α, mammalian, inhibited by, 223-224
DNA polymerase-β, mammalian, no inhibition of, 223-224
terminal deoxynucleotidyl transferase, effected by, 225
Ara-A-5'-valerate, 220, 225
adenosine deaminase and, 225
antiviral activity of, 220
structure of, *264*
synthesis of, 220
9-β-D-Arabinofuranosyladenine, *see* Ara-A
9-β-D-Arabinofuranosyladenine 5'-valerate, *see* Ara-A-5'-valerate
1-β-D-Arabinofuranosylthymine, *see* Ara-T
Arabinosylnucleosides, 216-236
Ara-C, antitumor effect, increased with

deoxycoformycin, 224
combination therapy with 5-azacytidine, 144-145
pyrazofurin, potentiated by, 286
Ara-Hx, deoxyadenosine antagonism of, 271-272
excretion of, 232
formation of, 218, 232
Ara-HxMP, antiviral effects of, 226
Ara-T (Spongothymidine,) 233-236
antibacterial activity of, 233
antiviral activity of, 233-236, *233*, *234*, *236*
biochemical properties of, 233-236
DNA synthesis, viral, inhibited by, 233-234
excretion of, 235
herpes simplex virus inhibited by, 233-236, *233*, *234*, *236*
isolation of, 233
metabolism of, 235-236, *236*
phosphorylation of, 234, 235-236
structure of, *217*, 233
toxicity of, 233, 235
tumor cells not effected by, 235
ARA-TMP, antiviral activity of, 235
Ara-TTP-, dTMP competitively inhibited by, 236
DNA polymerase-α, inhibited by, 236
Ara-Tubercidin, synthesis of, 221
Aristeromycin, 147-149
AMP synthesis inhibited by, 147
biochemical properties of, 147-149
3′, 6′-cyclic phosphate of, 149
cytotoxicity of, 149
derivatives of, 149
DNA synthesis induced by, 147
H. Ep. #2 cells effected by, 147
mammalian cells in culture effected by, 147
phosphorylation of, 147
plants inhibited by, 148
rice disease inhibited by, 148
structure of, *147*
toxicity of, reversal by adenosine, 147, *148*
transmethylases inhibited by, 147-148
S-Aristeromycinyl-L-homocysteine (SAmH), analogs of, 147-148
"capping" reaction of mRNA inhibited by, 147-148
transmethylases inhibited by, 147-148
Arthritis, bredinin activity against, 151
Arthrobacter oxamicetus, oxamicetin from, 81
Aspergillus nidulans, 3′-aminoadenosine from, 146
cordycepin from, 118
Aspiculamycin, *see* Gougerotin
Asteromycin, *see* Gougerotin
5-Azacytidine, 135-145
S-adenosylmethionine decarboxylase effected by, 141
antileukemic effects of, 135, 139, *140*, 143-144
antitumor activity of, 137, 144-145
ascites cells inhibited by, 139
biochemical properties of, 136-145
cell cycle, effected by, 141
chemical properties of, 135
combination therapy with, 143-146
cytotoxic analog of cytidine, 135
deamination of, 135, 136, 137
2′-deoxy derivative of, *135*, 141-142
dihydroanalog of, 142-143
DNA, incorporation of, 136, 143
DNA synthesis inhibited by, 135, 137-138, 139, *140*
Ehrlich ascites cells effected by, 136, 139, 142
HeLa cells effected by, 137, 142
L-1210 cells inhibited by, 136, 137, 139, *140*
leukemia, human, inhibited by, 136, 143-144
metabolism of, 136
mRNA, incorporation of, 136
Novikoff hepatoma effected by, 137
orotidine 5′-phosphate decarboxylase inhibited by, 136
phosphorylation of, 136
physical properties of, 135
polyamine biosynthesis effected by, 141
polysome degradation by, 136
protein synthesis inhibited by, 135, 136, 137
reduction to deoxynucleotide, 136
RNA incorporation into, 136, 143
RNA synthesis inhibited by, 135, 137, 139, *140*
rRNA maturation, inhibited by, 137, *138*
structure of, *135*
synthesis of, 135
thymidine and thymidylate kinase effected by, 139
toxicity of, 136
tryptophan oxygenase effected by, 141
tyrosine aminotransferase, sensitivity to, 141
uridine kinase effected by, 145
5-Aza-2′-deoxycytidine, dCMP incorporation

deamination of, 141-142
DNA, incorporation of, 141-142
inhibited by, 141-142
synthesis of, 135
5-Azacytidine 5′-triphosphate into RNA, 136
5-Aza-2′-deoxycytidine 5′-triphosphate into DNA, 136
8-Azaguanine, biosynthesis of, 160
5-Azauracil, 136, 142
p-Azidopuromycin, structure of, *97*

Bacillus thuringiensis, RNA polymerase from, 191-192
thuringiensin produced by, 183, 184, 185, *185*, 188
Bamicetin (amicetin C), 73, *74*
peptide bond formation inhibited by, 73, 86-87, *86*, *87*
structure of, *74*
6-Benzylamino-7-β-D-glucopyranosylpurine, 313, *315*
Biosynthesis of, 3′-aminoadenosine, 146
ara-A, 221
8-azaguanine, 160
blasticidins S and H, 75-77, *76*
C-nucleosides, *56*
cordycepin, 118
decoyinine, 280
5,6-dihydro-5-azathymidine, 247-248, *248*
eritadenine, 301
formycin, *156*, 172
octosyl acids, 32-35, *33*, *34*, *35*
oxazinomycin (minimycin), 238-240, *241*
polyoximic acid, 34, *34*
polyoxins, 32-35, *33*, *34*, *35*
psicofuranine, 280
puromycin, 96
pyrazofurin, 56, 283, *286*, *287*
pyrrolopyrimidine nucleoside analogs, 159-161, *160*
showdomycin, 55-56, *56*
sinefungin, 21-22, *22*
thuringiensin, 188
Blasticidin H, 73-77
biosynthesis of, 75-77, *76*
chemical properties of, 74
discovery, production, isolation, 73-74
growth inhibited by, 75
molecular formula of, 74
nmr spectrum of, 74-75, *75*
pentopyranamine D and, 154
structure of, 74-74, *74*
Blasticidin S, biosynthesis of, 75-77, *76*
nmr spectrum of, 74-75, *75*
peptide bond formation inhibited, 86-87, *86*
structure of, 74-75, *74*
Bredinin, 149-154
adenine, reversal of, 152
aglycon, cytotoxicity of, 152
antiarthritic activity of, 152
antileukemic activity of, 153
antiviral activity of, 150-151
biochemical properties of, 151-154
chemical properties of, 150
chemical synthesis of, 151, *151*
chromosomal breakage induced by, 152
cytostatic activity of, 153
discovery, production, isolation, 150
DNA synthesis inhibited by, 153
enzymatic synthesis of, 151
GMP synthesis inhibited by, 152
growth inhibited by, 150-151
immunosuppressive activity, 150, 152
L5178Y cells, cytotoxicity of, 150, 152, 153-154, *153*
molecular formula of, 150
nmr spectra of, 150
phosphorylation of, 154
physical properties of, 150
protein synthesis not inhibited, 152
RNA synthesis inhibited by, 152, 153
structure of, *149*

Cancer chemotherapy, ara-A in, 231-232
"Capping" reaction, *S*-aristeromycinyl-L-homocysteine inhibition of, 147-148
S-tubercidinylhomocysteine inhibition of, 162-163
9-(3-Carboxypropyl) adenine, isolation of, 299, 300-301
structure of, *300*
synthesis, 301
Cell wall synthesis, fungi, tunicamycin inhibition of, 13
Chain terminator, ara-A as, 222
cordycepin as, 118, 120, 124, 129, 135
"Chemical proofreading," 131-133, *132*, 178
Chitin synthesis, nikkomycin and polyoxins, 31, 35-36, *37*, 44-45
Chondroitin sulfate, tunicamycin and, 16-18, *17*, 18t

Cholesterol, eritadenine and, 298, 302-308, 305t
Chromosome breaks, due to ara-A, 218, 224
  5-aza-2′-deoxycytidine, 139, 141
  bredinin, 152
Clindamycin, formation from lincomycin, 104
Clindamycin ribonucleotides, 104-107
  biochemical properties of, 107
  chemical properties of, 105, 106t
  discovery, production, isolation, 104-105
  growth inhibited by, 107
  nmr spectrum of, 105
  physical properties of, 105, 106
  structure of, *105*
Clitidine, 292-294
  biochemical properties of, 293-294
  chemical properties and synthesis of, 292-293
  discovery, production, isolation, 292
  hyperemic effect of, 292
  hyperesthetic effect of, 292
  molecular formula of, 292
  physical properties of, 292-293
  spectra, infrared, mass, nmr, 293, *293*
  structure of, 293
Clitocybe acromelalga, clitidine from, 292
C-nucleosides, biosynthesis of, 55-56, *56*
  ezomycins B_1, B_2, C_1, C_2, D_1, D_2, 23
Coformycin (CF), 259-260, 262-274
  adenosine deaminase inhibited by, 262, 267-268
  5′-AMP deaminase inhibited by, 271, 272t
  ara-A, increased potency, 226
    virus inhibited by, 227-228, *228*
  ara-A toxicity increased by, 266, 267
  biochemical properties of, 262-274
  biosynthesis of, 260, *261*
  chemical synthesis of, 260
  DNA synthesis inhibited by, 268
  erythrocytes, nucleotide pool of, 174-177
  excretion of, 272
  formycin activity increased by, 263-264
  immunosuppressive activity of, 274
  isolation of, 259
  pharmacology and toxicity of, 269, 272-273
  phosphorylation of, 226, 263
  structure of, *259*
  tumor grafts and, 274
Cordycepin (3′-deoxyadenosine), 118-135
  adenosine deaminase inhibitors with, 266
  ADP-ribosylation and, 129-131, *131*
  antitumor activity of, 125, 266, 269-270, *269*
  biochemical properties of, 118-135
  biosynthesis of, 118
  chain terminator, RNA, and, 118, 120, 129, 135
  coformycin and, 266
  deamination of, 125, 133, 181
  deoxycoformycin and DNA synthesis, 269-270, *269*
  DNA polymerases and, 119-120
  EHNA plus, RNA synthesis and, 125
    virus inhibited by, 270-271, *270*
  globin mRNA synthesis inhibited, 124
  HeLa cells affected by, 119, 120-121, *121*, 123, 124
  HnRNA synthesis not inhibited by, 120
  inhibition as nucleoside, 227
  L1210 and L5178Y cells affected by, 122, 124, 126-127, *126*
  2′-*O*-methylation of cap and, 121-122
  mRNA formation inhibited by, 121, *122*
  mRNA transport inhibited by, 120
  phosphorylation of, 118, 129
  plant tissue effected by, 127-128, 129, *130*
  poly (A) inhibited, 120-121, 120t, *121*
  protein kinase inhibited by, 129
  protein synthesis affected by, 119, 123, 129, *130*
  P388 ascites inhibited by, 125
  RNA polymerase inhibited by, 122-123, 128
  RNA synthesis, inhibited by, 118, 119-121, 122-123, 125-126, 127-128, 133-135, *134*
  rRNA synthesis inhibited by, 126-127, *126*
  Sarcoma 180 inhibited by, 124
  structure of, *119*
  3′-terminus of RNA, incorporation into, 122-123, 132-133, *132*
  toxicity of, 118
  viruses affected by, 128-129
  virus-specific RNA synthesis inhibited by, 128
Cordycepin 5′-monophosphate, purine biosynthesis inhibited by, 118
Cordycepin 5′-triphosphate (3′-dATP), DNA synthesis, in toluene-treated *E. coli*, inhibited by, 133, *134*
  poly (A) polymerases inhibited by, 122-123
  poly (A) synthesis, free and chromatin-associated with, 123
  poly (A) synthesis, mitochondrial, 124
  RNA, incorporation of, 118, 133, 134, 135

RNA synthesis inhibited by, 123
RNA synthesis, in toluene-treated *E. coli*, inhibited by, 133, *134*
terminal deoxynucleotidyl transferase inhibited by, 123-124
Cordyceps militaris, 3′-aminoadenosine from, 146
cordycepin from, 118
homocitrullyaminoadenosine from, 103
lysylaminoadenosine from, 103
Covidarabine, *see* 2′-Deoxycoformycin
Crotonoside (isoguanosine), 60-62
biochemical properties of, 61-62
structure of, 60
Croton tiglium L, crotonoside from, 60
Crown gall, Ti plasmid in, 49-51, *51*
C-substance, pentopyranic acid and, 157
Cyclic AMP, bredinin and, 153, *153*
crotonoside and, 61
octosyl acids and, 297
3′,5′-Cyclic ara-AMP, 219
Cytokinin activity, 180
Cytosine arabinoside, *see* Ara-C

Deamination of, ara-A, 218, 221, 225, 226
5-azacytidine, 135, 136, 137
cordycepin, 125, 181
formycin, 172, 174, 181
thuringiensin, 188
7-Deazanebularine, phosphorylation of, 244
Decoyinine, 279-291
biochemical properties of, 280-281
biosynthesis of, 280
derivatives of, 280
sporulation induction by, 280-281, *282*
structure ot, *280*
Dehydrogenases, NAD analogs, effect with, 129-131, *131*, 165, 179-180
2′-Deoxyadenosine, ara-A and ara-Hx, 271-272
toxic with deoxycoformycin, 268
3′-Deoxyadenosine, *see* Cordycepin
2′-Deoxycoformycin (dCF, 2′-dCF, CoV, covidarabine, pentostatin, Co-ara-A), 259, 260-278
adenosine analogs, effect increased, 266-267, 273-274
antiviral effect increased with, 262, 266-267, 267t, 273-274
adenosine deaminase, inactivation and reactivation, 176, 262, 264, 267-269
5′-AMP deaminase inhibited by, 271, 272t
ara-ATP increased by, 262
ara-C, effect increased, 273-274
biochemical properties of, 262-278
chemical properties of, 260, 262
chemical synthesis of, 262
2′-deoxyadenosine toxicity increased, 268
discovery, production, isolation, 260
excretion of, 273-274
formycin, antiviral effect increased, 175, 266
immunosuppressive activity of, 274
mammalian cells in culture, effect by, 263-267
molecular formula of, 260
nucleoside transport and, 271
pharmacology of, 272-273
physical properties of, 260, 262
spectra, infrared, ^{13}Cnmr, nmr, 262
structure of, 259
tumor-bearing animals effected by, 266-267
tumor grafts and, 274
viruses effected by, 227-228, *228*, 229-231, 266-267
Deoxyeritadenine, chemical synthesis of, 301
isolation of, 299
nmr spectrum of, 301
structure of, *299*
2′-Deoxy and 3′-deoxy NAD$^+$, in dehydrogenase binding, 129-131
DNA synthesis inhibited by, 129-131, *131*
in poly (ADP-ribosylation), 129-131, *131*
synthesis of, 129-131
Deoxyribonucleic acid, *see* DNA
3′-Deoxyuridine, synthesis of, 135
3′-Deoxyuridine 5′-triphosphate, 135
DHMPR, and adenosine deaminase, 225, 262, 263t, 268
structure of, *264*
5,6-Dichloro-β-D-ribofuranosylbenzimidazole (DRB), 133
2′,3′-Dideoxyadenosine, protein synthesis, 129
5,6-Dihydro-5-azacytidine, properties of, 142-143
synthesis of, 142-143
5,6-Dihydro-5-azathymidine, 245-250
biochemical properties of, 249-250
biosynthesis of, 247-248, *248*
chemical properties of, 247
discovery, production, isolation, 246
growth inhibition of, 247
immunosuppressive activity of, 249

physical properties of, 247
resistance to, 249-250
spectra, infrared, mass, nmr, 247
structure of, *246*
toxicity of, 247, 249
viruses, DNA, inhibited by, 249
1,6-Dihydro-6-hydroxymethylpurine ribonucleoside, *see* DHMPR
Diphtheria toxin, 165
Diumycin, 5
DNA, *S*-adenosylhomocysteine stimulates, 163
agrocin 84 inhibition of, 49-50, *50*
2′-aminoguanosine effect on, 95-96, *96*
antibiotic A201A effect on, 91, *91*
ara-A inhibition of, 218, 227
α-ara-A inhibition of, 223-224
aristeromycin induction of, 147
5-azacytidine inhibition of, 136, 139, *140*, 141
bredinin effect on, 150, 152
cordycepin effect on, 119, 122-123, 133, *134*
cordycepin plus deoxycoformycin, 269-270, *269*
3′-deoxyuridine 5′-triphosphate, 135
EHNA effect on, 176
formycin inhibition of, 172
incorporation of, α-ara-ATP, 223-224
5-azacytidine, 136, 143
5-aza-2′-deoxycytidine, 142
sangivamycin, 169
tubercidin, 162
N^6-methyladenosine inhibition of, 225
NAD and analogs, inhibition of, 129-131, *131*
nebularine inhibition of, 245, *246*
pyrazofurin inhibition of, 285, *288*
thuringiensin, no inhibition of, 188-189
toyocamycin inhibition of, 167
tubercidin effect on, 131, 161, *161*, 165
S-tubercidinyl homocysteine, 164-165
viral, EHNA inhibits, 227, 228
viral infected cells, ara-T inhibits, 233
virazole inhibition of, 286-287
DNA polymerase, ara-ATP and, 218, 222-224, *223*
ara-TTP inhibition of, 236
cordycepin effect on, 123
DNA viruses, ara-A inhibition of, 222-224, *223*, 227-231, *228*, *229*
with coformycin, 227-228, *228*
5,6-dihydro-5-azathymidine and, 247
with EHNA, 227-228, *229*
streptovirudin inhibits, 12
DRB, *see* 5,6-Dichloro-β-D-ribofuranosylbenzimidazole

Ecdysone, molting hormone, 192
mRNA and rRNA synthesis stimulated by, 192
EHNA [*Erythro*-9-(2-hydroxyl-3-nonyl) adenine], 225
adenosine deaminase inhibits, 267-268
antiviral activity of, 227-228, 270, *270*
ara-A, antitumor increased, 266, 268
ara-A and viruses inhibited, 229, *229*
ara-A deamination inhibited, 262
cordycepin plus, RNA synthesis and viruses, 125, 270
DNA synthesis, effected by, 175-176
viral, inhibited by, 227
nucleoside transport and, 271
protein synthesis inhibited by, 268
structure of, *264*
Ehrlich ascites cell effected by, 5-azacytidine, 136, 139, 142
EHNA plus ara-A, 266
formycin B, 181
nebularine, 244
oxazinomycin, 240
pentopyranines, 158, 158t
toyocamycin, 166
Eritadenine (lentysine, lentinacin), 298-319
analogs of, 301, 304, 306
biochemical properties of, 302-308
biosynthesis of, 301
chemical properties and synthesis of, 300, 301
cholesterol and, 302-303, 303t, 304-307
discovery, production, isolation, 299
esters, effects of, 306-307, *307*
fatty livers and, 308
hypocholesterolemia, 303, 303t, 304-307, 305t
hypolipidemic effect, 298, 304, 306
molecular formula of, 300
physical properties of, 300
spectra, infrared, mass, nmr, 300
structure of, *299*
toxicity of, 303, 303t
VLDL reduced by, 304, 305t
Erythrocytes, adenosine deaminase deficiency and, 175, *176*
ara-A and, 263-264, *265*

coformycin and, 262-264, *265*
formycin and, 262-264, *265*
nucleotide pool of, 262
Erythro-9-(2-hydroxy-3-nonyl)adenosine, *see* EHNA
Escherichia coli, 84, 86, 87, 94, 137, 141-142
ara-A and ara-Hx inhibits, 217
cordycepin deamination in, 125
crotonoside and, 60, 61
5,6-dihydro-5-azathymidine and, 247, 249-250
RNA and DNA polymerases, ara-ATP effect on, 224
showdomycin and, 52, 57
thuringiensin inhibits, 193-194, *193*
3-Ethylidene-L-azetidine-2-carboxylic acid, *see* Polyoximic acid
β-Exotoxin, *see* Thuringiensin
Ezomycins, 23-30
3′-amino sugar similar to, 23
biochemical properties of, 29-30
chemical properties of, 24, 25t
C-nucleosides B_1, B_2, C_1, C_2, D_1, D_2, 23, *24*
discovery, production, isolation, 23-24
growth inhibition of, 29
molecular formula of, 24
N-nucleosides, A_1, A_2, 23, *24*
octosyl acids similar to, 23
physical properties of, 24, 25t
polyoxins similar to, 23
pyrimidine nucleosides similar to, 23
spectra, CD, infrared, mass, ^{13}Cnmr, nmr, 24, 27, *29*

Formycin (Formycin A), 169-181
adenosine analog, 169, 177
AMP nucleosidase inhibited by, 174
anhydroanalogs of, 170, 172, 174, 178-179, *179*
biochemical properties of, 172-181
biosynthesis of, 172
cell division inhibited by, 172
chemical synthesis of, 170, *171*
^{13}Cnmr of, 170
coformycin and, 263-264, *265*
cytotoxicity of, 169, 173
deamination of, 172, 174, 180
DNA synthesis inhibited by, 172
erythrocytes and, 174
fluorescence of, 172
NAD analog of, 179-180
N-methyl analogs, 173
DNA synthesis inhibited by, 181-182
phosphorylation of, 172
platelet aggregation, 180-181
polymers of, 172, 174
protein synthesis inhibited, 172
purine nucleoside phosphorylase and, 177
purine synthesis inhibited, 172
RNA synthesis not inhibited, 172
structure of, *169*
tautomers of, 170, 172
tRNA, editing by, 177-179
xanthine oxidase inhibited by, 183
Formycin B (laurusin), biochemical properties of, 181-182
conversion of formycin, 172
cytotoxicity of, 173
Ehrlich ascites and L5178Y cells affected, 181
N^6-oxide of, 182
oxoformycin B from, 182
poly (ADP-ribose) polymerase inhibited by, 181-182
proinsulin synthesis, 182
purine nucleoside phosphorylase and, 182
pyrazofurin, 4-amino derivative, 182
structure of, *169*
xanthine oxidase inhibited by, 183
Formycin cyclic 3′,5′-monophosphate, 180
Formycin 3′,5′-diphosphate, 181
Formycin 5′-phosphate, AMP nucleosidase and, 174
Formycin 5′-triphosphate, adenosine deaminase deficiency and, 174-177, *176*
RNA incorporation into, 172-173
RNA polymerase and, 172
tRNA incorporation into, 173
Fungi inhibited by, amipurimycin, 40, 42
antibiotic A201A and A201B, 91
ezomycins, 23, 29
mildomycin, 80
polyoxins, 30, 31, 35
septacidin, 46
sinefungin, A9145A, A9145C, 19, 20-21
thraustomycin, 38, 40, *40*
tunicamycin, 5, 7, 12-13

Globin mRNA, cordycepin and, 124
5′-*O*-Glucosyl-*ribo*-nucleosides, 312-313
chemical synthesis of, 313
structures of, *313*

Glycolipids, tunicamycin inhibits, 5
Glycoproteins, tunicamycin and streptovirudins inhibit, 5, 7, *7*, 8, 11
Glycosaminoglycan, tunicamycin inhibits, 17, *17*
Glycosylation of interferon and serum proteins, tunicamycin inhibits, 10, 12
Gougerotin (aspiculamycin, asteromycin, moroyamycin), analog of, 87
  aspiculamycin, same as, 84, *85*
  binding to ribosomes, 88-89, *88*
  peptide bond formation inhibited by, 86, *86*
  peptide chain termination, 89
  structure of, 74
  synthesis of, 84, *85*
Guanine, nucleoside Q precursor of, 159
Guanosine analogs, 2′-aminoguanosine, 93
  isoguanosine, 61

Hadacidin, effects of, 281
HeLa cells, 2′-aminoguanosine inhibit, 94
  ara-ATP effect on, 222, *222*
  5-azacytidine effect on, 137, 142
  cordycepin effect on, 119, 123, 124
  EHNA effect on, 228
  pyrazofurin effect on, 281
Helminthosporium, 3′-aminoadenosine from, 146
H. Ep.#.2 cells, aristeromycin and, 147
Heparan sulfate, tunicamycin effect on, 17, *17*, 18t
Herbicidins A and B, 311-312
  biochemical properties of, 312
  chemical properties of, 312, 312t
  discovery, production, isolation, 311
  herbicidal activity of, 312
  physical properties of, 312, 312t
  toxicity of, 311
Herpes simplex encephalitis and keratitis in humans, ara-A and ara-AMP effect on, 219, 230-231
Herpes simplex virus, inhibition by, ara-T, 233-234, 235-236, *236*
  5,6-dihydro-5-azathymidine, 249
  EHNA, 270-271, *270*
  EHNA plus cordycepin, 270-271, *270*
  1-methylpseudouridine, 243
  pyrazofurin, 286
Hikizimycin (anthelmycin), 77-79
  anthelmintic activity of, 78-79
  chemical properties of, 77-78
  discovery, production, isolation, 77
  growth inhibited by, 78-79
  molecular formula of, 77
  physical properties of, 77-78
  spectra, ^{13}Cnmr, infrared, mass, nmr, 78
  structure of, *77*
HnRNA, *see* RNA, heterogeneous nuclear
Homocitrullylaminoadenosine, 103-104
Hyaluronic acid, tunicamycin effect on, 17-18, *17*, 18t
N-Hydroxy-*N*-acetyl-2-aminofluorine, 59-60, *60*
3-Hydroxy-3-methylglutaryl CoA reductase, inhibition by compactin and ML236A, B, 304
Hypocholesterolemic agents, eritadenine, compactin and ML236A, B, 302-304

Immunoglobulins, tumicamycin and, 16
Immunosuppressive agents, bredinin, 150, 151-152
  coformycin, 274
  deoxycoformycin, 274
  5,6-dihydro-5-azathymidine, 249
Insects, thuringiensin and, 184, 186-187, 195
Interferon, induction, by poly (formycin B), 182
  poly (tubercidin) inhibition of, 165
  toyocamycin inhibition of, 167
  tunicamycin and, 12
Isoguanosine, *see* Crotonoside

KB cells, tubercidin inhibition of, 161, *161*
Keratan sulfate, tunicamycin and, 17, *17*, 18t

L1210 cells, ara-A plus coformycin, 273-274
  ara-A plus deoxycoformycin, 267
  ara-C plus deoxycoformycin, 273
  5-azacytidine, 139, *140*
  cordycepin, 124, 126
  cordycepin plus deoxycoformycin, 269-270, *269*
  methylformycins, effect on, 173
L5178Y cells and α-ara-A, 224
  bredinin, 150, 152, 153, *153*
  coformycin, 263, 268
  cordycepin, 122
  formycin B, 181
  pyrazofurin, 285
L-cells, ara-AMP and, 226

2′-deoxyadenosine and deoxycoformycin in, 268
pyrazofurin inhibition of, 281
toyocamycin effect on, 166
Lentinus edodes, 9-(3-carboxypropyl)adenine from, 299
deoxyeritadenine from, 299
eritadenine from, 298, 299, 300
Lentinacin, *see* Eritadenine
Lentysine, *see* Eritadenine
Leukemia, 5-azacytidine activity against, 137, 139
bredinin activity against, 153
human, ara-A inhibition of, 232-233
5-azacytidine inhibition of, 137, 145, 319
pyrazofurin effect on, 221
resistant to 5-azacytidine, 145
sangivamycin effect on, 169
Linocomycin, 133
Lipid-pyrophosphoryl-*N*-acetylmuramyl-pentapeptide synthesis, tunicamycin inhibits, 13-14, *14*
Lmyphoma, α-ara-A effect on, 222
3′ dATP effect on, 123
Lysylaminoadenosine, 103-104

Macarbomycin, similar to tunicamycin, 5
Maleimycin, structure of, *52*
Mammalian cells, ara-AMP and, 219
3′,5′-C-ara-AMP and, 219
bredinin inhibition of, 152
deoxycoformycin effect on, 266-267
tunicamycin inhibition of, 5
Mass spectrum of, clitidine, 293
5,6-dihydro-5-azathymidine, 247
eritadenine, 300
ezomycins, 24
hikizimycin, 78
1-methylpseudouridine, 242
nikkomycin, 43
norplicacetin, 81
octosyl acids, B, C, 296
oxamicetin, 82
pyrazofurin, 283
thraustomycin, 38-39, *39*
Mediocidin, similar to oxazinomycin, 237
Melanoma, 5-azacytidine effect on, 145
N6-Methyladenosine, adenosine deaminase and, 225
Methylation, RNA, cap formation, 162-165
cordycepin and, 121-122
nuclear, 121-122, *122*
toyocamycin and, 166, 167
tubercidin and, 162-165
1-Methylpseudouridine, 241-245
biochemical properties of, 243-245
discovery, production, isolation, 242
growth inhibition by, 243
molecular formula of, 242
physical properties of, 242
spectra, ^{13}Cnmr, mass, and nmr, 242
structure of, *241*
5′-Methylthiotubercidin, 163
Methyltransferases, cordycepin inhibition of, 121-122
sinefungin inhibition of, 22-23, *22*
Miharamycin A, B, similar to nikkomycin, 42
Mildiomycin, 79-80
discovery, production, isolation, 79
fungi inhibited by, 80
growth inhibited by, 80
molecular formula of, 80
physical properties of, 80
spectra, CD, ^{13}Cnmr, and nmr, 80
structure of, *79*
Minimycin, *see* Oxazinomycin
Mitotic spindle formation, thuringiensin inhibition of, 192-193
Moenomycin, similar to tunicamycin, 5
Moroyamycin, *see* Gougerotin
mRNA, *see* RNA, messenger
Mucopolysaccharides, tunicamycin inhibits, 16-18, *17*, 18t
Mushrooms, clitidine from, 293
eritadenine from, 298, 299
nebularine from, 244
Mycospocidin, similar to tunicamycin, 5

NAD$^+$, *see* Nicotinamide adenine dinucleotide
Nebularine, 244-246
biochemical properties of, 244-245
7-deazanebularine, 244
DNA synthesis inhibited by, 245, *246*
LD_{50} of, 245
phosphorylation of, 245
structure of, *244*
toxicity of, 245
virus inhibited by, 244
Newcastle disease virus, tunicamycin inhibits, 10, *10*
Nicotinamide adenine dinucleotide (NAD$^+$), analogs of, 129-131, *131*, 165, 179-180

ADP-ribosylation, 129-131, *131*, 179-180
DNA synthesis affected by, 129-131, *131*, 179-180
in dehydrogenase reactions, 129-131, 165, 179-180
Nikkomycin, 43-45
biochemical properties of, 44-45
chemical properties of, 43-44
discovery, production, isolation, 43
growth inhibition of, 44
mass spectrum of, 43
molecular formula of, 43
physical properties of, 43-44
Streptomyces tendae Tü 901, 43
structure of, *43*
Nmr spectra, 2′-aminoguanosine, 93-94
amipurimycin, 41
antibiotic A201A, 90
blasticidins H and S, 74, *75*
bredinin, 150
clindamycin, 105
clitidine, 293
2′-deoxycoformycin, 262
deoxyeritadenine, 300
5,6-dihydro-5-azathymidine, 247
eritadenine, 300
ezomycins, 24
hikizimycin, 78
1-methylpseudouridine, 242
mildiomycin, 80
octosyl acids, A, B, C, 296
oxamicetin, 82
oxazinomycin, 237
pentopyranines A and C, 155, *156*
sinefungin, 20
thuringiensin, 185
^{13}Cnmr spectra, 2′-aminoguanosine, 93-94
amipurimycin, 41
2′-deoxycoformycin, 262
ezomycins, 24, 27, *29*
hikizimycin, 78
1-methylpseudouridine, 242
mildiomycin, 80
showdomycin, 55, *55*
sinefungin, 20
thuringiensin, 185
Nocardia interforma, coformycin, formycin, formycin B, oxoformycin from, 169, 259
Norplicacetin (amicetin B), 80-81
chemical properties of, 81
discovery, production, isolation, 80-81
growth inhibited by, 81
mass spectrum of, 81
molecular formula of, 81
physical properties of, 80-81
structure of, *81*
Novikoff hepatoma, 5-azacytidine and, 137
deoxyuridine and, 287, *288*
pyrazofurin and, 285
toyocamycin and, 166
Nuclear magnetic resonance, *see* Nmr spectra
Nucleocidin, 102-103
biochemical properties of, 102-103
structure of, 103
Nucleoside Q, guanine into, 159, 160-161
pyrrolopyrimidine ring in tRNA, 159
structure of, 159
Nucleoside transport, crotonoside and, 61
deoxycoformycin and, 370
EHNA and, 370
showdomycin and, 57

Octosyl acid, 295-297
biochemical properties of, 297
biosynthesis of, 30-37, 297
chemical properties and synthesis, 295-296
discovery, production, isolation, 295
ezomycins compared to, 295
physical properties of, 295-296
structure of, *296*
Octosyl acid A, adenine derivative of, 296, 297
cCMP analog of, 297
molecular formula of, 295
structure of, *296*
transglycosylation of, 297
Octosyl acid B, 296
molecular formula of, 296
spectra, mass and nmr, 296
structure of, *296*
Octosyl acid C, 296
molecular formula of, 296
spectra, mass and nmr, 296
structure of, *296*
Optical rotatory dispersion (ORD) spectrum, thuringiensin, 185
Orotidine 5′-phosphate decarboxylase, 5-azacytidine and, 137
Osteogenic sarcoma, 5-azacytidine and, 145
Ovalbumin, tunicamycin inhibits, 7, 17
Oxamicetin, 81-84
chemical properties of, 82

discovery, production, isolation, 81-82
growth inhibited by, 84
molecular formula of, 82
peptidyl transferase inhibited by, 84
physical properties of, 82
spectra, infrared, mass, nmr, 82
structure of, 82-83, *82*, *83*
toxicity of, 84
Oxazinomycin (minimycin), 236-241
antibacterial activity of, 240
antitumor activity of, 240
biochemical properties of, 240
biosynthesis of, 238, 240
chemical properties and synthesis, 237, *239*
discovery, production, isolation, 237
Ehrlich ascites inhibited by, 240
LD_{50} of, 240
mediocidin similar to, 237
molecular formula of, 237
physical properties of, 237
spectra, mass, nmr, ultraviolet, 237, 238t
structure of, *237*
x-ray analysis of, 237
Oxoformycin B, 182-183
formation of, 169
3-methyl, 183
structure of, *169*

Pactamycin, 98
PAN, *see* Puromycin aminonucleoside
Parasites, sinefungin inhibits, 19, 20-21
Pentopyranic acid, 77
chemical synthesis of, 157
C-substance and, 157
isolation of, 154
structure of, 155
Pentopyranine, 154-158
biochemical properties of, 157-158
chemical properties-synthesis, 154, 155t, 157
discovery, production, isolation, 154
physical properties of, 154, 155t
RNA synthesis inhibited by, 157-158, 158t
structure of, 155
Pentopyranine A, chemical synthesis of, 157
nmr spectrum of, 156, *156*
structure of, *155*
Pentopyranine C, chemical synthesis of, 157
nmr parameters of, 156, *156*
structure of, 155
Pentopyranamine D, blasticidin H and, 77
chemical synthesis of, 157
structure of, *155*
Pentostatin, *see* 2′-Deoxycoformycin
Peptide bond formation, inhibition by, 97
nucleocidin, 102-103
pyrimidine nucleoside analogs, 86-87, *86*
Peptide chain termination, amicetin effect on, 86-87, *86*
gougerotin effect on, 86-87, *86*
Peptidoglycan, tunicamycin inhibits, 5, 13, *14*
Peptidyl-puromycin, 88, 97
Peptidyltransferase, 84, 86, 98, 178, *179*
Phage T3 RNA polymerase, thuringiensin effect on, 194-195, *194*
Pharmacology, 5-azacytidine, 143
coformycin, 272-273
deoxycoformycin, 272-273
Phosphatase, acid, tunicamycin inhibition of, 13
Phosphofructokinase, tubercidin 5′-triphosphate and, 162
Physarum polycephalum RNA, cordycepin inhibits, 120-121, *120*
Plants, aristeromycin inhibition of, 148
cordycepin effect on, 123, 127-128, 129, *130*
tunicamycin inhibition of, 5, 16
Plasmid, in crown gall, 50-51, *51*
Platelet aggregation, effect of analogs, 181
Platenocidin, 46-47
biochemical properties of, 47
chemical properties of, 46-47
discovery, production, isolation, 46
physical properties of, 46-47
polyoxins similar to, 47
structure of, *47*
Plicacetin, peptide bond formation inhibited by, 86-87, *86*
structure of, *74*
Poly (A), cordycepin inhibition of, 120-121, 120t, *121*, 124
5,6-dichloro-β-D-ribofuranosyl-benzimidazole effect on, 133
in mRNA synthesis, 120
toyocamycin inhibition of, 167, *168*
Poly (A) polymerases, ara-ATP effect on, 224
cordycepin and 3′dATP as inhibitors and substrates for, 122-123
Polyadenylation, ara-ATP inhibition of, 224
toyocamycin prevention of, 166
Poly (adenylic acid), *see* Poly (A)
Poly (ADP-ribose) polymerase, ara-ATP effect on, 224

formycin B inhibition of, 182
nucleoside analogs effect on, 130-131
showdomycin inhibition of, 182
Poly (ADP-ribosylation), NAD analogs effect on, 165
Polyamine biosynthesis, 163
5-azacytidine effect on, 141
Poly(formycin), 172
Poly(formycin B), 181
Polynucleotide phosphorylase, aristeromycin in study of, 148
thuringiensin and, 188
Polyoxamic acid (2-amino-2-deoxy-L-xylonic acid), 34, *35*
Polyoximic acid, biosynthesis of, 34, *34*
Polyoxin, 30-37
analogs of, 32
biochemical properties of, 35-37
biosynthesis of, 32-35
chitin synthetase, inhibited by, 30, 32, 35-37
ezomycins similar to, 23
5-fluorouracil incorporation into, 32, 37
fungi inhibited by, 30, 31, 35-37
nikkomycin similar to, 43
platenocidin similar to, 47
rice diseases inhibited by, 35
Streptomyces cacaoi, 32, 33, 34
structures of, A-N, *30*
transnucleosidation of, 32, 36
UDP-*N*-acetylglucosamine analog, 4, 26-27, *37*
Polyoxin D, chitin synthetase inhibited by, 36-37
Polyoxin J, chemical synthesis of, 31-32, *31*
Polyribosome "run-off", antibiotic A201A, 92-93, *92*
puromycin, 92-93, *92*
Polysome, degradation by 5-azacytidine, 137
Poly(tubercidin), interferon production inhibited by, 165
Prasinomycin, TM similar to, 5
Procollagen, tunicamycin inhibits, 16
Proinsulin synthesis, formycin B in study of, 182
Protein, affected by, ara-A, 227
bredinin, 152
eritadenine, 308
thuringiensin, 188-189, 192
toyocamycin, 167
chemical proofreading, 131-133, *132*
inhibited by, agrocin-84, 49-50, *50*
2′-aminoguanosine, 94-96, *96*
antibiotic A201A, 91-92, *91*
5-azacytidine, 136, 137, 139
cordycepin, 120, 122
EHNA, 268
formycin, 172
homocitrullyaminoadenosine, 103
lysylaminoadenosine, 103
nebularine, 245
nikkomycin, 44-45
nucleocidin, 102-103
puromycin, 96-97
tubercidin, 161-162, *161*
plants, cordycepin effect on, 127-128, *130*
stimulation by cordycepin, 129, *130*
2′,3′-dideoxyadenosine, 129
Protein kinase, aristeromycin 3′,6′-cyclic phosphate effect on, 149
cordycepin inhibition of, 129
eritadenine esters effect on, 306-307, *307*
tubercidin effect on, 162
Protozoa, tunicamycin inhibition of, 5
Pseudouridine, biosynthesis of, 244
Psicofuranine, 279-281
biochemical properties of, 280-281
biosynthesis of, 280
derivatives of, 280
GMP biosynthesis inhibited by, 281
purine biosynthesis inhibited by, 281
structure of, *280*
Purine biosynthesis inhibited by, cordycepin 5′-monophosphate, 118
decoyinine, 281
nebularine, 244
psicofuranine, 281
Purine nucleoside analogs, 89-104
Purine nucleoside phosphorylase, formycin and formycin B inhibition of, 177
Puromycin, 96-102
analogs of, 101-102
biochemical properties of, 96-102
biosynthesis of, 96
peptidyl transferase and, 99-100, *102*
photoaffinity labeling of ribosomes, 99, *101*
protein synthesis inhibition by, 96, 99-100
"reversed" nucleoside synthesis of, 96
structure of, *90*, *97*
requirements for activity, 101-102
Puromycin aminonucleoside (PAN), 195
demethylated analog, 195
phosphorylation of, 195

5′-deoxy PAN, 195
nephrotic effect of, 195
RNA synthesis inhibited by, 195
structure of, *195*
Puromycin reaction, 97-98, *98*
Pyrazofurin (Pyrazomycin), 281-289
analogs of, 182-183, 281
antitumor activity of, 285-286
ara-C increased effect with, 285-286
biochemical properties of, 285-289
biosynthesis of, 283, *286-287*
chemical synthesis of, 283, *284*
2′-deoxypyrazofurin, synthesis of, 281
DNA synthesis inhibited by, 285-287, *288*
HeLa, L5178Y and L-cells inhibited by, 281
leukemia affected by, 285
mass spectrum of, 283
Novikoff hepatoma inhibited by, 281, 285-287, *288*
phosphorylation of, 285
pyrimidine nucleotide synthesis inhibited by, 285
structure of, *283*
toxicity of, 285, 289
viruses inhibited by, 286-289
Pyrazolopyrimidine nucleoside analogs, 169-183. *See also* Formycin; Formycin B; *and* Oxoformycin B
Pyrazomycin, *see* Pyrazofurin
Pyridine nucleosides, clitidine, 292-294
nikkomycin, 43-45
Pyrimidine deoxyribonucleoside kinase (dPyK), ara-T and, 233-234, *234*
Pyrimidine nucleoside analogs, *see* Aminoacylaminohexosylcytosine analogs
Pyrimidine nucleosides, ezomycins similar to, 23
Pyrimidine nucleotide synthesis, pyrazofurin inhibition of, 286-287
Pyrrolopyrimidine nucleoside analogs, 158-169. *See also* Sangivamycin; Toyocamycin; *and* Tubercidin
Pyrrolopyrimidine ring, nucleoside Q in tRNA, 159, *159*, 160-161

Q base, guanine incorporation into, 161
Q nucleoside, *see* Nucleoside Q
Quenine, *see* Q base
Queuosine, *see* Nucleoside Q

Raphanatin, 313, *315*
Ribavirin, *see* Virazole
Ribonucleic acid, *see* RNA
Ribonucleotide reductase, ara-ADP and ara-ATP and, 218
tubercidin and, 162
Ribosomal binding, aminoacylaminohexosylcytosine analogs and, 86-87, *87*
antibiotic A201A effect on, 92
nucleocidin in, 103
puromycin in, 99-100, *101*
Ribosomes, biosynthesis of cordycepin inhibition of, 125
gougerotin binding to, 88-89, *88*
photoaffinity labeling of, 99-100, *101*
Rice diseases inhibited by, amipurimycin, 42-43
aristeromycin, 148
polyoxins, 35
RNA, affected by, eritadenine, 308
sangivamycin, 169
toyocamycin, 167
heterogeneous nuclear (HnRNA), cordycepin no inhibition of, 120
subclasses of, 133
synthesis of, toyocamycin no effect on, 167, *168*
incorporation of, ara-ATP, 224
5-azacytidine, 136-137, 143
5-azauracil, 136
cordycepin, 122-123
formycin, 174
maturation, thuringiensin effect on, 185, 192
messenger (mRNA), 5-azacytidine incorporation into, 137
formation, cordycepin inhibits, 120-121, *121*
globin synthesis, cordycepin inhibits, 123, 124
steps in processing of, 121, *122*
synthesis of, ecdysone stimulation of, 192
transport, cordycepin inhibits, 121, *122*
methylation of, *S*-tubercidinyl homocysteine, inhibition of, 163
nucleolar, thuringiensin inhibits, 194-195, *194*
pre-rRNA processing, toyocamycin inhibits, 166
ribosomal (rRNA), 5-azacytidine inhibits maturation, 137-138, *138*
processing of, 120
synthesis of, cordycepin inhibits, 120, 126-127, *126*

ecdysone stimulates, 192
thuringiensin inhibits, 190, *191*
toyocamycin 5′-triphosphate inhibits, 166-167
synthesis of, agrocin 84 inhibits, 49, *50*
2′-aminoguanosine effect on, 94-96, *96*, 145
antibiotic A201A effect on, 91-92, *91*
α-ara-A effect on, 224
5-azacytidine inhibition of, 139, *140*
bredinin effect on, 152
cordycepin, inhibition of, 118, 120-121, *121*, 122-123, 124, 124-125, 127-128, *130*
cordycepin plus deoxycoformycin, inhibition of, 270-271, *270*
3′dATP effect on *E. coli*, toluene-treated, 133, *134*
3′dATP inhibition of, 123
3′-deoxyuridine 5′-triphosphate inhibition of, 135
elongation, cordycepin effect on, 133-135, *134*
formycin no effect on, 172
initiation, cordycepin effect on, 133-135, *134*
nebularine inhibition of, 245, *246*
nikkomycin inhibition of. 44-45
nucleocidin effect on, 102-103
pentopyranines inhibition of, 157-158, 158t
plants, cordycepin effect on, 127-128, *130*
poly(A) sequence for, 121
puromycin aminonucleoside inhibits, 195
thuringiensin inhibition of, 188, 189, 192
toyocamycin inhibition of, 166-167
tubercidin inhibition of, 161-162, *161*
viral, tunicamycin effect on, 10
virazole inhibition of, 286-287
virus-specific, cordycepin inhibition of, 129
transfer (tRNA), adenosine analogs and, 132-133, *132*
3′-aminoadenosine incorporation into, 146
ara-ATP incorporation into 3′-terminus, 224
5-azacytidine incorporation into, 137
2′- or 3′-deoxyadenosine in, 178
formycin and, 178
methylation of, *S*-tubercidinylhomo-cysteine inhibition of, 163
misacylation/deacylation of, 133
mischarging of, 133-135, *134*
peptide bond, formycin and, 178
precursors of, formycin inhibits, 178
processing, formycin and, 178
pyrrolopyrimidine ring in, 160-161
sangivamycin incorporation into, 169
3′-terminus, ATP analogs and, 224
RNA polymerase, 3′-aminoadenosine inhibition of, 146
ara-ATP effect on, 224
cordycepin and 3′dATP inhibition of, 122-123
5,6-dichloro-β-D-ribofuranosylbenzimidazole inhibits, 133
E. coli, ara-ATP effect on, 224
ecdysone-stimulated, thuringiensin inhibition of, 192
formycin 5′-triphosphate and, 173
nucleolar, thuringiensin and, 189
nucleoplasmic, thuringiensin and, 189
phage T3, thuringiensin and, 194, *194*
thuringiensin inhibition of, 189, 191, 192, 194, *194*
RNA polymerase I, nucleolar, thuringiensin and, 189, *190*
RNA polymerase II, cordycepin effect on, 133-135, *134*
nucleoplasmic, thuringiensin and, 189, *190*
RNA viruses, non-oncogenic, ara-A and, 218, 227-228, *228*
streptovirudins inhibition of, 12
tumor, ara-A inhibition of, 218, 277-228, *228*
Rous sarcoma virus, toyocamycin and, 167, *168*

Sangivamycin, 169
biochemical properties of, 169
leukemia inhibited by, 169
structure of, *158*
Sarcoma 180, 2′-aminoguanosine inhibition of, 93
cordycepin effect on, 124
Schistosoma mansoni, tubercidin effect on, 165
Semlicki forest virus, membrane synthesis, 11
Septacidin, 45-46
structure of, *46*
Shiitake, eritadenine from, 298, 299
Showdomycin, 52-60

analogs of, 53, *54*
biochemical properties of, 56-60
biosynthesis of, 55-56, *56*
carcinogenesis and, 59-60
chemical synthesis of, 53, *54*
^{13}Cnmr spectrum of, 55, *55*
cytotoxicity of, 57
isoshowdomycin, 53
maleimycin, 52
oxidative phosphorylation of, 59-60
phosphorylation of, 53
poly(ADP-ribose) polymerase inhibition of, 181
structure of, *52*
thymidylate synthetase, 59, *59*
transport, 57
Sigma factor, thuringiensin effect on, 191-192
Sinefungin (A9145), A9145A, C, E, H, 19-23
S-adenosylmethionine analog of, 19
biochemical properties of, 22-23
biosynthesis of, 21-22, *22*
chemical properties of, 20
discovery, production, isolation, 20
fungi inhibited by, 20-21
growth inhibited by, 20-21
LD_{50} of, 21
methyltransferase inhibited by, *22*, 23
molecular formula of, 20
physical properties of, 20
spectra, ^{13}Cnmr, infrared, and nmr, 20
structure, *19*
toxicity of, 21
viruses inhibited by, 21, 23
Spacer region, mRNA synthesis and, 121, *122*
Sparsomycin, 98
Spectinomycin, 107
Spermidine, biosynthesis of, 162
Splicing process, mRNA synthesis, 119
Spongothymidine, *see* Ara-T
Streptomyces sp., production of, amipurimycin, 41
antibiotic A201A, 89
ara-A, 217
aristeromycin, 147
5-azacytidine, 135
blasticidin H and S, 73
coformycin, 259
decoyinine, 279
2′-deoxycoformycin, 260
5,6-dihydro-5-azathymidine, 245, 246
ezomycins, 23-24
formycin, 169
formycin B, 169
gougerotin, 73
herbicidins A and B, 311
hikizimycin, 77
maleimycin, 52, 56
1-methylpseudouridine, 241, 242
nebularine, 244
nikkomycin, 43
norplicacetin, 80
nucleocidin, 102
octosyl acids, 295
oxamicetin, 81
oxazinomycin, 237
oxoformycin B, 169
pentopyranines, 154
platenocidin, 46
polyoxins, 30, 31, 32-35
psicofuranine, 279
puromycin, 96
pyrazofurin, 281
sangivamycin, 159
showdomycin, 52, 56
sinefungin, 19, 20
thraustomycin, 38
toyocamycin, 159
tubercidin, 159
tunicamycin, 4, 5, 6
S. alboniger (puromycin), 96
S. antibioticus (ara-A, 2′-deoxycoformycin), 217, 260
S. armentosus (clindamycin sulfoxide), 104
S. A-5 (hikizimycin), 77
S. cacaoi var. *asoensis* (polyoxins, octosyl acids), 30, 31, 32-35, 295
S. candidus (pyrazofurin), 281
S. capreolus (A201A, A201B), 89
S. citricolor (aristeromycin), 147
S. clavus (nucleocidin), 102
S. exfoliatus (thraustomycin), 38
S. griseochromogenes (blasticidin S, pentopyranines), 73, 154
S. griseolus (sinefungin, A9145A, A9145C), 19, 20
S. gunmeances (formycins), 169
S. H 273 N-SYZ (platenocidin), 46
S. hygroscopicus (oxazinomycin, psicofuranine, decoyinine), 237, 279
S. kaniharensis (coformycin), 259
S. ladakanus (5-azacytidine), 135
S. lavendulae (formycins), 169

S. longissimus (hikizimycin), 77
S. lysosuperificus (tunicamycin), 4, 5-6
S. mediocidicus (oxazinomycin), 237
S. novoguineensis (amipurimycin), 41
S. oxamicetus (oxamicetin), 81
S. piomogenus (polyoxins L, M), 30
S. platensis (1-methylpseudouridine, 5,6-dihydro-5-azathymidine), 241, 242, 245, 246
S. plicacetus (amicetin, bamicetin, plicacetin, norplicacetin), 81
S. punipalus (1-demethylclindamycin), 104
S. rochei (lincomycin), 104
S. saganonensis (herbicidins A and B), 311
S. showdoensis (showdomycin), 52, 56
S. tanesashinensis (oxazinomycin), 237
S. tendae Tü 901 (nikkomycin), 43
S. toyocaensis (aspiculamycin), 84
S. yokosukanensis (nebularine), 244
Streptomycin, 107
Streptoverticillium rimofaciens (mildiomycin), 79
Streptovirudins, 5, 12
  structure, different from tunicamycin, 12
"Suicide inactivation," by ara-A, 227
  by cordycepin, 227
SV-40, tubercidin inhibition of, 166

Teichoic acid, tunicamycin inhibits, 15, *15*
Terminal deoxynucleotidyl transferase (TdT), adenosine deaminase deficiency and, 176-177
  α- and β-ara-ATP effects on, 225
  3′dATP inhibition of, 123-124, 225
Tetrahydrouridine, deaminase inhibitor, 137, 142-143
Thraustomycin, 38-40
  biochemical properties of, 40
  chemical properties of, 38
  discovery, production, isolation, 38
  molecular formula of, 38
  physical properties of, 38
  spectra, mass and ultraviolet, 38, *39*
  Streptomyces exfoliatus, 38
  structure of, *38*
Thuringiensin (β-exotoxin), 183-185
  analogs of, 186
  Bacillus thuringiensis, 183, 188
  biochemical properties of, 188-195
  biosynthesis of, 188
  chemical properties and synthesis of, 185, 186, *187*
  dephosphorylation of, 184, 185, 188
  discovery, production, isolation, 184-185, *185*
  DNA synthesis, not inhibited, 188
  growth inhibited by, 186-187
  insect control with, 186-187, 195
  mitotic spindle formation inhibited by, 192-193
  molecular formula of, 185
  N^1-oxide of, 189
  physical properties of, 185, 186, *187*
  protein synthesis affected by, 188, 192
  rice disease control by, 186-187
  RNA maturation affected by, 184, 189-191, 192, 194
  RNA polymerase inhibited by, 184, 189, 191-192
  rRNA synthesis inhibited by, 190, *191*
  spectra, ^{13}Cnmr, infrared, ORD, nmr, ^{31}Pnmr, 185, 186
  structure of, *183*
  toxicity of, 183, 184, 188
Tight binders, adenosine deaminase and, 267-268
Toyocamycin, 166-169
  adenosine replaced by, 166
  analogs of, 169
    Ehrlich ascites cells inhibited by, 166
    biochemical properties of, 166-169
  biosynthesis of, 149-151, *160*
  DNA synthesis inhibited by, 167
  HnRNA not inhibited by, 167
  interferon inhibited by, 167
  methylation and, 166, 167
  Novikoff hepatoma inhibited by, 166
  phosphorylation of, 166
  poly(A) addition inhibited by, 167, *168*
  pre-rRNA processing inhibited by, 166
  protein synthesis not inhibited by, 167
  RNA, appearance in cytoplasm, inhibited by, 167
  RNA incorporation of, 166
  RNA synthesis inhibited by, 166, 167
  rRNA synthesis inhibited by, 166-167
  structure of, *158*
  viruses inhibited by, 167, *168*
tRNA, *see* RNA, transfer
Translocase, tunicamycin inhibits, 13
Transmethylases, aristeromycin inhibits, 147-148
Transmethylation, *S*-adenosylmethionine in, 162-165, *164*

Tryptophan oxygenase, 5-azacytidine and, 141
Tubercidin, 158-166
  analogs of, 163, 166
    S-adenosylhomocysteine analog of, 162-165, *164*
  anthelmintic properties of, 165
  ara-tubercidin formation, 221
  ATP competitive inhibitor of, 162
  biochemical properties of, 161-166
  clinical application of, 165
  3′,5′-cyclic monophosphate of, 165
  DNA synthesis, 161-162, *161*
  NAD analog of, 131, 165
  pharmacology of, 165
  phosphorylation of, 162
  poly(tubercidin), 165
  protein kinase affected by, 162
  protein synthesis inhibited by, 161-162, *161*
  purine synthesis inhibited by, 161
  ribonucleotide reductase affected by, 162
  RNA synthesis inhibited by, 161-162, *161*
  structure of, *158*
  toxicity of, 162
S-Tubercidinyl-L-homocysteine, 147-148, 162-165, *164*
  5′-cap inhibition of, 163
  DNA synthesis inhibited by, 163
  methylases inhibited by, 162-165, *164*
  RNA methylation inhibited by, 163
  tRNA methylation inhibited by, 163
Tumors, affected by, ara-A, 218
  5-azacytidine, 136, 144-145
  cordycepin plus 2′-deoxycoformycin, 125
  2′-deoxycoformycin in animals, 266-267
  toyocamycin, 167
Tunicamine, 5, 6
Tunicaminyl uracil, 5, 6, *6*
Tunicamycins A, B, C, D, 4-19, *62*
  N-acetylglucosaminyl-lipids inhibited by, 5, 8, 16
  N-acetylglucosaminyl pyrophosphate inhibited by, 8, *8*
  antiviral activity reversal of, 9-12
  biochemical properties of, 7-19, *62*
  cell wall synthesis inhibited by, 12-13
  chemical properties of, 6
  collagen inhibited by, 16
  discovery, production, isolation, 5-6
  dolichyl-N-acetylglucosaminyl pyrophosphate, 4, 13
  fungi inhibited by, 5, 7
  glucosamine incorporation inhibited by, 8, 13
  glycolipid synthesis inhibited by, 5
  glycoprotein inhibited by, 7-9, *7*, 11
  glycosaminoglycans inhibited by, 16-19, *17*, 18t
  glycosylation inhibited by, 4, 11, 16-19
  growth inhibited by, 7
  immunoglobulin secretion inhibited by, 16
  interferon glycosylation inhibited by, 12
  lipid-pyrophosphoryl-N-acetyl-muramyl-pentapeptide
    inhibited by, 13-15, *14*
  mammalian cells in culture inhibited by, 5
  mannolipid effected by, 8, *8*
  Newcastle disease inhibited by, 4, 9-10, *10*
  ovalbumin inhibited by, 8-9, *9*
  peptide synthesis inhibited by, 14
  peptidoglycans inhibited by, 13-15, *14*
  physical properties of, 6
  plants inhibited by, 16
  polyoma-transformed cells, sensitivity to, 11
  RNA synthesis, viral, effected by, 9-12
  saccharide-lipids and, 9
  serum proteins, inhibited by, 16
  spectra, infrared, nmr, 6
  structure of, *4*
  SV-40 transformed cells, 10-11
  teichoic acid inhibited by, 15, *15*
  UDP-N-acetylglucosamine, analog of, 4
  viral coat formation inhibited by, 9-12, *10*
  viral envelope inhibited by, 5
  VSV assembly inhibited by, 11-12
  yeast, glycoprotein, inhibited by, 13
Tyrosine aminotransferase, 5-azacytidine and, 141

UDP-N-acetylglucosamine, polyoxins and, 36-37, *37*
Undecaprenyl-N-acetylglucosaminyl lipid, tunicamycin and, 13-15
Uridine kinase, 5-azacytidine and, 136, 145

Vaccinia virus, bredinin inhibition of, 150
  poly(A) in, 119-120
Vesicular stomatitis virus (VSV) assembly, inhibited by tunicamycin, 11-12
Viral coat formation, tunicamycin inhibits, 5, 10, *10*
Virazole (Ribavirin), 286-287
Viruses, inhibited by, ara-A, 218-219, 221-224, 227-228
  ara-AMP, 226
  ara-A-5′-valerate, 221

ara-HxMP, 226
ara-T, 233-236
bredinin, 150
5-bromotubercidin, 166
cordycepin, 128-129
deoxycoformycin, 227-228, 266-267
5,6-dihydro-5-azathymidine, 249
EHNA, 227-228
EHNA plus cordycepin, 270-271
nebularine, 244-245
pyrazofurin, 281, 286-289
sinefungin, 23
toyocamycin, 167-168
tunicamycin, 5
virazole, 286-287
see also specific type

Xanthine oxidase, formycins and, 183

Yeast, affected by, ezomycins, 29
platenocidin, 47
toyocamycin, 166
tunicamycin, 4, 5, 13